Recent Developments in Forward Osmosis Processes

Recent Developments in Forward Osmosis Processes

Edited by

Rodrigo Valladares Linares, Zhenyu Li, Menachem Elimelech, Gary Amy and Hans Vrouwenvelder

IWA PUBLISHING

Published by **IWA Publishing**
Alliance House
12 Caxton Street
London SW1H 0QS, UK
Telephone: +44 (0)20 7654 5500
Fax: +44 (0)20 7654 5555
Email: publications@iwap.co.uk
Web: www.iwapublishing.com

First published 2017

British Library Cataloguing in Publication Data
A CIP catalogue record for this book is available from the British Library

ISBN 9781780408118 (Paperback)
ISBN 9781780408125 (eBook)

Cover image: Ana Mabel López Villanueva

Contents

Additional Image credits

Part 1 – Spring-fed pool, "Ojo de Agua" ranch, Villa Unión, Coahuila, Mexico. Photograph taken by Rodrigo Valladares Linares

Part 2 – Recycle for clean water, Photographer: serrnovik. Photo ID: csp9783511 http://www.canstockphoto.com/recycle-for-clean-water-9783511.html

Part 3 – Scanning electron microscopy image of a biofouled membrane showing bacterial cells and exopolymeric substances. Photograph taken by Szilard Bucs and Rodrigo Valladares Linares

Part 4 – 3-D representation of Bisphenol A and Paracetamol molecules. Paracetamol molecule – Author: Ben Mills – Benjah-bmm27. Public Domain. Available from https://en.wikipedia.org/wiki/Paracetamol#/media/File:Paracetamol-from-xtal-3D-balls.png, Author: Bisphenol A molecule – Author: Edgar181. Public Domain. Available from https://commons.wikimedia.org/wiki/File:Bisphenol_A.png

Part 5 – Scanning electron microscopy cross-section image of a clean cellulose triacetate forward osmosis membrane. Photograph taken by Szilard Bucs and Rodrigo Valladares Linares

Part 6 – Visualization of membrane and feed spacer fouling process using computational fluid dynamics software. Simulation image created by Szilard Bucs

Part 7 – Water. Photographer: ifong. Photo ID: csp4605553 http://www.canstockphoto.com/water-4605553.html

List of Contributors

We want to acknowledge the contribution of our co-authors in the various chapters.

Muhannad Abu-Ghdaib
King Abdullah University of Science and Technology, Saudi Arabia.

Gary L. Amy
King Abdullah University of Science and Technology, Saudi Arabia.

Cyril Aubry
King Abdullah University of Science and Technology, Saudi Arabia.

Szilard S. Bucs
King Abdullah University of Science and Technology, Saudi Arabia.

Hongmin Chen
University of Missouri-Kansas City, United States.

Tai-Shung Neal Chung
National University of Singapore, Singapore.

Menachem Elimelech
Yale University, New Haven, United States.

Qingchun Ge
National University of Singapore, Singapore.

Noreddine Ghaffour
King Abdullah University of Science and Technology, Saudi Arabia.

Laura A. Hoover
Yale University, New Haven, United States.

Yan-Ching Jean
University of Missouri-Kansas City, United States.

TorOve Leiknes
King Abdullah University of Science and Technology, Saudi Arabia.

Qingyu Li
King Abdullah University of Science and Technology, Saudi Arabia.

Zhenyu Li
King Abdullah University of Science and Technology, Saudi Arabia.

Rodrigo Valladares Linares
King Abdullah University of Science and Technology, Saudi Arabia.

Mingming Ling
National University of Singapore, Singapore.

Jeffrey R. McCutcheon
Yale University, New Haven, United States.

Yong-Gyun Park
GS Engineering and Construction, Seoul, South Korea.

William A. Phillip
University of Notre Dame, United States.

Cristian Picioreanu
Delft University of Technology, The Netherlands.

Sarper Sarp
GS Engineering and Construction, Seoul, South Korea.

Alberto Tiraferri
Yale University, New Haven, United States.

Johannes S. Vrouwenvelder
King Abdullah University of Science and Technology, Saudi Arabia; Delft University of Technology, The Netherlands; Wetsus, The Netherlands.

Kai Yu Wang
National University of Singapore, Singapore.

Chunhai Wei
King Abdullah University of Science and Technology, Saudi Arabia.

Victor A. Yangali-Quintanilla
King Abdullah University of Science and Technology, Saudi Arabia; Grundfos Holding A/S, Bjerringbro, Denmark.

Ngai Yin Yip
Yale University, New Haven, United States.

Tong Zhan
King Abdullah University of Science and Technology, Saudi Arabia.

Sui Zhang
National University of Singapore, Singapore.

Preface

A constant and reliable source of fresh water is essential for the development of modern economies and settlements around the world. It has been projected that by 2025, between 2.4 billion and 3.2 billion people could live under water-scarce or water-stressed conditions, a four-fold increase in the number of people that lived under these situations at the beginning of the century. Research has identified the potential of forward osmosis (FO) systems to produce high quality water from seawater, while recovering impaired water from wastewater sources.

The study of FO has increased strongly in the past five years, compared to the previous fifteen years, as indicated by a more than tripling of the number of scientific papers. However, no single source has given an updated overview of FO.

This book gives a complete and comprehensive overview of all aspects of forward osmosis: (i) introduction, (ii) water recovery, (iii) fouling, (iv) rejection of micropollutants, (v) draw solutions and membranes, (vi) numerical modelling, and an (vii) outlook with an overview on low-energy applications, a life cycle cost assessment, and the possible niches in the water industry for forward osmosis processes and hybrids.

Rodrigo Valladares Linares
rodrigo.valladares@sinacqua.com.mx
rodrigo.valladares@kaust.edu.sa

Zhenyu Li
zhenyu_li@hks.harvard.edu
lizhenyu@nwafu.edu.cn

Menachem Elimelech
menachem.elimelech@yale.edu

Gary L. Amy
gamy@clemson.edu
cheagl@nus.edu.sg

Johannes S. Vrouwenvelder
j.s.vrouwenvelder@tudelft.com
johannes.vrouwenvelder@kaust.edu.sa

Summary

Since more than 97% of the water in the world is seawater, desalination technologies have the potential to solve the fresh water crisis. The most used desalination technology nowadays is seawater reverse osmosis (SWRO), where a membrane is used as a physical barrier to separate the salts from the water, using high hydraulic pressure as the driving force. However, the use of high hydraulic pressure imposes a high cost on operation of these systems, in addition to the known persistent fouling problems associated with reverse osmosis (RO) membrane filtration systems.

Forward osmosis (FO) is an alternative membrane process that uses an osmotic pressure difference as the driving force. FO uses a concentrated draw solution to generate high osmotic pressure, which extracts water across a semi-permeable membrane from a feed solution. Afterwards, fresh water can be obtained when the diluted draw solution is regenerated in a second treatment step, e.g., low pressure reverse osmosis (LPRO). Research has identified the potential for hybrid forward osmosis/low-pressure reverse osmosis (FO/LPRO) systems for several applications, including seawater desalination, and to reduce the cost and fouling propensity of producing fresh water from impaired-quality water sources, compared to conventional high pressure RO systems.

One of the main advantages of FO is the limited amount of external energy required to extract water from the feed solution, using only a very low amount of energy to recirculate the draw solution on one side of the membrane, while the feed solution is passively in contact with the other side of the membrane.

The objective of this book is the evaluation of the recent developments in FO processes for wastewater recovery and seawater desalination, characterization of membrane fouling and cleaning, pollutant rejection capabilities of FO membranes, state-of-the-art FO membranes and draw solutions, modeling the effect of concentration polarization, reverse salt flux and fouling on water flux through the FO membrane, and the analysis of FO applications/niches and the life cycle cost of a large-scale system compared to conventional desalination and water recovery alternatives.

WATER RECOVERY

Forward osmosis membrane processes can be adapted to the existing urban runoff collection system in coastal regions for the management of urban runoff (Chapter 2.1). During testing,

FO process provided a sustainable permeate flux for the recovery of fresh water, and reliable and stable rejection of urban runoff-derived contaminants in a simulated osmotic detention pond. A nearly complete rejection of trace metals and phosphorus was achieved while the rejection rates of nitrate and total nitrogen (TN) were moderate to high in the lab scale testing. By using FO as a bridge, urban runoff management, seawater desalination can be integrated to achieve a sustainable solution to the water-energy nexus in coastal regions.

Another study revealed the ability of a FO process to integrate seawater desalination and municipal wastewater treatment for drinking water production (Chapter 2.2). The FO process showed a high rejection for chemical oxygen demand, phosphate and trace metals, and moderate rejection for ammonia and TN. Organic carbon analysis revealed that the membrane tested was unable to reject low molecular weight acids and low molecular weight neutral compounds, such as sodium acetate and urea. Biopolymer-like substances played a dominant role in the formation of fouling on the membrane surface. The study showed that FO is a reliable barrier to effectively reject most wastewater contaminants and salts from either the wastewater as feed solution or seawater as draw solution while allowing clean water to pass through, providing a possible significant energy-saving strategy to integrate municipal wastewater treatment and seawater desalination to further promote sustainable urban water management and water reuse in coastal cities.

Experiments with FO membranes using a secondary wastewater effluent as a feed water and seawater as a draw solution demonstrated that the technology is promising (Chapter 2.3). Indirect desalination with a hybrid system FO/LPRO was implemented. The system consumes only 50% of the energy used for normal high pressure RO desalination, and produces a good quality water extracted from the impaired feed water.

FOULING

Flux patterns and membrane fouling propensity during seawater FO desalination were investigated (Chapter 3.1). Flux decreased significantly in each batch, which could be mainly attributed to external concentration polarization (ECP), fouling related resistance, and fouling related concentrative internal concentration polarization (ICP). The membrane surface was covered by a fouling layer after 18–30 day periods. The fouling layer was made up of scaling-like foulants surrounded by biopolymeric substances. The membrane scaling was caused by the polymerization of dissolved silica. The polymerized silica cluster facilitated the accumulation and deposition of natural organic matter (NOM), especially biopolymers. A higher crossflow velocity could improve the flux by reducing ECP and fouling. Most NOM foulants in the fouling matrix could be removed by water flushing while silica scaling was difficult to hydraulically clean.

Characterization of the organic foulants in both wastewater and seawater was performed (Chapter 3.2). Organic carbon analysis (liquid chromatography coupled with organic carbon detection (LC-OCD) and three-dimensional fluorescence excitation emission matrices (3-D FEEM)) suggest that biopolymers and protein-like substances, present in the feed water, form a fouling layer on feed side of the FO membrane, reducing the water flux and thus, affecting the efficiency of the seawater dilution process. Transparent exopolymer particles (TEP) were identified in the support layer of the FO membrane in contact with the seawater, which contains a significant amount of these particles, reducing the flux of the FO membrane.

A novel FO calcium carbonate ($CaCO_3$) scaling mechanism induced by the interaction of feed solute and draw solute was explored (Chapter 3.3). This scaling can occur at the beginning of the FO process with a low concentration of paired cation and anion from feed and draw solution, respectively, and cause a dramatic and irreversible flux decline. Membrane surface morphology and roughness is crucial in the reversibility of $CaCO_3$ scaling reported. The study adds one more important criterion for the selection of DS in the FO process. In addition to the characteristics of a favorable DS, such as high osmotic pressure, non-toxic, low cost and easy separation/recycle, the risk of induced scaling should also be considered. A deeper understanding of selective ion transport in the FO process and possible impact on scaling formation could contribute to the development of FO-based water treatment processes.

The study on the influence of feed spacer thickness (28, 31 and 46 mil, 1 mil = 0.0254 mm) on performance and biofouling development on the feed side of FO membranes (Chapter 3.4) led to the following conclusions: (i) the biomass amount alone does not determine the flux decline: the same amount of biomass was found for all spacer thicknesses after the same run time at the same feed flow, while the flux reduction decreased with thicker spacer; (ii) the flux decline caused by biomass accumulation can be reduced by using a thicker spacer; (iii) spatial distribution of the biofilm differed with feed spacer thickness. Findings are in agreement with reported data for high pressure reverse osmosis cross-flow systems: thicker spacers reduce the impact of biofouling on performance. This result clearly contradicts observations obtained with particulate and colloidal fouling in forward osmosis.

Physical and chemical methods were used and compared in an effort to set an effective protocol for FO membrane cleaning (Chapter 3.5). NOM fouling showed high hydraulic reversibility, up to 90% when *in-situ* air scouring for 15 minutes was used as a cleaning technique. Chemical cleaning with a mixture of Alconox, an industrial detergent containing phosphates, and sodium ethylenediaminetetraacetic acid (EDTA) showed to improve the reversibility further (93.6%). Osmotic backwashing using a 4% NaCl solution and deionized water proved to be ineffective to recover the flux due to the salt diffusion phenomena occurring at the active layer (the membrane separation layer). The same detergent solution used to clean the active layer was used to clean the support layer; 95% of flux was recovered, showing that the chemically irreversible fouling of the FO membrane is in the order of 5.5%, which might be associated with the adsorption of biopolymers on the active layer and some TEP residuals on the support layer. Physical cleaning (air scouring) proved to be the most effective way to control organic fouling.

REJECTION OF POLLUTANTS

Applying practical conditions of water reuse applications, FO membranes were able to reject most of the organic micropollutants spiked in the feed water (Chapter 4.1). Rejections were moderate for hydrophilic neutral compounds (44–95%), moderate for hydrophobic neutral contaminants (48–92%), and high for the hydrophilic ionic micropollutants (96–99%). FO coupled with LPRO was effective in rejecting low molecular weight hydrophilic neutral micropollutants, with rejections exceeding 89%. For the rest of the compounds, rejections were greater than 99%. A hybrid FO/LPRO system serves as a double barrier against micropollutants, including pharmaceutically active compounds, hormones and other pollutants.

Experiments on boron rejection by CTA and TFC FO membranes (Chapter 4.2) have shown that: (i) membrane characteristics play a role in boron flux, (ii) at low and high water flux, the new generation TFC FO membranes show a lower boron flux than CTA FO membranes, (iii) boron rejection is not affected by reverse draw solute flux, nor the draw solute charge, and (iv) NOM fouling layer on the membrane has no significant impact on boron flux. This demonstrates that the new generation TFC FO membrane is potentially suitable for desalination, particularly in terms of solute rejection and a reduced boron flux.

DRAW SOLUTION AND MEMBRANES

A comprehensive review on the progress of draw solution for FO processes is presented (Chapter 5.1). Ideally, the FO semipermeable membrane performs as a barrier that allows only water to pass through but rejects all others. However, in reality, depending on draw solute's chemistry property and physical structure, the reverse flux of draw solutes may take place across FO membranes which not only results in a lower effective osmotic driving force but also facilitates fouling. In addition, the asymmetric structure of FO membranes and the transport resistance of draw solutes within the FO membranes cause concentration polarization and lower the water flux. Furthermore, the regeneration of draw solutes from diluted draw solutions and the production of clean water might be energy intensive if inappropriate draw solutes and recycle processes are utilized. An assessment on the advantages and limitations of the existing draw solutes were made, presenting various FO integrated processes for water production and draw solute regeneration methods suitable for different applications.

Membranes with an ultra-thin selective layer and a fully porous support have been fabricated which showed high performance in FO processes (Chapter 5.2). The following conclusions could be made from this work: (i) the interaction between polymers and the casting substrate can be a determinant factor for the structure of the bottom layer of the membrane formed during phase inversion; (ii) the thickness of the ultra-thin bottom selective layer was found to be around 95 nm; meanwhile, an open-cell, fully porous structure was formed in the middle of the membrane during phase inversion; (iii) seawater desalination performance via FO process of the double dense-layer membrane showed low salt leakage and mitigated ICP as compared to previous reports; taking into consideration both ECP and ICP in a model, a structural parameter of 54 μm was found, which is much smaller than those reported in literature, (iv) by introducing an intermediate immersion in an organic/water phase mixed coagulant bath prior to complete phase transformation in water, the top layer of the membrane could be made fairly porous and the structural parameter was further reduced to 51 μm, leading to even lower ICP during the FO process; and (v) the double dense-layer membrane showed good resistance to the irreversible fouling of large particles in the novel FO – membrane bioreactor integrated system as compared to the single-dense-layer membranes using nanoparticles of 200 nm as the model foulant.

MODELING

The impact of external and internal concentration polarization on the osmotic driving force for both symmetric and asymmetric membranes was quantified through modeling (Chapter 6.1). Permeate flux through both of these membranes was modeled for a variety of draw and feed concentrations at a set of specified experimental conditions. Water flux through asymmetric membranes was modeled in both FO and pressure retarded osmosis (PRO) modes. It was determined that both ECP and ICP played major roles in the reduction of the osmotic driving force in both FO and PRO modes, though, in general, ICP impacted permeate water flux more. With asymmetric membrane flux modeling, the value of the solute resistance to diffusion, K, was varied to determine the impact of membrane design on flux performance. Overall, reducing K was found to significantly improve permeate water flux performance in either the PRO or FO mode. In both PRO and FO, reduced K also allowed for higher salinity feed waters to be used, indicating the possibility of higher feed water recovery. Improving the membrane would allow for better flux performance, higher feed water recoveries, and lower energy use of the draw solute recovery thereby improving the economic viability and utility of a variety of osmotically driven membrane processes.

The impact of biofouling on membrane performance in a cross-flow forward osmosis (FO) system was evaluated with numerical simulations supported by experimental data (Chapter 6.2). The two-dimensional numerical model couples liquid flow with solute transport in the FO feed and draw channels, in the FO membrane support layer and in the biofilm developed on one or both sides of the membrane. The developed model was tested against experimental measurements at various osmotic pressure differences and in batch operation without and with the presence of biofilm on the membrane active layer. Numerical studies explored the effect of biofilm properties (thickness, hydraulic permeability and porosity), biofilm membrane surface coverage, and biofilm location on salt external concentration polarization and on the permeation flux. The numerical simulations revealed that (i) when biofouling occurs, external concentration polarization became important, (ii) the biofilm hydraulic permeability and membrane surface coverage have the highest impact on water flux, and (iii) the biofilm formed in the draw channel impacts the process performance more than when formed in the feed channel. The proposed mathematical model helps to understand the impact of biofouling in FO membrane systems and to develop possible strategies to reduce and control biofouling.

OUTLOOK

FO applications described in this chapter (Chapter 7.1) avoid draw solution regeneration and thus can perform while using very little energy. These low-energy applications have the potential to enhance the energy efficiency of existing technologies, minimize pollution of aquatic and marine environments, utilize wastewater as a resource, or alleviate water shortages in agriculture. Unique to these resource-efficient FO applications is the exposure of the membranes to feed and draw solution streams with high fouling potential. Hence, there is a pressing need to develop membranes with low propensity for irreversible fouling in order to enable continuous operation without the need for costly pretreatment of the feed and draw solutions. Furthermore, membrane module configurations that maximize mixing for the continuous removal of foulants from the membrane surface need to be developed. Adapting the advances of the past few decades in membrane technology and materials science to FO can put these technological achievements within our reach.

FO/LPRO has been depicted as a near horizon low-energy desalination technology considering the use of a hybrid system for desalination and wastewater recovery, using the principle of osmotic concentration/dilution (Chapter 7.2). Based on the economic analysis of water treatment systems producing 100,000 $m^3 \cdot d^{-1}$ of water described, it can be concluded that: (i) a hybrid FO/LPRO system has lower costs for producing water compared to conventional seawater desalination by RO; (ii) the sensitivity analysis showed that the most critical aspect in terms of economic feasibility for these hybrid FO/LPRO systems is the FO module cost; (iii) the proposed hybrid FO/LPRO system has a comparable cost to wastewater treatment and recovery system; and (iv) additional advantages of hybrid FO/LPRO systems include the reduction in wastewater volume to be post-treated, recoverable biogas production based on anaerobic post-treatment of concentrated wastewater effluent, and the reduction of greenhouse gas emissions compared to conventional high-energy desalination technologies.

Forward osmosis (FO) is an emerging membrane technology with a range of possible water treatment applications (desalination and wastewater recovery). An overview of applications, advantages, challenges, costs and knowledge gaps is given (Chapter 7.3). With current commercial technology, hybrid FO systems for both desalination and water recovery applications have proven to have higher capital cost compared to conventional technologies. Nevertheless, due to the demonstrated lower operational costs of hybrid FO systems, the unit cost for each m^3 of fresh water produced with the FO system are lower than conventional desalination/water recovery technologies (i.e. ultrafiltration/RO systems).

There are key benefits of using FO hybrid systems compared to RO: (i) chemical storage and feed systems may be reduced for capital, operational and maintenance cost savings, (ii) reduced process piping costs, (iii) more flexible treatment units, and (iv) higher overall sustainability of the desalination process, while producing high quality water.

The major challenges of FO to be a commercially viable technology are: (i) developing a higher flux membrane, capable of maintaining an elevated salt rejection and a reduced ICP effect, (ii) the availability of appropriate draw solutions, which can be recirculated via an efficient recovery process, (iii) better understanding of fouling and biofouling occurrence, (iv) assuring the high quality of the water produced, and (v) hybridization with other technologies that can increase the benefits of FO use (i.e. water recovery, energy production, etc.).

Additional research and development as well as process demonstration at the pilot-scale must be completed to permit successful implementation and commercialization. Doing so will also give important information on the direction that should be taken to develop robust low cost water treatment hybrid systems to produce high quality water.

Part 1
Introduction

Chapter 1.1

Population distribution and water scarcity

Globally, the level of urbanization has significantly increased during the last six decades. In 2010, urban areas comprised 3.5 billion people, or 50.5 per cent of the world's population (United Nations, 2011). Due to the development of coastal regions in many countries, two-fifths of cities with populations of 1 million to 10 million people are located near coastlines (Tibbetts, 2002). Moreover, 14 of the largest 17 cities in the world are situated along coasts (Figure 1.1.1) (Creel, 2003b).

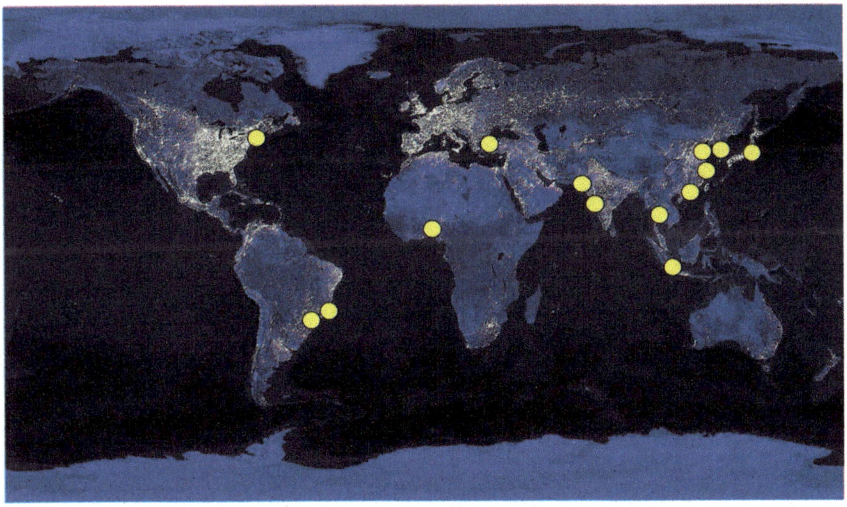

Figure 1.1.1 Fourteen of the world's largest cities are located along the coast (indicated in orange), which translates into an opportunity to integrate drinking water and wastewater management, especially in water-stressed areas (Creel, 2003) (image adapted from: NASA (2000)).

With increasing population growth, the demand for clean water is also increasing. Under an average economic growth scenario and when no efficiency gains are assumed, global fresh water demand will increase 53% from 2009 to 2030, from 4.5 trillion m^3 to 6.9 trillion m^3 (The Barila Group *et al.* 2009), with a proportionally greater deficit in many developing countries. This situation will leave between 2.4 billion and 3.2 billion people under water-scarce or water-stressed conditions by year 2025, four-folding the number of people that lived in areas with limited fresh water availability at the beginning of the 20th century. Water shortage is likely to grow especially acute in the Middle East and much of Africa (Engelman *et al.* 2000), as well as vast areas in Australia, China, India and Mexico (Figure 1.1.2), directly associated with the physical absence of additional fresh water resources to cover their demand. Besides the amount of water itself, the quality must comply with the minimum standards set by each country's regulations to be considered as potable; diarrheal disease alone is responsible for the death of 1.8 million people every year, and it was estimated that 88% of these cases are attributable to unsafe water supply, sanitation and hygiene (Prüss-Üstün & Corvalá, 2004).

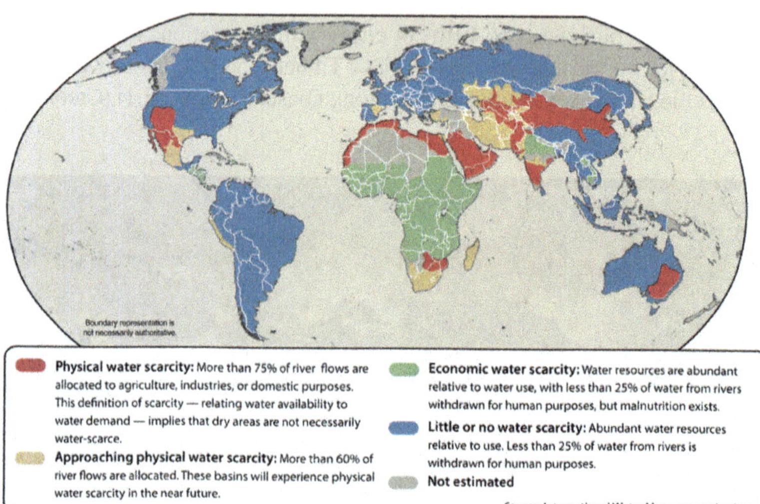

Figure 1.1.2 Projected global water scarcity by 2025 (international water management institute).

Human population distribution along the coasts represents a great challenge in terms of water management, due to the contamination of surface and ground water (basins, aquifers, rivers, etc.), limited fresh water sources in some areas, and more stringent environmental regulations that restrict the use of certain water resources. Therefore, a comprehensive strategy for the management of water resources is crucial for the sustainable development of these areas (Li *et al.* 2014).

Since more than 97% of the water in the world is seawater (Figure 1.1.3), desalination technologies have the potential to solve the fresh water crisis, particularly in coastal areas. The most used desalination technique nowadays is reverse osmosis, where a membrane is used as a physical barrier to separate the salts from the water, using high hydraulic pressure (Amjad, 1993). Nevertheless, the use of pressure imposes a high cost on operation of these systems, besides the known persistent fouling problems associated with membrane filtration systems (Ridgway & Flemming, 1996; Shannon *et al.* 2008; Vrouwenvelder *et al.* 2008).

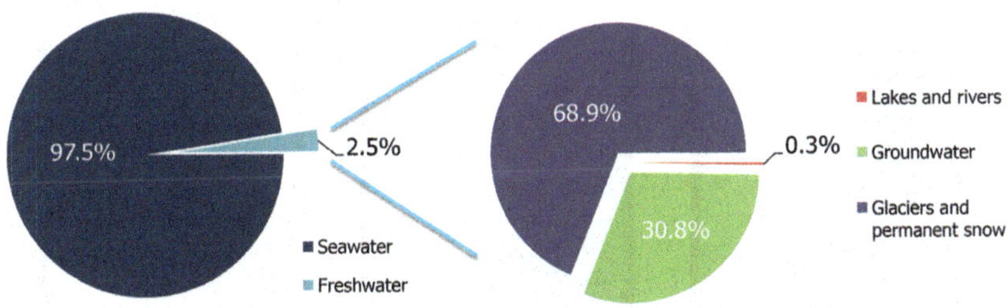

Figure 1.1.3 Total water distribution in the world (UNESCO, 1999).

Research has identified the potential for hybrid forward osmosis/reverse osmosis (FO/RO) systems for several applications, including sweater desalination (Choi *et al.* 2009), to reduce the cost and fouling propensity of producing fresh water from impaired-quality water sources (Chang *et al.* 2002; Achilli *et al.* 2009; Boo *et al.* 2013). Recently, studies have shown the potential of these systems to produce low cost high quality fresh water using low pressure desalination, while simultaneously recovering impaired water from a recycled feed water (Cath *et al.* 2009; Yangali-Quintanilla *et al.* 2011). Nevertheless, there are concerns about the use of FO membranes as a barrier for rejecting micropollutants and nutrients from the wastewater, besides the inevitable fouling problems that can occur during the filtration process, when the membrane is submerged in the recycled feed water, resulting in a poor water flux and an increase in the operational cost due to membrane cleaning.

1.1.1 OSMOTIC MEMBRANE PROCESSES

Osmosis is defined as the transport of water through a semipermeable membrane caused by a difference in osmotic pressure for the solutions on both sides of the membrane. The osmotic pressure is related to the concentration of dissolved ions in solution and the temperature (Cath *et al.* 2009).

Osmotic pressure (π) can be calculated using the van't Hoff equation (van't Hoff, 1887):

$$\pi = R\,T\,\Sigma i\,M \tag{1.1.1}$$

where i is the dimensionless van't Hoff factor for the specific ion, M is the molarity of the specific ion, R is the gas constant ($0.08206\ \mathrm{L \cdot atm \cdot mol^{-1} \cdot K^{-1}}$), and T is the temperature in Kelvin.

The most common applied membrane filtration processes are reverse osmosis (RO), pressure retarded osmosis (PRO) and forward osmosis (FO), also known as osmosis or direct osmosis, are shown in Figure 1.1.4.

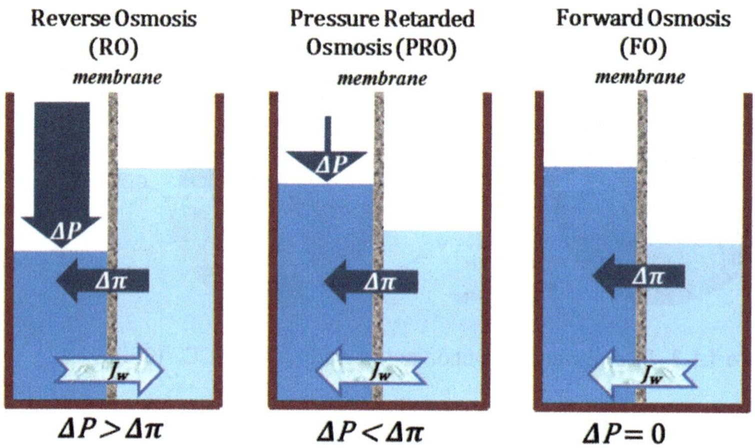

Figure 1.1.4 Osmotic processes in membrane filtration. ΔP – applied hydraulic pressure; $\Delta\pi$ – osmotic pressure difference between the two solutions; J_w – water flux.

RO occurs when the osmotic pressure difference between the two solutions is overcome by a hydrostatic pressure applied to the high concentration solution, reversing the osmosis process. PRO is an osmosis process in which there is a hydraulic pressure applied to the high concentration solution, but the osmotic pressure difference is bigger, so the water flux is opposite to the flux in RO process. When there is no hydraulic pressure applied, the osmotic process is called forward osmosis (FO), which is the natural flow of water from the low solute concentration to the high solute concentration solution to equilibrate the osmotic pressure difference.

1.1.2 FORWARD OSMOSIS (FO)

Forward osmosis (FO) uses a concentrated draw solution to generate high osmotic pressure, which extracts water across a semi-permeable membrane from the feed solution (Mi & Elimelech, 2008). As a result, the feed stream reduces in volume, increasing the salt concentration, which results in a decrease in permeate flux to the draw solution side (Choi *et al.* 2009).

According to Lee *et al.* (1981), the general equation describing water transport through the RO membrane is:

$$J_w = A\,(\sigma\Delta\pi - \Delta P) \tag{1.1.2}$$

where J_w is the water flux, A the water permeability coefficient of the membrane, $\sigma\Delta\pi$ the effective osmotic pressure difference in reverse osmosis, being σ the reflection coefficient, and ΔP the applied pressure; for FO, $\Delta P = 0$; for RO, $\Delta P > \Delta\pi$ (Cath *et al.* 2006). This equation is not suitable for FO processes because the parameter A and the reflection coefficient are calculated considering that pressure is being applied to the brine; besides, the driving force considered is the difference between osmotic pressure and the applied hydraulic pressure (ΔP).

For FO, it is necessary to refer to the Lee's equation for low concentration solutions in the porous substructure side of the membrane (Lee *et al.* 1981):

$$\frac{J_1}{A\pi_{\mathrm{Hi}}} = \frac{1 - \left(\frac{C_{\mathrm{Low}}}{C_{\mathrm{Hi}}}\right)\exp(J_1K)}{1 + \frac{B}{J_1}(\exp(J_1K) - 1)} \tag{1.1.3}$$

where A is the water permeation constant for RO and B is the solute permeability coefficient for the RO, J_1 is the flux measured during the experiment, C the concentration of the solutions, π_{Hi} the osmotic pressure of the high concentration solution, and K is the solute resistivity of the membrane.

Instead, Loeb *et al.* derived a formula (Equation 1.1.4) from Lee's equation (Equation 1.1.3) that can be applied to a case in which two different solutions, one with higher osmotic pressure, are kept in contact through a semipermeable membrane, making several important considerations for FO. The π_{Low} is always considerably higher than zero; with this, $A\pi_{\mathrm{Low}}$ and, therefore, $A\pi_{\mathrm{Hi}}$ are much higher than B or J; the relation between the concentrations can be equal to the relation between osmotic pressures ($C_{\mathrm{Hi}}/C_{\mathrm{Low}} = \pi_{\mathrm{Hi}}/\pi_{\mathrm{Low}}$) when restricting the condition of appreciable osmotic pressure on both sides of the membrane.

$$K = \frac{1}{J}\ln\left(\frac{\pi_{\mathrm{Hi}}}{\pi_{\mathrm{Low}}}\right) \tag{1.1.4}$$

where K is the solute resistivity of the membrane, $\pi_{\mathrm{Hi}}/\pi_{\mathrm{Low}}$ the ratio between the osmotic pressure of the high concentration solution and the low concentration solution, and J the flux. The driving force in the case of FO keeps a logarithm ratio between the osmotic pressures, unlike the RO, where the driving force is the only the difference between the osmotic pressures and the pressure generated by the applied force (Loeb *et al.* 1997).

The main advantage of using FO is the low energy required to extract pure water from a wastewater or recycled feed, being only the energy used to recirculate the draw solution (Adham *et al.* 2007). A major limiting factor of FO systems performance is an eventual flux decline due to concentration polarization (McCutcheon *et al.* 2005).

Many FO membrane potential applications have been studied, including desalination (Low, 2009), concentration of dilute industrial wastewater, concentration of landfill leachate, direct potable reuse for advanced life support systems, food processing, pharmaceutical industry processes (Cath *et al.* 2006) and concentration of digested sludge liquids (Holloway *et al.* 2007). Because forward osmosis is gaining interest as a potential technology to reduce the cost of producing fresh water and treating wastewater effluents, a rapid increase (factor of 3) in the number of peer reviewed publications was observed in the last 5 years (Figure 1.1.5).

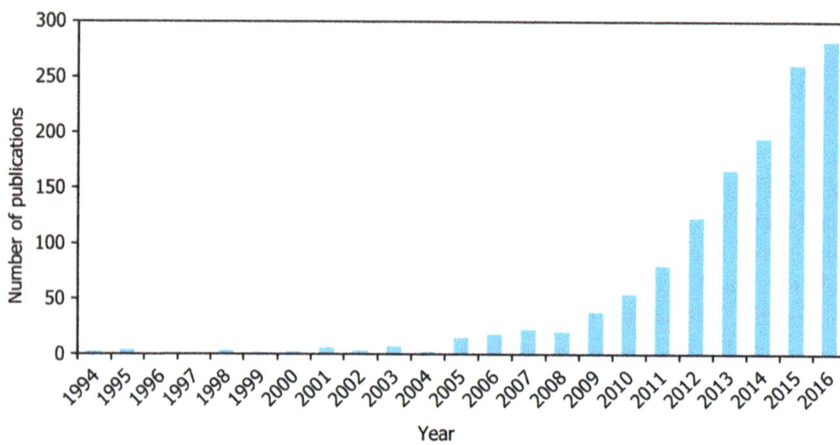

Figure 1.1.5 Development of annual peer reviewed publications on "forward osmosis" as a keyword in the period 1994–2016 (scopus database on: december 2016).

1.1.3 FO SYSTEM FOR WASTEWATER RECOVERY AND SEAWATER DESALINATION

The studies comprised in this work were based on the layout of a hybrid forward osmosis/low pressure reverse osmosis (FO/LPRO) system shown in Figure 1.1.6. Wastewater was used as feed water for the FO process; once concentrated, the wastewater is ready for a post-treatment step to recover energy in form of biogas. The draw solution for the FO process is seawater; once diluted, the seawater is fed into a LPRO unit to remove the remaining dissolved salts and produce fresh water.

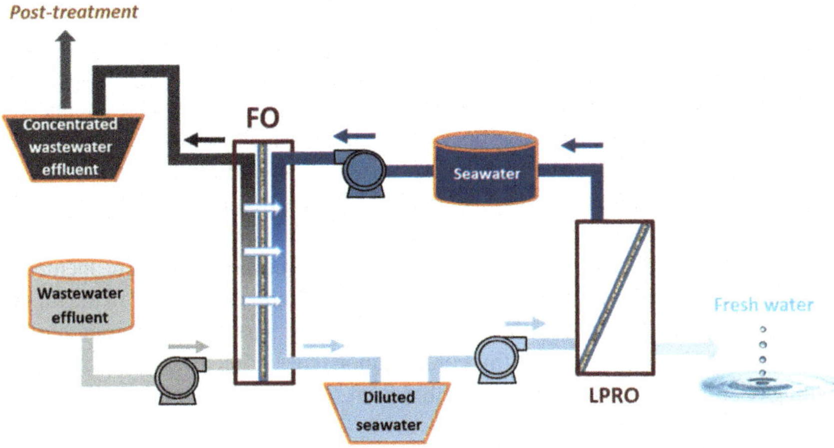

Figure 1.1.6 FO-LPRO system layout combining wastewater recovery and seawater desalination.

To better understand the process, an example of FO filtration is given: using a municipal primary wastewater effluent (conductivity ≈ 2000 μS · cm^{-1}) as feed solution and Red seawater (conductivity ≈ 52,000 μS · cm^{-1}) as draw solution, a 60% seawater dilution can be achieved, resulting in partially desalinated seawater (conductivity ≈ 21,000 μS · cm^{-1}), which can further be treated with a LPRO system (feed pressure ≈ 20 bar), to produce fresh water (conductivity ≈ 250 μS · cm^{-1}) at a lower energy consumption than a traditional RO system (feed pressure ≈ 60 bar). The concentrated wastewater effluent can be further treated in an anaerobic bioreactor to recover biogas.

1.1.4 CONCENTRATION POLARIZATION IN FO MEMBRANES

Concentration polarization (CP) occurs when the difference in salt concentration across the active later is different than the difference in concentration in the bulk solutions itself (McCutcheon & Elimelech, 2006). CP can refer to the formation of a concentration layer at the membrane surface, defined as external concentration polarization (ECP), or in the porous structure of asymmetric membranes, defined as internal concentration polarization (ICP); both reduce the driving force of the osmotic process, and thus, the water flow through the membrane decreases.

ECP can be controlled by several hydrodynamic techniques such as increasing the cross-flow velocity (Sablani *et al.* 2001). ICP, which is one of the major drawbacks in FO, occurs within the membrane support layer (Gray *et al.* 2006), and thus, it is difficult to mitigate, being mainly dependent on the porous support structure.

The effects of ICP for a commercial FO membrane tested with two different orientations, one in which the active layer (dense and thin separation layer) is facing the draw solution (AL-DS), and the other one where the active layer is in contact with the feed solution (AL-FS) have been studied. After several tests, flux decline was up to 15% lower for AL-DS in comparison with the AL-FS orientation (Cath *et al.* 2006). ICP is very minimally affected by changing the hydraulic conditions in the membrane cell, but it is affected by the orientation of the dense layer of the membrane (McCutcheon & Elimelech, 2006). Other studies refer to the overall performance of the membranes and concludes that the optimal configuration to prevent membrane damage is with the active layer facing the feed solution, reducing the ICP in the support layer, because the direction of the water flow is opposite to the concentration of the bulk draw solution into this layer; this configuration is also effective against membrane fouling (Cornelissen *et al.* 2008; Mi & Elimelech, 2008; Wang *et al.* 2010c).

1.1.5 FO MEMBRANE FOULING

Fouling is a severe problem in membranes, reducing the flux considerably, and affecting the overall performance. Fouling supposes that membrane elements have to be cleaned and eventually replaced. Biofouling has been identified as one of the major problems in spiral wound nanofiltration and reverse osmosis membrane module operation (Figure 1.1.7) (Winters & Isquith, 1979; Paul, 1991; Tasaka *et al.* 1994; Flemming *et al.* 1997; Khedr, 2000; Saeed *et al.* 2000; Vrouwenvelder *et al.* 2008; van Loosdrecht *et al.* 2012).

Figure 1.1.7 Biofouling affecting a reverse osmosis spiral wound module used in a water filtration system (photo by johannes vrouwenvelder).

In FO membrane filtration, water flux decline due to fouling is less severe than in RO, because the FO process itself does not induce suspended solids and other organic contaminants into the membrane (Holloway *et al.* 2007). Both reversible and irreversible membrane fouling were found to be negligible when using a FO membrane submerged in secondary wastewater effluent (Cornelissen *et al.* 2008). In a different study with a FO/RO hybrid system, using impaired water as feed solution, results show the low fouling propensity of the FO process, and the ability to treat large volumes of water with almost no need for physical or chemical cleaning (Cath *et al.* 2009). One of the few studies on biofouling in FO membrane filtration suggests that the effect of the biofilm layer is less severe than for RO processes in the same hydrodynamic cross-flow channel conditions (Yoon *et al.* 2013).

Fouling can change the surface characteristics of the membrane, either to improve or degrade the rejection capability and flux. Several studies show both positive and negative impacts for NF and RO membranes (Kimura *et al.* 2003; Bellona *et al.* 2004; Xu *et al.* 2006), but few studies have given insight on FO membranes (Cath *et al.* 2011; Xie *et al.* 2012).

1.1.6 ENERGY DEMAND IN DESALINATION AND WATER TREATMENT PROCESSES

High energy demand for current desalination technologies has limited its use in several regions, representing more than 70% of the operating costs of the plant and up to 50% of the cost of the final product (potable water). For water treatment processes, more than 50% of the energy required is used for the activated sludge aeration; another significant fraction of the energy is used for the pumping system in all the water treatment stages (Metcalf & Eddy, 2003). There are several recommendations that require implementing operational changes and retrofitting equipment to reduce the energy consumption in wastewater treatment plants as well as in desalination facilities; however, all these measures represent a significant expenditure and cannot be applied in many cases (Farooque *et al.* 2008).

One of the main advantages of FO is the limited amount of external energy required to extract water from the feed, only using a very low hydraulic pressure to recirculate the DS on one side of the membrane, while the feed is passively in contact with membrane on the other side (Cath *et al.* 2006). However, it should be noted that special attention must be taken into the water quality, because the dilution might contaminate the water and affect the downstream RO process (Shannon *et al.* 2008). This might turn into an energy intensive solution, having to add an advanced pretreatment process, which turns the research on contaminant removal with FO membranes critical.

1.1.7 SCOPE AND OUTLINE

The objective of this book is the assessment of the recent developments in FO processes, including the performance evaluation of hybrid FO/LPRO systems for wastewater recovery and seawater desalination, characterization of membrane fouling and cleaning, pollutant rejection capabilities of FO membranes, state-of-the-art FO membranes and draw solutions, modeling the effect of concentration polarization and fouling on water flux through the membrane, and the analysis of FO applications/niches and the life cycle cost of a large-scale system compared to conventional desalination and water recovery alternatives.

The book is structured in five themes studying forward osmosis: (i) water recovery, (ii) fouling, (iii) rejection of pollutants, (iv) draw solution and membranes, and (v) modeling. A layout of the structure is shown in Figure 1.1.8.

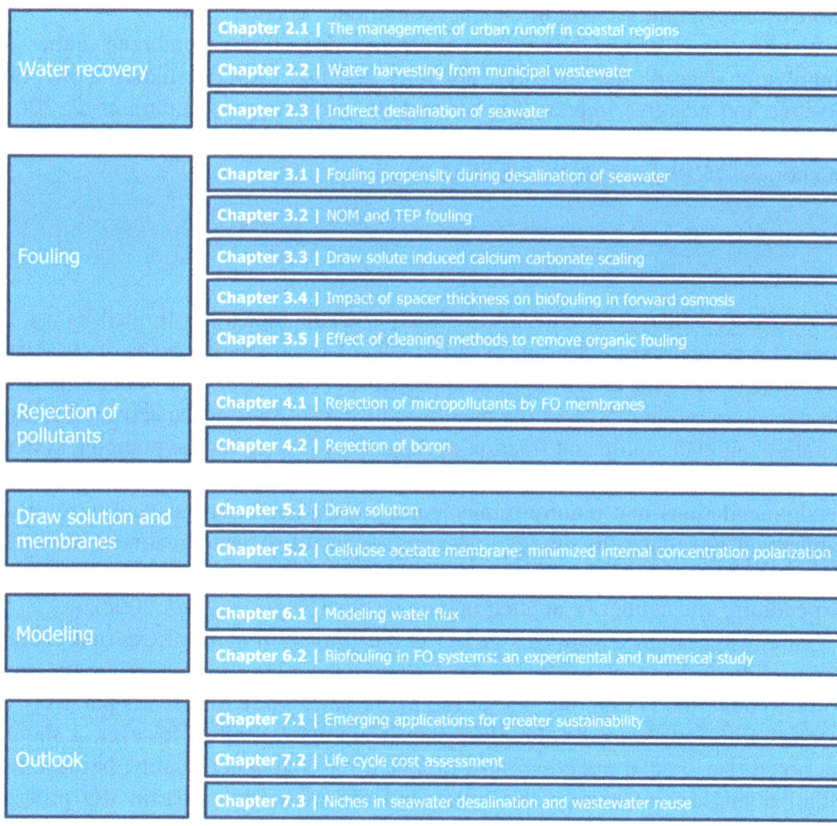

Figure 1.1.8 Book structure.

The first theme (water recovery) describes the general method and systematic studies for the assessment of a combined wastewater concentration and seawater dilution system based on a submerged FO membrane cell. Chapter 2.1 focuses on the use of urban runoff as feed water, Chapter 2.2 explores a municipal primary wastewater effluent as feed water and Chapter 2.3 utilized real seawater as draw solution for the hybrid FO/LPRO system.

The second theme (fouling) gives attention to fouling propensity during seawater desalination, particularly polymerized silica fouling (Chapter 3.1) and NOM and TEP fouling characterization (Chapter 3.2) which was found to be responsible for the reduction of system performance. A novel draw solution induced calcium carbonate scaling mechanism is reported in Chapter 3.3. Chapter 3.4 focuses on biofouling and its effects on forward osmosis, as well as describing the influence of feed spacer on biofouling within a cross-flow cell. Physical and chemical cleaning methods were used and compared to try to set a protocol for FO membrane cleaning (Chapter 3.5).

The third theme (rejection of pollutants) reports the rejection capabilities of different FO membranes to organic micropollutants (Chapter 4.1) and boron (Chapter 4.2).

The fourth theme (draw solution and membranes) comprises a review on the progress of draw solution for FO processes (Chapter 5.1) and the fabrication method for FO membranes with an ultra-thin selective layer and a fully porous support with a high performance in FO processes (Chapter 5.2).

The last theme (modeling) focuses on the impact of external and internal concentration polarization and the value of solute resistance to diffusion on the osmotic driving force, and hence the effective water flux through the FO membrane (Chapter 6.1). Chapter 6.2 describes impact of (i) biofilm properties (thickness, hydraulic permeability, and porosity), biofilm membrane surface coverage, and biofilm location (feed channel, draw channel and both) on FO membrane performance, and (ii) biofouling on external concentration polarization in different FO operation conditions.

An overview is given on low-energy FO applications (Chapter 7.1), including a life cycle cost assessment of a hybrid FO/LPRO large-scale system for seawater desalination and wastewater recovery (Chapter 7.2), and the possible niches in the water industry for forward osmosis processes (Chapter 7.3).

The book is structured as a compilation of published manuscripts, i.e. it consists of a number of scientific articles, except for the introduction chapter. Some repetitions are consequently unavoidable in individual chapters. Small adaptations have been made to improve the chapters.

1.1.8 REFERENCES

Achilli A., Cath T. Y., Marchand E. A. and Childress A. E. (2009). The forward osmosis membrane bioreactor: a low fouling alternative to MBR processes. *Desalination*, **239**, 10–21.

Adham S., Oppenheimer J., Liu L. and Kumar M. (2007). WateReuse Dewatering Reverse Osmosis Concentrate from Water Reuse Applications using Forward Osmosis. WateReuse Foundation, Alexandria, VA, USA.

Amjad Z. (1993). Reverse Osmosis: Membrane Technology, Water Chemistry, and Industrial Applications. International Thomson Publishing, New York.

Bellona C., Drewes J. E., Xu P. and Amy G. (2004). Factors affecting the rejection of organic solutes during NF/RO treatment – a literature review. *Water Research*, **38**, 2795–2809.

Boo C., Elimelech M. and Hong S. (2013). Fouling control in a forward osmosis process integrating seawater desalination and wastewater reclamation. *Journal of Membrane Science*, **444**, 148–156.

Cath T. Y., Childress A. E. and Elimelech M. (2006). Forward osmosis: principles, applications, and recent developments. *Journal of Membrane Science*, **281**, 70–87.

Cath T. Y., Drewes J. E. and Lundin C. D. (2009). A Novel Hybrid Forward Osmosis Process for Drinking Water Augmentation using Impaired Water and Saline Water Sources. Water Research Foundation, Denver, CO.

Cath T. Y., Hancock N., Xu P. and Heil D. (2011). A comprehensive study of micropollutants rejection by forward osmosis and hybrid FO-RO. In 2012 Membrane Technology Conference American Water Works Association, Glendale, Arizona.

Chang I.-S., Le Clech P., Jefferson B. and Judd S. (2002). Membrane fouling in membrane bioreactors for wastewater treatment. *Journal of Environmental Engineering*, **128**(11), 1018–1029.

Choi Y.-J., Choi J.-S., Oh H.-J., Lee S., Yang D. R. and Kim J. H. (2009). Toward a combined system of forward osmosis and reverse osmosis for seawater desalination. *Desalination*, **247**, 239–246.

Cornelissen E. R., Harmsen D., de Korte K. F., Ruiken C. J., Qin J.-J., Oo H. and Wessels L. P. (2008). Membrane fouling and process performance of forward osmosis membranes on activated sludge. *Journal of Membrane Science*, **319**, 158–168.

Creel L. (2003). Ripple Effects: Population and Coastal Regions. Population Reference Bureau, Washington, D.C.

Engelman R., Cincotta R. P., Dye B., Gardner-Outlaw T. and Wisnewski J. (2000). People in the Balance: Population and Natural Resources at the Turn of the Millennium.

Farooque A. M., Jamaluddin A. T. M., Al-Reweli A. R., Jalaluddin P. A. M., Al-Marwani S. M., Al-Mobayed A. A. and Qasim A. H. (2008). Parametric analyses of energy consumption and losses in SWCC SWRO plants utilizing energy recovery devices. *Desalination*, **219**, 137–159.

Flemming H. C., Schaule G., Griebe T., Schmitt J. and Tamachkiarowa A. (1997). Biofouling – the Achilles heel of membrane processes. *Desalination*, **113**, 215–225.

Gray G. T., McCutcheon J. R. and Elimelech M. (2006). Internal concentration polarization in forward osmosis: role of membrane orientation. *Desalination*, **197**, 1–8.

Holloway R. W., Childress A. E., Dennett K. E. and Cath T. Y. (2007). Forward osmosis for concentration of anaerobic digester centrate. *Water Research*, **41**, 4005–4014.

Khedr M. G. (2000). Membrane fouling problems in reverse-osmosis desalination applications. *International Desalination and Water-Reuse Quarterly*, **10**, 8–17.

Kimura K., Amy G., Drewes J. E., Heberer T., Kim T.-U. and Watanabe Y. (2003). Rejection of organic micropollutants (disinfection by-products, endocrine disrupting compounds, and pharmaceutically active compounds) by NF/RO membranes. *Journal of Membrane Science*, **227**, 113–121.

Lee K. L., Baker R. W. and Lonsdale H. K. (1981). Membranes for power generation by pressure-retarded osmosis. *Journal of Membrane Science*, **8**, 141–171.

Li Z., Valladares Linares R., Abu-Ghdaib M., Zhan T., Yangali-Quintanilla V. and Amy G. (2014). Osmotically driven membrane process for the management of urban runoff in coastal regions. *Water Research*, **48**, 200–209.

Loeb S., Titelman L., Korngold E. and Freiman J. (1997). Effect of porous support fabric on osmosis through a Loeb-Sourirajan type asymmetric membrane. *Journal of Membrane Science*, **129**, 243–249.

Low S. C. (2009). Preliminary studies of seawater desalination using forward osmosis. *Desalination and Water Treatment*, **7**(1-3), 41–46.

McCutcheon J. R. and Elimelech M. (2006). Influence of concentrative and dilutive internal concentration polarization on flux behavior in forward osmosis. *Journal of Membrane Science*, **284**, 237–247.

McCutcheon J. R., McGinnis R. L. and Elimelech M. (2005). A novel ammonia – carbon dioxide forward (direct) osmosis desalination process. *Desalination*, **174**, 1–11.

Metcalf and Eddy. (2003). Wastewater Engineering Treatment and Reuse, 4th edn., McGraw Hill, Columbus, OH.

Mi B. and Elimelech M. (2008). Chemical and physical aspects of organic fouling of forward osmosis membranes. *Journal of Membrane Science*, **320**, 292–302.

NASA. (2000). Astronomy Picture of the Day. http://apod.nasa.gov/apod/ap001127.html

Paul D. (1991). Reverse osmosis: scaling, fouling and chemical attack. *Desalination Water Reuse*, **1**, 8–11.

Prüss-Üstün A. and Corvalá C. (2004). Preventing Disease through Healthy Environments towards an Estimate of the Environmental Burden of Disease. World Health Organization, Geneva, Switzerland.

Ridgway H. F. and Flemming H.-C. (1996). Membrane Biofouling in Water Treatment Membrane Processes. AWWA Research Foundation, Water Research Commission, South Africa.

Sablani S. S., Goosen M. F. A., Al-Belushi R. and Wilf M. (2001). Concentration polarization in ultrafiltration and reverse osmosis: a critical review. *Desalination*, **141**, 269–289.

Saeed M. O., Jamaluddin A. T., Tisan I. A., Lawrence D. A., Al-Amri M. M. and Chida K. (2000). Biofouling in a seawater reverse osmosis plant on the Red Sea coast, Saudi Arabia. *Desalination*, **128**, 177–190.

Shannon M. A., Bohn P. W., Elimelech M., Georgiadis J. G., Marinas B. J. and Mayes A. M. (2008). Science and technology for water purification in the coming decades. *Nature*, **452**, 301–310.

Tasaka K., Katsura T., Iwahori H. and Kamiyama Y. (1994). Analysis of RO elements operated at more than 80 plants in Japan. *Desalination*, **96**, 259–272.

The Barila Group, The Coca-Cola Company, The International Finance Corporation, McKinsey & Company, Nestle S.A., New Holland Agriculture, SABMiller plc, Standard Chartered Bank, and Syngenta AG. (2009). Charting Our Water Future, Economic Frameworks to Inform Decision-Making. 2030 Water Resources Group, Washington, DC.

Tibbetts J. (2002). Coastal cities: living on the edge. *Environmental Health Perspectives*, **110**, A674–A681.

United Nations (2011). Population Distribution, Urbanization, Internal Migration and Development: An International Perspective. Department of Economic and Social Affairs, Population Division, New York.

van Loosdrecht M. C., Bereschenko L., Radu A., Kruithof J. C., Picioreanu C., Johns M. L. and Vrouwenvelder H. S. (2012). New approaches to characterizing and understanding biofouling of spiral wound membrane systems. *Water Science and Technology*, **66**, 88.

van't Hoff J. H. (1887). The function of osmotic pressure in the analogy between solutions and gases. *Proceedings of the Physical Society of London*, **9**, 307.

Vrouwenvelder J. S., Manolarakis S. A., van der Hoek J. P., van Paassen J. A. M., van der Meer W. G. J., van Agtmaal J. M. C., Prummel H. D. M., Kruithof J. C. and van Loosdrecht M. C. M. (2008). Quantitative biofouling diagnosis in full scale nanofiltration and reverse osmosis installations. *Water Research*, **42**, 4856–4868.

Wang Y., Wicaksana F., Tang C. Y. and Fane A. G. (2010). Direct microscopic observation of forward osmosis membrane fouling. *Environmental Science & Technology*, **44**, 7102–7109.

Winters H. and Isquith I. R. (1979). In-plant microfouling in desalination. *Desalination*, **30**, 387–399.

Xie M., Price W. E. and Nghiem L. D. (2012). Rejection of pharmaceutically active compounds by forward osmosis: role of solution pH and membrane orientation. *Separation and Purification Technology*, **93**, 107–114.

Xu P., Drewes J. E., Kim T.-U., Bellona C. and Amy G. (2006). Effect of membrane fouling on transport of organic contaminants in NF/RO membrane applications. *Journal of Membrane Science*, **279**, 165–175.

Yangali-Quintanilla V., Li Z., Valladares R., Li Q. and Amy G. (2011). Indirect desalination of Red Sea water with forward osmosis and low pressure reverse osmosis for water reuse. *Desalination*, **280**, 160–166.

Yoon H., Baek Y., Yu J. and Yoon J. (2013). Biofouling occurrence process and its control in the forward osmosis. *Desalination*, **325**, 30–36.

Part 2
Water Recovery

Chapter 2.1

The management of urban runoff in coastal regions*

2.1.1 INTRODUCTION

Globally, the level of urbanization has significantly increased during the last five decades and coastal regions are being developed in many countries (Tibbetts, 2002; Creel, 2003). Thus, there is a convergence of urban and coastal development reflected in megacities, such as Manila and Jakarta. The concentrated population in coastal regions has placed great stress on the interaction of water management and the local coastal environment, including the scarcity of fresh/clean water, contamination of surface and/or ground water, and input of pollutants and nutrients to the ocean. Therefore, a comprehensive strategy for the management of water resource is crucial for the sustainable development of these areas.

Accompanying rapid urbanization, urban runoff has been one of the leading causes of quality impairment of surface waters (rivers, lakes and wetlands) (Boller, 1997; EPA, 2000). Urban runoff has also been shown to be a main contributor to coastal water pollution (Novotny, 1995). When precipitation occurs, part of it becomes runoff originating from streets, parking lots, airports, construction sites, roofs and lawns, and contains a wide range of chemical and biological contaminants including organics (e.g., oil and grease, humic substances, hydrocarbons), inorganics (e.g., trace metals, silica and nutrients) and pathogens, which often cause severe water quality problems that result in detrimental impacts on the ecology and biodiversity of receiving water bodies (Schiff *et al.* 2003; Walsh, 2004). Management of urban runoff should be not only to attenuate excessive flow (i.e., flooding) during storm events but also to sustain

* This chapter is based, with permission from the copyright holder, on the previously published paper. (Li Z., Valladares Linares R., Abu-Ghdaib M., Zhan T., Yangali-Quintanilla V. and Amy G. (2014). Osmotically driven membrane process for the management of urban runoff in coastal regions. *Water Research*, **48**, 200–209. doi:10.1016/j.watres.2013.09.028)

water resources (EPA, 2000). For example, Singapore plans to collect storm water from 2/3 of the island's area, store it in reservoirs like the new Marina Barrage, and membrane filter it before sending it to drinking water supply (Public Utilities Board of Singapore, 2007).

Although the treatment and disposal of urban runoff have received increasing attention, methods and practices for the treatment of urban runoff are still limited since many urban runoff drainage systems are normally separate from the sewage system and are intended to divert runoff water straight to natural water bodies (such as rivers, lakes and the ocean) to avoid area flooding. Several methods have been employed in the treatment of urban runoff, especially for the removal of trace metals (Table 2.1.1). Potentially, large treatment facilities with large amounts of absorbents and chemicals are required due to the large volume of runoff water, although flow equalization can be practiced. The removal rates of contaminants vary according to site and season, and are low to moderate in most cases (Table 2.1.1).

Membrane technology has been widely applied in water treatment. However, it has not previously been employed to remove contaminants from urban runoff in large-scale applications even though microfiltration (MF) and ultrafiltration (UF) can remove most particles, bacteria and macromolecules while nanofiltration (NF) and reverse osmosis (RO) are reliable barriers to many trace metals and organic pollutants. The major reason is that the cost of energy in these pressure driven membrane processes is still high.

Forward osmosis (FO) is an emerging membrane process which has been explored for novel applications (Cath *et al.* 2006; Chung *et al.* 2012). FO or FO hybrids have demonstrated abilities to treat impaired quality waters as well as desalinate seawater and brackish water (Choi *et al.* 2009; Hancock *et al.* 2011; Li *et al.* 2012; Zhao *et al.* 2012), potentially showing a competitive separation performance to that of NF or RO. By utilizing the osmotic pressure (osmotic energy) as the driving force, FO has unique attributes including less fouling, especially irreversible fouling, minimized energy input and low operating and maintenance cost. However, it has not been previously reported to employ FO in the treatment of urban runoff.

The present study proposes a novel strategy for the management of urban runoff in coastal regions by using FO as a barrier to reject runoff-derived contaminants (Figure 2.1.1). The main components of the system include an urban runoff drainage system which has already been developed in many cities, a constructed osmotic detention pond, submerged FO membrane cassettes and a seawater recirculation system. Seawater is employed as the draw solution (DS) to provide a natural osmotic pressure created by the salinity in seawater. Urban runoff, the feed solution (FS), is collected in an osmotic detention pond by the runoff drainage system and then concentrated in the pond. The fresh water is extracted from urban runoff to seawater by osmotic pressure. The concentrated urban runoff with contaminants can be collected for further treatment and the diluted seawater for further utilization.

Table 2.1.1 Removal efficiency of pollutants and nutrients from urban runoff by various methods.

Methods	Removal of Target Solutes (%)									References
	Zn	Pb	Cu	Cd	Ni	Cr	Mn	P	N	
Wetland	13–100	0–94	0–92	0–100	17–52	43–99	n.a.	0–89	0–46	(Scholes *et al.* 1998; Carleton *et al.* 2001)
Bioretention/ Biofiltration	64–95	70–95	43–97	n.a.	n.a.	n.a.	n.a.	50–80	30–50	(Davis *et al.* 2003; Bratieres *et al.* 2008; Aryal *et al.* 2010)
Settling basin	18–42	58–70	20–50	n.a.	n.a.	n.a.	n.a.	n.a.	n.a.	(Berbee *et al.* 1999;
Sand/gravel filtration	14–60	51–83	32–70	n.a.	n.a.	n.a.	n.a.	40–70	30–50	Berbee *et al.* 1999; Hatt *et al.* 2007)
Zero-valent iron	75–90	n.a.	90	n.a.	n.a.	n.a.	n.a.	n.a.	n.a.	(Rangsivek & Jekel, 2005)
Reactive filter	27–49	11–31	6–51	0–41	0–11	0–27	53–84	n.a.	n.a.	(Renman & Hallberg, 2007)

n.a., not available.

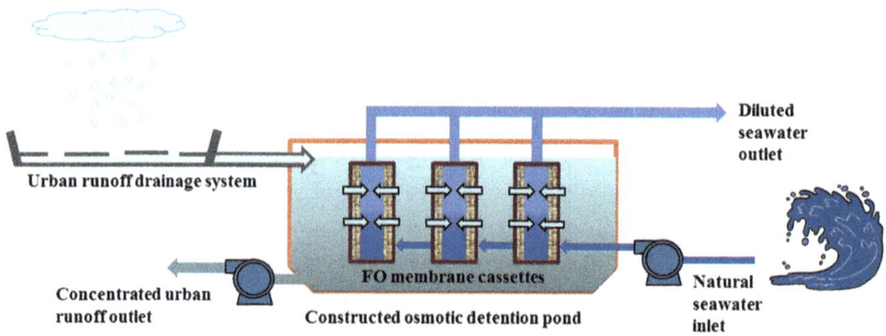

Figure 2.1.1 Diagram of the osmotic detention pond for the treatment of urban runoff in coastal regions.

In lab scale testing, a simulated osmotic detention pond is employed to identify the performance of the process. This research investigated the sustainability of the process and removal efficiency of trace metals and nutrients from synthetic urban runoff. The experimental data and simulated osmotic pond can potentially be scaled up and operated with current urban runoff drainage systems and reservoirs (ponds or basins) developed in many temperate or tropical coastal regions such as Europe and Southeast Asia.

2.1.2 MATERIALS AND METHODS

2.1.2.1 Synthetic urban runoff and seawater

Synthetic urban runoff (SUR) was prepared by a method reported previously with some modifications (Davis *et al.* 2003), and used as the feed water/solution (FS). The recipe of SUR is described in Table 2.1.2. The target contaminants include trace metals and nutrients. The DS was a synthetic seawater (SSW) corresponding to a 4% NaCl solution.

2.1.2.2 FO membrane and simulated osmotic detention pond

A commercial flat sheet FO membrane was used in this study. The membrane is made of cellulose triacetate embedded about a polyester screen mesh (HTI, Scottsdale, AZ). The active layer is the shiny side of the membrane (Li *et al.* 2012). Before using, the membrane was soaked in deionized water for 24 hours to remove the glycerin which was used to replace the water during shipment.

The membrane cell was a custom-made plate and frame assembly. The cell sandwiched 2 pieces of the flat-sheet membranes with a total area of 140 cm^2 in a plate and frame configuration. The active layer of the membrane faced the FS, and the support layer faced the DS.

Table 2.1.2 Recipe of synthetic urban runoff.

Target Pollutants	Chemicals	Concentration ($mg \cdot L^{-1}$)
Trace metals		
Cu	$CuSO_4$	0.08
Zn	$ZnCl_2$	0.6
Cd	$CdCl_2$	0.001
Pb	$PbCl_2$	0.08
Nutrients		
Nitrate	$NaNO_3$	2 (as N)
Organic nitrogen	Glycine	4 (as N)
Phosphorus	Na_2HPO_4	0.6 (as P)
Dissolved solids	$CaCl_2$	120

Note: 1 M HCl was used to maintain pH of 5.5 in acidic condition testing. 1 $mg \cdot L^{-1}$ (as DOC) humic acid (HA) was added in NOM fouling condition testing.

The experimental set-up of a simulated osmotic detention pond is shown in Figure 2.1.2. The membrane cell was immersed in a feed tank containing SUR and was connected to a DS reservoir containing SSW. A gear pump (Coleparmer) was used to continuously recirculate DS inside the lumen space of the cell, extracting water from SUR into DS reservoir. In order to homogenize the feed water in the feed tank, SUR in the feed tank was circulated by a gear pump as well at a flow rate of 30 $L \cdot h^{-1}$. The temperature of both the DS reservoir and feed tank was controlled at 21 ± 1 °C by heat exchange coils connected to the digital recirculating temperature controller (1156D, VWR). During the FO process, a low flow rate is preferred to reduce the energy consumption. Therefore the crossflow velocity of SSW was maintained at 4.17×10^{-4} $m \cdot s^{-1}$. Process parameters including conductivity, pH, temperature, flow rate and pressure were controlled and measured *in situ*.

Three processes were tested to model different urban runoff conditions. The neutral condition was performed at pH of 6.9 ± 0.1, containing compositions shown in Table 2.1.2. The acidic condition was performed at pH of 5.5 ± 0.1 by adding 1 M HCl to SUR to simulate more acidic rainfall. The natural organic matter (NOM) fouling condition was performed by adding and adjusting the initial dissolved humic acid (HA) (Sigma-Aldrich, Germany) concentration to 1 $mg \cdot L^{-1}$ (as dissolved organic carbon, DOC) in SUR with the pH maintained at 6.9 ± 0.1. The experiments were conducted in a semi-batch mode. Each process was initiated with 20 L SUR in the feed tank and 1 L SSW in DS reservoir. Each batch was interrupted when 2 L of permeate was extracted from SUR to SSW. Then, diluted SSW was replaced by fresh aliquots while concentrated SUR was continuously used to the next batch until 7 batches were completed. In other words, 14 L permeate was extracted from SUR and the recovery rate was 70% in each process.

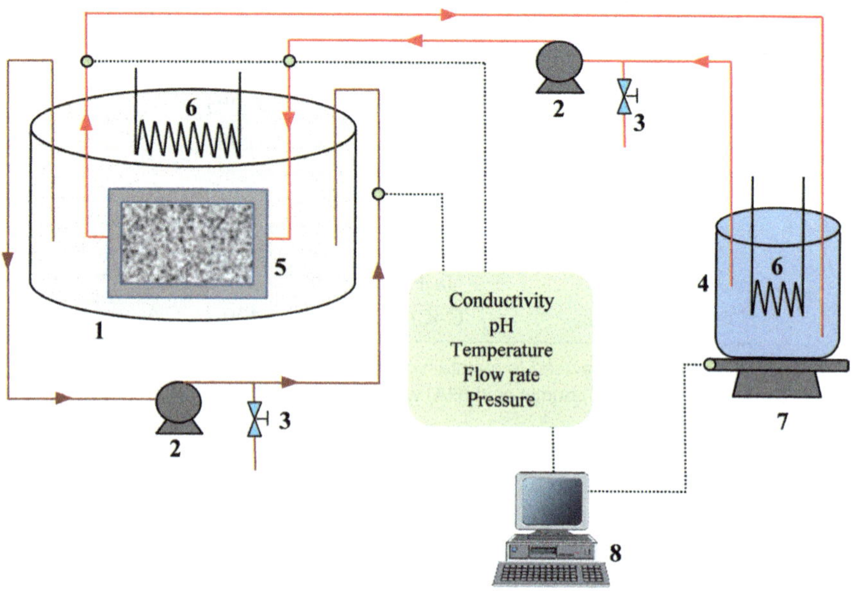

Figure 2.1.2 Experimental set-up of a simulated osmotic detention pond (1, Feed tank; 2, Pump; 3, By-pass valve; 4, Draw solution reservoir; 5, Membrane cell; 6, Heat exchange coil; 7, Digital balance; 8, Data acquisition system).

2.1.2.3 Analytical methods

The variation of DS volume was measured by a digital balance (Mettler Toledo) and converted to the permeate flux. The first hour of each batch was used to stabilize the system, then the initial flux (J_{ini}) was detected. The average flux (J_{ave}) was calculated from the volume of water collected in the DS reservoir and the time of each batch. The salt leakage from the DS side to SUR was estimated by measuring the conductivity in the DS reservoir and converted to the salt amount. The loss of salt in each batch was considered to be the salt leakage to the feed tank. The trace metals were analyzed by Inductively Coupled Plasma/Mass Spectrometry (ICP-MS 7500; Agilent, US). DOC and total nitrogen (TN) was measured by an Organic Carbon Analyzer (TOC-VCPH, Shimadzu, Japan). Phosphorus and nitrate were analysed by Hach TNT reagents with a spectrophotometer (DR, 2800; Hach). The NOM fouled membrane was visualized by a Scanning Electron Microscopy (SEM, Magellan, FEI).

2.1.3 RESULTS AND DISCUSSION

2.1.3.1 Effects of feed water condition on flux patterns

Flux patterns were investigated under three feed conditions (i.e., neutral, acidic and NOM fouling) with a semi-batch mode. The period of the FO process varied from 360 to 410 hours under each feed condition to complete 7 batches for a total recovery rate of 70% (Figure 2.1.3). The flux in each batch of all feed conditions decreased rapidly. The rate of flux decline varied from 48% to 67%, mainly due to the loss of osmotic pressure by diluting SSW continuously in each batch.

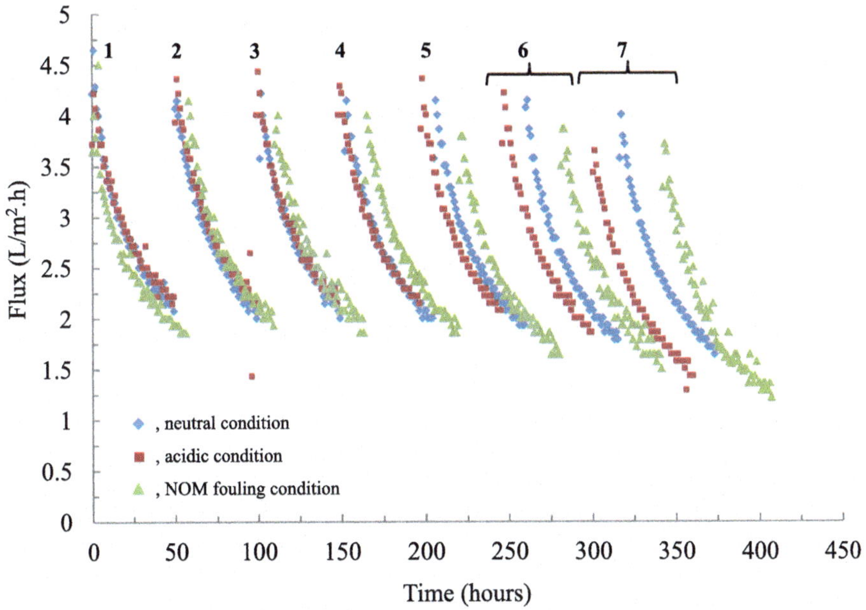

Figure 2.1.3 Flux patterns in the simulated osmotic detention pond under different feed water conditions (1–7, the operating batch number).

During a large scale application with natural seawater as the DS, it would be possible to maintain a constant salinity at the DS side by continuously supplying seawater into the system. Therefore, the average flux (J_{ave}) and initial flux (J_{ini}) of each batch in this study are more meaningful to prove the sustainability of the system (Figure 2.1.4). The conductivity of feed water was monitored to indicate the ionic concentration. The variation of conductivity in feed water was more significant under the acidic and NOM fouling condition than the neutral condition. The initial conductivity of feed water was about

0.2 mS · cm^{-1} before FO process and increased to 4.2, 5.3 and 7.8 mS · cm^{-1} after 7 batches of FO process under the neutral, acidic and NOM fouling condition, respectively.

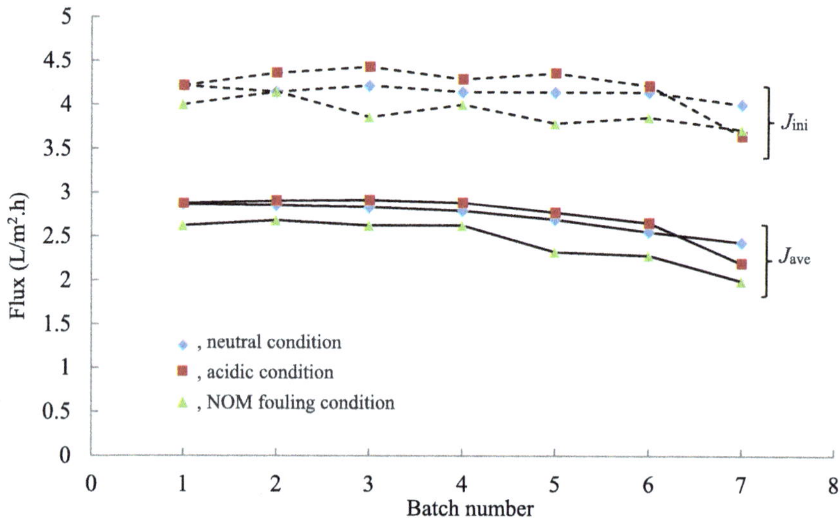

Figure 2.1.4 Effect of feed water condition on the average (J_{ave}) and initial (J_{ini}) flux.

The decline of J_{ave} and J_{ini} were much lower than those of flux versus time in each batch. The decline of J_{ave} between the first batch and the last batch was 15% under the neutral condition, 23.7% under the acidic condition and 24% under NOM fouling condition. J_{ini} was more stable. The decline of J_{ini} between the first and last batch was 4.9% under the neutral condition, 13.5% under the acidic condition and 7.3% under NOM fouling condition. During the FO process, apart from the intrinsic resistance of membrane, J_{ini} should be dominated by the ionic concentration (or salt concentration) of a feed water and DS while J_{ave} is limited by fouling, the internal concentration polarization (ICP) inside the membrane and external concentration polarization (ECP) near the membrane surface at the DS and FS side. Adding acid to SUR increased the ionic concentration in the feed water while HA contributed to fouling of the membrane. Therefore, J_{ave} and J_{ini} were more stable under the neutral condition than the acidic and NOM fouling conditions. Although the comparatively low flux (up to about 2.9 L · m^{-2} · h^{-1} for J_{ave} and 4.4 L · m^{-2} · h^{-1} for J_{ini}) was observed in this study, the increase of flux can be expected by further optimizing the dilution rate of DS, changing the membrane orientation (e.g. active layer facing the DS) and developing better FO membrane. The remained osmotic pressure in concentrated urban runoff after 7 batches of FO process was 2.0, 2.5 and

3.5 bar under the neutral, acidic and NOM fouling condition, respectively. Compared to the osmotic pressure of 26.1 bar in natural seawater (Table 2.1.3), there is enough driving force (osmotic pressure) for FO process to further increase the recovery rate to 80–90%.

Table 2.1.3 The linkage of osmotic pressure and energy for desalination between the osmotic detention pond and SWRO.

	Natural Seawater	Osmotically Diluted Seawater	Original Urban Runoff	Osmotically Concentrated Urban Runoff
TDS (ppm)[a]	35,627	11,876	161.61	1616.11
Osmotic pressure (bar)[b]	26.14	8.71	0.13	1.25
Operational pressure for SWRO (bar)[c]				
Number of elements: 1	52.43	23.34	–	–
Number of elements: 7	32.82	11.62	–	–
Specific energy for SWRO ($kWh \cdot m^{-3}$)[c]				
Number of elements: 1	12.23	5.49	–	–
Number of elements: 7	7.74	2.83	–	–

[a]The calculations are made using a recovery rate of 90% and rejection rate of 100% in the osmotic detention pond, and a dilution factor of 3 for seawater in each batch.
[b]The osmotic pressure is evaluated by Lenntech Osmotic Pressure Calculator (www.lenntech.com/calculators/osmotic/osmotic-pressure.htm).
[c]The calculations are made by DOW Reverse Osmosis System Analysis (ROSA) using membrane SW 30–4040 from DOW Filmtec with a water production of 4.50 $m^3 \cdot day^{-1}$, 1 stage and 1 vessel operation.

2.1.3.2 Salt leakage and NOM fouling

The membrane provided a reliable barrier to prevent NaCl leakage from the DS to FS through the process. The leakage rate of NaCl from the DS to FS was less than 3% under all conditions tested (Figure 2.1.5). The neutral condition led to a relatively higher salt leakage rate (varied from 1.7% to 2.4%) than acidic and NOM fouling conditions. The reverse salt flux was 1.35 $g \cdot m^{-2} \cdot h^{-1}$ in the first batch and 1.42 $g \cdot m^{-2} \cdot h^{-1}$ in the last batch of FO process under the neutral condition. The salt leakage was only observed in the early stages of FO under the acidic and NOM fouling conditions, then became negligible or not detectable in following batches. Increased negative charge of membrane surface under either acidic or NOM (HA in this study) fouling condition may have resulted in a higher rejection of NaCl by enhanced electrostatic repulsion and electroneutrality (Childress & Deshmukh, 1998; Fan *et al.* 2001; Shim *et al.* 2002).

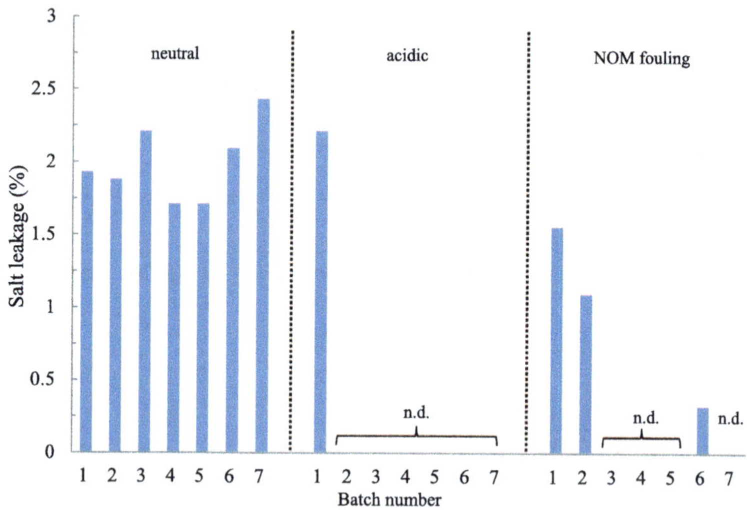

Figure 2.1.5 Effect of feed water condition on the salt leakage (n.d., not detectable).

The formation of a HA fouling layer on the membrane in the simulated osmotic detention pond is revealed by SEM images. Although there was not a hydraulic pressure applied during the FO process, HA fouling layer still formed on the membrane surface at the FS side. The magnified image suggests that this HA fouling layer was very porous (Figure 2.1.6). The HA layer did not cause a significant resistance to the permeation of water indicated by J_{ini} in Figure 2.1.4. Only 7.3% reduction of J_{ini} was observed between the first and the last batch of FO process under NOM fouling condition. The minor decline of J_{ini} under NOM fouling condition could be explained by a critical flux concept in FO process. The critical flux concept has been applied in pressure-driven membrane process, stating that the impact of fouling is negligible when the permeate flux is lower than a critical value (Bacchin *et al.* 2006). Recent researches of FO process revealed that this concept is also applicable in FO process (Zou *et al.* 2011; Zou *et al.* 2013). In this study, the fluxes were low and likely below the critical flux. The deposit of HA on the membrane surface was mainly attributed to the adsorption rather than the drag force induced by the flux. The lack of strong crossflow (shear stress) near the membrane surface in the submerged FO process could facilitate the adsorption and deposit of HA on the membrane. However the porous structure of this fouling layer may obstruct the diffusion of ions (salts) from membrane surface into the feed bulk during FO process. The accumulation of ions in the fouling layer resulted in not only a lower apparent slat leakage from DS to FS comparing to neutral condition, but also an elevated osmotic pressure near the membrane surface which led to a drop in the net driving force (Boo *et al.* 2012). This caused relatively lower flux under NOM fouling condition comparing to neutral and acidic conditions as shown in the flux patterns (Figure 2.1.3 and 2.1.4).

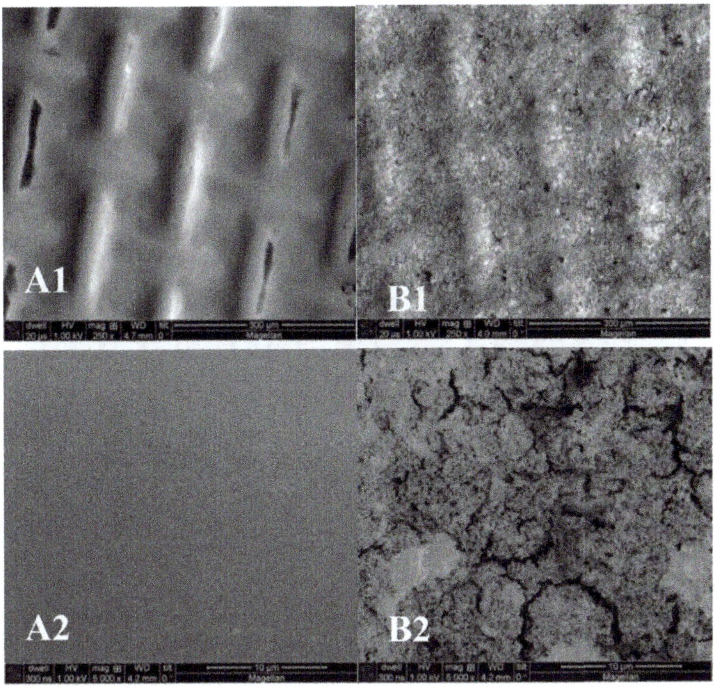

Figure 2.1.6 Membrane fouling by humic acid (A1 and A2, the original clean membrane; B1 and B2, fouled membrane; scale bar 300 μm in A1 and B1, 10 μm in A2 and B2).

2.1.3.3 Rejection of trace metals and nutrients

The simulated osmotic detention pond provided a stable and excellent capacity to reject all trace metals (Cu, Pb, Cd and Zn) tested in this study (Figure 2.1.7). The different feed conditions did not affect the rejection rate of trace metals. A nearly complete rejection of trace metals was observed through all testing conditions. The rejection rate showed a minor variation from 98% to 100%. Furthermore, it has been proven that FO membrane is also able to reject most of trace organic contaminants (TrOCs). Rejections were moderate for hydrophilic neutral and hydrophobic neutral TrOCs, and high for the hydrophilic ionic TrOCs (Valladares Linares *et al.* 2011; Xie *et al.* 2012).

The rejection of nutrients was evaluated by determining phosphorus, nitrate and TN in FS and DS (Figure 2.1.8). The highest rejection was obtained for phosphorus. Nearly 100% rejection of phosphorus was reached under all feed water conditions and operational batches, which can be explained by the steric hindrance and electrostatic repulsion based on the molecular size and negative charge of phosphate (Leo *et al.* 2011).

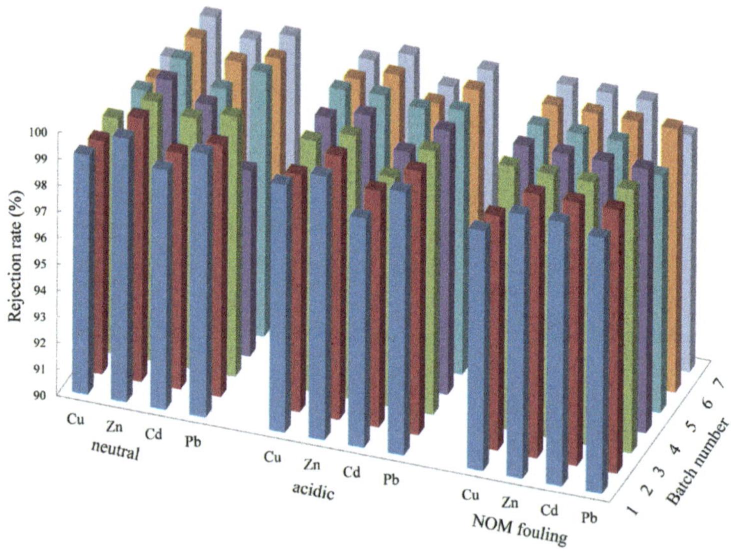

Figure 2.1.7 Rejection of trace metals in the simulated osmotic detention pond.

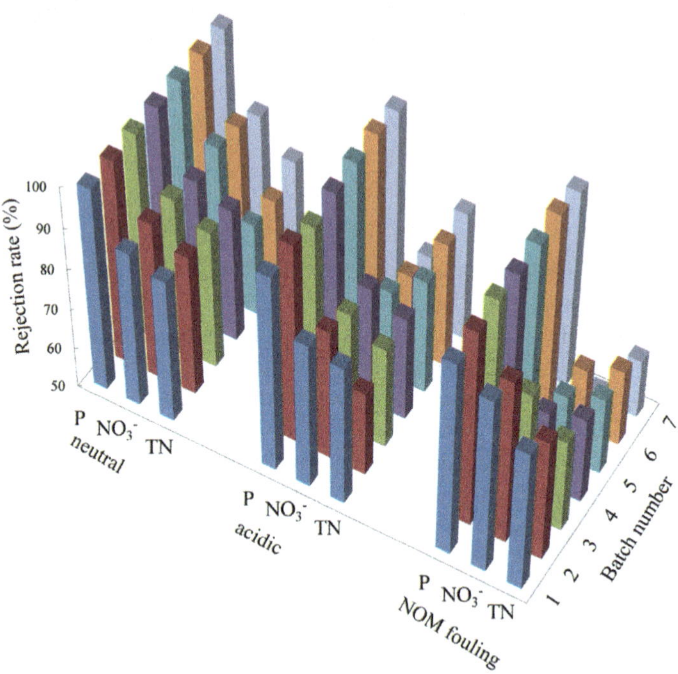

Figure 2.1.8 Rejection of nutrients in the simulated osmotic detention pond.

The rejection of nitrate was more susceptible to the change of pH and formation of a fouling layer on the membrane. The average rejection rate of nitrate in each batch decreased with batch sequence under acidic and NOM fouling conditions while it was relatively stable under neutral condition. The average rejection of nitrate was 88.4%, 86.2% and 93.9% in the first batch of neutral, acidic and NOM fouling condition, respectively, and reduced to 82.9%, 66.9% and 52.1% in the last batch of each condition. Adding acid enhanced the negative charge of the membrane surface while concentrative ECP at the FS side, promoted by the presence of HA in the feed water, can increase the concentration of nitrate at the membrane surface. Therefore, the transport of nitrate was enhanced under the acidic and NOM fouling conditions (Paugam *et al.* 2004).

In the case of TN, the rejection rates varied from 85.4% to 71.9% under the neutral condition, from 84.7% to 71.1% under the acidic condition, and from 85.1% to 64.7% under NOM fouling condition. The rejection rate of TN decreased with batch number under NOM fouling condition while there were no clear correlations between rejection rate and batch number under neutral and acidic conditions. TN comprised the nitrogen from both nitrate and glycine (used as a model organic nitrogen compound in SUR). Glycine is an amino acid which has the smallest side-chain (a hydrogen) and has much less steric hindrance than the other amino acids. A change of pH in the feed did not significantly impact the rejection of glycine since the predominant form of glycine in the pH range (5.5 and 6.8) used in this study was a zwitterion which was not affected by electrostatic interactions (Gotoh *et al.* 2004). In addition, it has been proven that the FO membrane used in this study provides a high rejection of humic-like substance (Li *et al.* 2012). Thus the transport of TN to the DS was also mainly attributed to nitrate and glycine under the NOM fouling condition with HA in the feed water. If diluted DS is used for agriculture, such as for salt tolerant crops, nitrogen is an attribute.

2.1.3.4 Environmental and economic implications

During the management of urban runoff, a sustainable solution to the water-energy nexus can be created by integrating the osmotic detention pond with post-treatment of concentrated urban runoff and diluted seawater (Figure 2.1.9). The variation of TDS and the osmotic pressure in the osmotic detention pond, and specific energy required for subsequent seawater reverse osmosis (SWRO) process were evaluated in Table 2.1.3.

The contaminants in urban runoff can be enriched in the pond and finally cost-effectively treated by simple methods such as lime precipitation. The process utilizes natural osmotic energy, minimizes the amounts of chemicals and absorbents, and avoids secondary contamination to wetlands if used (Scholes *et al.* 1998; Carleton *et al.* 2001). In addition, the process efficiency and sustainability can be significantly improved by developing novel FO membranes. For example, a newly developed 3-bilayer FO membrane has an impressive pure water flux of 28 $L \cdot m^{-2} \cdot h^{-1}$ using 0.5 M $MgCl_2$ as the DS (Cui *et al.* 2013). Novel thin-film composite hollow fiber FO membranes can achieve pure water flux up to about $40 \text{ L} \cdot \text{m}^{-2} \cdot \text{h}^{-1}$ using 2 M NaCl as DS with different membrane orientations (Wang *et al.* 2010; Sukitpaneenit & Chung, 2012).

Figure 2.1.9 An integrated urban runoff management involving FO hybrids in coastal regions for a sustainable solution to the water-energy nexus (1, Osmotic detention pond; 2, Seawater collection basin; 3, Urban runoff collection basin; 4, Low pressure RO (LPRO) plant for post-treatment of diluted DS; 5, Pressure retarded osmosis (PRO) plant; 6, Salt tolerant crops irrigation; 7, Diluted seawater outlet (optional); 8, Seawater intake; 9, Ocean; 10, Urban runoff intake; 11, Urban runoff drainage system).

The diluted seawater permits the use of a low-pressure RO membrane to significantly reduce the energy cost in SWRO process. It is also possible to recover energy from the DS side by a pressure retarded osmosis process (PRO) (Hoover *et al.* 2011). In addition to being the feed in SWRO, the diluted seawater can also be used to irrigate salt-tolerant plants/crops in semiarid coastal areas, called saline farming or saline agriculture (Hillel, 2000; Zhu, 2001; Fedoroff *et al.* 2010). A 40% seawater irrigation solution was safe to ensure economic yield for the 100 day growth period of sunflower (Tang *et al.* 2004). Some selected glycophytes such as *Chamaerops*, *Olea*, *Vetiveria* and *Elytrigia*, can have good growth with the irrigation by 40–60% seawater (Sardo, 2005).

2.1.4 SUMMARY

An osmotic detention pond was proposed for the management of urban runoff in coastal regions. Forward osmosis was employed as a bridge to utilize natural osmotic energy from seawater for concentrating and reusing urban runoff water, and as a barrier to reject runoff-derived contaminants. The process was demonstrated by a lab scale testing using synthetic urban runoff (as the feed solution) and synthetic seawater (as the draw solution). The submerged forward osmosis process was conducted under neutral, acidic and natural organic matter fouling condition, respectively. Forward osmosis flux decline was mainly attributed to the dilution of seawater during a semi-batch process in lab scale testing. However, it is possible to minimize flux decrease by maintaining a constant salinity at the draw solution side. Various changes in urban runoff water quality, including acidic conditions (acid rain) and natural organic matter presence, did not show significant effects on the rejection of trace metals and phosphorus, but influenced salt leakage and the rejection of nitrate and total nitrogen. Rejection of trace metals varied from 98% to 100%, phosphorus varied from 97% to 100, nitrate varied from 52% to 94% and total nitrogen varied from 65% to 85% under different feed water conditions. The work described in this study contributes to an integrated system of urban runoff management, seawater desalination and possible power generation in coastal regions to achieve a sustainable solution to the water-energy nexus.

2.1.5 REFERENCES

Aryal R., Vigneswaran S., Kandasamy J. and Naidu R. (2010). Urban stormwater quality and treatment. *Korean Journal of Chemical Engineering*, **27**, 1343–1359.

Bacchin P., Aimar P. and Field R. W. (2006). Critical and sustainable fluxes: theory, experiments and applications. *Journal of Membrane Science*, **281**, 42–69.

Berbee R., Rijs G., de Brouwer R. and van Velzen L. (1999). Characterization and treatment of runoff from highways in the Netherlands Paved with impervious and pervious asphalt. *Water Environment Research*, **71**, 183–190.

Boller M. (1997). Tracking heavy metals reveals sustainability deficits of urban drainage systems. *Water Science and Technology*, **35**, 77–87.

Boo C., Lee S., Elimelech M., Meng Z. and Hong S. (2012). Colloidal fouling in forward osmosis: role of reverse salt diffusion. *Journal of Membrane Science*, **390–391**, 277–284.

Bratieres K., Fletcher T. D., Deletic A. and Zinger Y. (2008). Nutrient and sediment removal by stormwater biofilters: a large-scale design optimisation study. *Water Research*, **42**, 3930–3940.

Carleton J. N., Grizzard T. J., Godrej A. N. and Post H. E. (2001). Factors affecting the performance of stormwater treatment wetlands. *Water Research*, **35**, 1552–1562.

Cath T. Y., Childress A. E. and Elimelech M. (2006). Forward osmosis: principles, applications, and recent developments. *Journal of Membrane Science*, **281**, 70–87.

Childress A. E. and Deshmukh S. S. (1998). Effect of humic substances and anionic surfactants on the surface charge and performance of reverse osmosis membranes. *Desalination*, **118**, 167–174.

Choi Y.-J., Choi J.-S., Oh H.-J., Lee S., Yang D. R. and Kim J. H. (2009). Toward a combined system of forward osmosis and reverse osmosis for seawater desalination. *Desalination*, **247**, 239–246.

Chung T.-S., Zhang S., Wang K. Y., Su J. and Ling M. M. (2012). Forward osmosis processes: yesterday, today and tomorrow. *Desalination*, **287**, 78–81.

Creel L. (2003). Ripple Effects: Population and Coastal Regions. Population Reference Bureau, Washington D.C.

Cui Y., Wang H., Wang H. and Chung T.-S. (2013). Micro-morphology and formation of layer-by-layer membranes and their performance in osmotically driven processes. *Chemical Engineering Science*, **101**, 13–26.

Davis A. P., Shokouhian M., Sharma H., Minami C. and Winogradoff D. (2003). Water quality improvement through bioretention: lead, copper, and zinc removal. *Water Environment Research*, **75**, 10.

EPA. (2000). National Water Quality Inventory: 2000 Report. US Environmental Protection Agency, Washington D.C.

Fan L., Harris J. L., Roddick F. A. and Booker N. A. (2001). Influence of the characteristics of natural organic matter on the fouling of microfiltration membranes. *Water Research*, **35**, 4455–4463.

Fedoroff N. V., Battisti D. S., Beachy R. N., Cooper P. J. M., Fischhoff D. A., Hodges C. N., Knauf V. C., Lobell D., Mazur B. J., Molden D., Reynolds M. P., Ronald P. C., Rosegrant M. W., Sanchez P. A., Vonshak A. and Zhu J.-K. (2010). Radically rethinking agriculture for the 21st century. *Science*, **327**, 833–834.

Gotoh T., Iguchi H. and Kikuchi K.-I. (2004). Separation of glutathione and its related amino acids by nanofiltration. *Biochemical Engineering Journal*, **19**, 165–170.

Hancock N. T., Xu P., Heil D. M., Bellona C. and Cath T. Y. (2011). Comprehensive bench- and pilot-scale investigation of trace organic compounds rejection by forward osmosis. *Environmental Science & Technology*, **45**, 8483–8490.

Hatt B. E., Fletcher T. D. and Deletic A. (2007). Treatment performance of gravel filter media: implications for design and application of stormwater infiltration systems. *Water Research*, **41**, 2513–2524.

Hillel D. (2000). Salinity Management for Sustainable Irrigation: Integrating Science, Environment, and Economics. World Bank Publications, Washinton D.C.

Hoover L. A., Phillip W. A., Tiraferri A., Yip N. Y. and Elimelech M. (2011). Forward with osmosis: emerging applications for greater sustainability. *Environmental Science & Technology*, **45**, 9824–9830.

Leo C. P., Chai W., Mohammad A., Qi Y., Hoedley A. and Chai S.-P. (2011). Phosphorus removal using nanofiltration membranes. *Water Science & Technology*, **64**, 199–205.

Li Z.-Y., Yangali-Quintanilla V., Valladares-Linares R., Li Q., Zhan T. and Amy G. (2012). Flux patterns and membrane fouling propensity during desalination of seawater by forward osmosis. *Water Research*, **46**, 195–204.

Novotny V. (1995). Water Quality Management Library – Volume 9/Nonpoint Pollution and Urban Stormwater Management, Lancaster, US. CRC Press, Basel, Switzerland.

Paugam L., Taha S., Dorange G., Jaouen P. and Quéméneur F. (2004). Mechanism of nitrate ions transfer in nanofiltration depending on pressure, pH, concentration and medium composition. *Journal of Membrane Science*, **231**, 37–46.

Public Utilities Board of Singapore. (2007). NEWater & Marina Barrage. www.pub.gov.sg.

Rangsivek R. and Jekel M. R. (2005). Removal of dissolved metals by zero-valent iron (ZVI): kinetics, equilibria, processes and implications for stormwater runoff treatment. *Water Research*, **39**, 4153–4163.

Renman G. and Hallberg M. (2007). Reactive filters for removal of dissolved metals in highway runoff. In: Highway and Urban Environment, G. Morrison and S. Rauch (eds), Springer, Netherlands, pp. 465–474.

Sardo V. (2005). Halophytes and salt-tolerant glycophytes a potential resource. In: The Use of Nonconventional Water Resources, A. Hamdy (ed.), CIHEAM/EU DG Research, Bari, pp. 87–98.

Schiff K., Bay S. and Diehl D. (2003). Stormwater toxicity in Chollas Creek and San Diego Bay, California. *Environmental Monitoring and Assessment*, **81**, 119–132.

Scholes L., Shutes R. B. E., Revitt D. M., Forshaw M. and Purchase D. (1998). The treatment of metals in urban runoff by constructed wetlands. *Science of the Total Environment*, **214**, 211–219.

Shim Y., Lee H.-J., Lee S., Moon S.-H. and Cho J. (2002). Effects of natural organic matter and ionic species on membrane surface charge. *Environmental Science & Technology*, **36**, 3864–3871.

Sukitpaneenit P. and Chung T.-S. (2012). High performance thin-film composite forward osmosis hollow fiber membranes with macrovoid-free and highly porous structure for sustainable water production. *Environmental Science & Technology*, **46**, 7358–7365.

Tang Q., Liu Z., Liu L., Zheng Q. and Chen M. (2004). Effects of irrigation with diluted seawater on 'Oleic sunflower G101'grown in salinized soils, Laizhou, China. *Acta Phytoecological Sinica*, **29**, 1000–1006.

Tibbetts J. (2002). Coastal cities: living on the edge. *Environmental Health Perspectives*, **110**, A674–A681.

Valladares Linares R., Yangali-Quintanilla V., Li Z. and Amy G. (2011). Rejection of micropollutants by clean and fouled forward osmosis membrane. *Water Research*, **45**, 6737–6744.

Walsh C. J. (2004). Protection of in-stream biota from urban impacts: minimise catchment imperviousness or improve drainage design? *Marine and Freshwater Research*, **55**, 317–326.

Wang R., Shi L., Tang C. Y., Chou S., Qiu C. and Fane A. G. (2010). Characterization of novel forward osmosis hollow fiber membranes. *Journal of Membrane Science*, **355**, 158–167.

Xie M., Nghiem L. D., Price W. E. and Elimelech M. (2012). Comparison of the removal of hydrophobic trace organic contaminants by forward osmosis and reverse osmosis. *Water Research*, **46**, 2683–2692.

Zhao S., Zou L. and Mulcahy D. (2012). Brackish water desalination by a hybrid forward osmosis–nanofiltration system using divalent draw solute. *Desalination*, **284**, 175–181.

Zhu J.-K. (2001). Plant salt tolerance. *Trends in Plant Science*, **6**, 66–71.

Zou S., Gu Y., Xiao D. and Tang C. Y. (2011). The role of physical and chemical parameters on forward osmosis membrane fouling during algae separation. *Journal of Membrane Science*, **366**, 356–362.

Zou S., Wang Y.-N., Wicaksana F., Aung T., Wong P. C. Y., Fane A. G. and Tang C. Y. (2013). Direct microscopic observation of forward osmosis membrane fouling by microalgae: critical flux and the role of operational conditions. *Journal of Membrane Science*, **436**, 174–185.

Chapter 2.2

Water harvesting from municipal wastewater*

2.2.1 INTRODUCTION

Under an average economic growth scenario and if no efficiency gains are assumed, global water demand will increase 53% by 2030, up to 6.9 trillion m^3 per year (The Barila Group *et al.* 2009). Along with the growing demand for fresh water, there is also an increase in the amount of wastewater that needs to be treated adequately to meet public health and environmental discharge regulations. This problem is more significant along coastlines, where today, more than 3 billion people live, and it is likely that by 2025 this number will double (Creel, 2003). As a result of these trends, over the last two decades, wastewater recovery for direct and indirect planned reuse has become a more common practice, with an increasing number of studies that compare various technologies that can treat and recover secondary wastewater effluent (Fujita *et al.* 1996; Madireddi *et al.* 1997; Drewes & Fox, 1999; Drewes *et al.* 2003; Melin *et al.* 2006). However, there are still effluents that are discharged into water bodies, which may contain contaminants that are not removed by conventional treatment processes (Levine & Asano, 2004). The disposal of wastewater effluents in coastal areas poses an additional concern because of the relatively high concentration of nutrients in the treated water, i.e., nitrogen and phosphorous, which enhance eutrophication and impact the ecosystem, endangering the food source of millions of people, and disrupting the economic activities associated with coastlines, besides the public health hazards involved (Areas & Council, 1993; Ng Yan Yan & Barron, 2000). In addition, a significant part of the anthropogenic emissions of trace metals ends up in wastewater. The sources of these trace metals include industrial processes and/or urban runoff in combined sewers (Karvelas *et al.* 2003). The presence of trace

* This chapter is based, with permission from the copyright holder, on the previously published paper. (Valladares Linares R., Li Z., Abu-Ghdaib M., Wei C.-H., Amy G. and Vrouwenvelder J. S. (2013). Water harvesting from municipal wastewater via osmotic gradient: an evaluation of process performance. *Journal of Membrane Science*, **447**, 50–56. doi:10.1016/j.memsci.2013.07.018)

metals in the influent to municipal wastewater treatment plants can adversely affect the performance of the treatment processes and quality of the effluent, and influence decisions regarding reuse (Johnson *et al.* 2008).

Several advanced methods have been studied for wastewater reclamation, but they all involve highly energy intensive processes, such as nanofiltration (NF) and reverse osmosis (RO) (Shannon *et al.* 2008; Yangali-Quintanilla *et al.* 2010). A membrane bioreactor (MBR) has been considered to be a sustainable wastewater treatment process to achieve high quality effluents (Günder & Krauth, 1999; Jefferson *et al.* 2001; Melin *et al.* 2006). Nevertheless, this process still involves a considerable amount of energy to extract the treated effluent through the membrane and to provide oxygen for biological degradation. To solve this problem, forward osmosis (FO) can contribute to increase water recovery at lower energy consumption, and therefore, considerable cost reduction is feasible (Cath *et al.* 2009; Bamaga *et al.* 2011). The forward osmosis (FO) process has been investigated for various applications in wastewater treatment by using different feed streams and process models including, for example, activated sludge (Cornelissen *et al.* 2008), anaerobic digester centrate (Holloway *et al.* 2007), membrane bioreactor effluent (Hancock *et al.* 2011), drilling mud and fracturing wastewater from oil and gas industry (Hickenbottom *et al.* 2013), primary effluent (Lutchmiah *et al.* 2011), secondary and tertiary treated effluent (Cath *et al.* 2010). By an osmotic dilution process, volume-minimizing treatment of wastewater and reduction of osmotic pressure of seawater prior to RO desalination can be achieved simultaneously (Cath *et al.* 2010). The osmotic membrane bioreactor (OsMBR) has demonstrated a sustainable flux with relatively low reverse transport of solutes from the draw solution (DS) into the mixed liquor (Achilli *et al.* 2009). However, the energy that needs to be used for aeration (intended for bacterial growth) and air stripping (intended for membrane cleaning) is still required, and it is about 10% of total energy estimate in OsMBR while about 15% in the conventional MBR (Bharwada, 2011).

On the other hand, wastewater has been considered as an energy source to produce biogas by an anaerobic process. The crucial factor for successful anaerobic treatment of wastewater with biogas production is the concentration of organic matter compounds including polysaccharides, lipids, protein, simple aromatics (Metcalf & Eddy, 2003). The soluble organic fraction in wastewater cannot be concentrated easily with a low cost process, consequently becoming a barrier to the direct anaerobic treatment of wastewater (McCarty *et al.* 2011). In the treatment of wastewater, the idea of osmotic concentration of raw wastewaters using FO membranes has not been studied to its full extent as an alternate method to provide high organic concentration for an anaerobic process, and avoid primary and aerobic treatment of wastewater.

The objective of this research is the assessment of a combined wastewater concentration and seawater dilution system based on a submerged FO membrane cell that concentrates a synthetic municipal wastewater using synthetic seawater as a draw solution (DS). Special attention is given to nutrients removal and high recovery rates intended for optimal feed concentrations for subsequent anaerobic treatment as well as membrane fouling analysis. On the other hand, high quality water can be extracted through the FO membrane, diluting the seawater which can be utilized to produce fresh water via low-pressure reverse osmosis (LPRO) (Yangali-Quintanilla *et al.* 2011). This study provides insight and details into the concept of using the FO process to avoid secondary aerobic treatment of a primary wastewater effluent that has not been reported in the literature, as well as additional information on the partial desalination of seawater, including: (i) the use of a submerged FO module which makes it possible to adapt the process to a primary clarifier tank, (ii) analysis of the quality of the diluted seawater for further post-treatment, (iii) FO rejection of nutrients and heavy metals in the wastewater, and (iv) analysis of fractional organic carbon composition in the fouling layer on the FO membrane.

2.2.2 EXPERIMENTAL

2.2.2.1 Feed water and draw solution

Synthetic municipal wastewater containing chemical compounds, food ingredients and trace metals was prepared by a method reported previously with some modifications (Nopens *et al.* 2001), and used as the feed water/solution (FS) (Table 2.2.1). The chemical oxygen demand (COD) of the synthetic municipal wastewater was estimated to be about 360 mg · L^{-1} before the FO process. For the DS, a solution of 4% NaCl was used to simulate Red Sea water, with a conductivity of 55 ± 0.5 mS · cm^{-1}.

Table 2.2.1 Recipe of synthetic municipal wastewater.

Solutes	mg · L^{-1}
Chemical Compounds	
Urea	91.74
NH_4Cl	12.75
Sodium Acetate	79.37
Peptone	17.41
$MgHPO_4 \cdot 3H_2O$	29.02
KH_2PO_4	23.40
$FeSO_4 \cdot 7H_2O$	5.80
Food Ingredients	
Starch	122.00
Skimmed Milk Power	116.19
Yeast extract	52.24
Trace Metals	
$Cr(NO_3)_3 \cdot 9H_2O$	0.770
$CuCl_2 \cdot 2H_2O$	0.536
$MnSO_4 \cdot H_2O$	0.108
$NiSO_4 \cdot 6H_2O$	0.336
$PbCl_2$	0.100
$ZnCl_2$	0.208

2.2.2.2 FO membrane and experimental set-up

The FO membrane was provided by Hydration Technology Innovations (HTI, Scottsdale, AZ). The FO membrane is a cellulose triacetate (CTA) layer embedded in a polyester support mesh. A layout of the experimental setup can be found in a previous publication (Valladares Linares *et al.* 2012). The membrane cell was a custom-made plate and frame assembly, accommodating flat-sheet membranes with a total area of 250 cm^2. Two orientations of the membrane were used in the experiments: active layer (AL) of the membrane facing the FS side (AL-FS mode) and active layer facing the DS side (AL-DS mode). The membrane cell was immersed in a feed tank containing FS, and connected to a receptacle containing DS. A gear pump (Cole-Parmer) was used to continuously recirculate the DS in the lumen space of the cell, extracting water from the FS into the DS reservoir. A digital balance (Sartorius AG, Germany) connected to a computer was used to measure the variation of DS weight and convert it to flux. The temperature of the FS and DS was kept constant at 20 ± 0.5 °C by using chiller/heater devices.

The experimental procedure started with 20 L of FS in the feed tank and 1 L of DS in the DS receptacle. The recirculation pump was started at a flow rate of 100 $mL \cdot min^{-1}$ and dilution of the DS started, meanwhile the conductivity and flux data acquisition were also started. Since the flow rate is low, the hydraulic pressure induced by the flow in the lumen space of the cell was not detectable by the pressure gauge (1008 Ashcroft, U.S.A.) and thought to be negligible in this study. The high osmotic pressure of the DS inside the channel resulted in an extraction of clean water (permeate) from the FS to DS while most contaminants or solutes in either FS or DS were rejected by the membrane. A stirrer was used to provide mixing in the feed tank with a global velocity gradient of 50 s^{-1}. The each cycle of FO process was performed until the final volume of DS was 2 L, i.e., 1 L of permeate was extracted from FS to DS. The cycle was repeated by replacing diluted DS with fresh aliquot (1 L of 4% NaCl solution). The osmotic concentration process was terminated after 10 cycles, when FS was concentrated 50% (10 L of permeate was extracted after 10 cycles).

Osmotic backwash was used as a cleaning technique to determine its impact on flux recovery. After 10 cycles of osmotic concentration, the cell was moved to a separate tank and soaked in 4% NaCl solution while deionized (DI) water was recirculated inside the lumen space of the cell at a flow rate of 100 $mL \cdot min^{-1}$. The cleaning process was performed for 12 hours. The permeability fluxes of the membrane before and after osmotic backwash were determined under the same operational condition of osmotic concentration process with DI water as feed and 4% NaCl solution as DS.

2.2.2.3 Analytical methods

The conductivity of the FS and DS was monitored with a conductivity meter (WTW, Germany) connected to a computer.

The salt leakage from the DS side to feed wastewater was estimated by measuring the conductivity in the DS reservoir and converting to salt amount. The loss of salt in each batch was considered to be the salt leakage to the feed tank (Equation 2.2.1).

$$\text{Salt leakage} = (1 - (C_{DS2} \cdot V_{DS2}/C_{DS1} \cdot V_{DS1})) \cdot 100 \tag{2.2.1}$$

where C_{DS1} and C_{DS2} are salt concentrations in the DS before and after each cycle, respectively. V_{DS1} and V_{DS2} are the volumes of DS before and after each cycle, respectively.

The trace metals (Cr, Mn, Ni, Cu, Zn and Pb) were analyzed by Inductively Coupled Plasma/Mass Spectrometry (ICP-MS 7500; Agilent, US). DOC and total nitrogen were measured by an Organic Carbon Analyzer (TOC-VCPH, Shimadzu, Japan). COD, ammonia and phosphate were analyzed by Hach TNT reagents with a spectrophotometer (DR, 2800; Hach). The proper dilution of sample was done to eliminate the possible interference caused by water salinity. Hach TNT reagent tests were performed previously with different water salinities in the samples to determine the accuracy of the method when used with concentrated FS containing leaked salts and diluted DS. All of the tests showed reliable results due to the fact that TNT reagents chosen have been adapted to be used with saline waters samples (Boyles, 1997; HACH, 2012).

After 10 cycles of the osmotic concentration process and osmotic backwash, membrane samples (2×2 cm^2) were cut and soaked in 15 mL DI water, and then treated by ultrasound for 30 minutes. The supernatant was collected for Liquid Chromatography – Organic Carbon Detection analysis (LC-OCD, Model 8, DOC Labor, Germany) to reveal membrane foulants. The FS and DS collected during the FO process as well as the original synthetic municipal wastewater were also analyzed by LC-OCD. All samples for LC-OCD analysis were pre-filtered by 0.45 μm pore size filter as required by the instrument.

2.2.3 RESULTS AND DISCUSSION

2.2.3.1 Flux patterns

The flux in each cycle decreased rapidly. Flux decrease is due to two reasons: (i) a decrease in the osmotic pressure difference caused by the dilution of DS and concentration polarization in each cycle during a batch operation in this study, (ii) fouling on the membrane surface. During a full scale application, it is possible to have a continuous supply and recirculation of DS. In such a system, the recovery rate can achieve up to 80%. The decrease of flux in the full scale application would mainly be associated with membrane fouling and fouling enhanced concentration polarization which can be demonstrated by the decrease of initial flux (J_{ini}) and average flux (J_{ave}) in each cycle of this study. Figure 2.2.1 shows the flux patterns for 10 cycles in terms of J_{ini} and J_{ave} for each mode tested. Both J_{ini} and J_{ave} decreased with batch number, which indicated that membrane fouling also caused a negative impact, limiting the flux in the FO process even though there was no hydraulic pressure applied. During the FO process, the fouling layer itself may not give a significant resistance to the water permeability since it was loose and easily removed by an osmotic backwash (proved in Section 2.2.3.3). However, fouling enhanced concentration polarization created by accumulating reversely diffused salt within fouling layer can reduce the effective osmotic driving force and, consequently, limit the water flux (Boo *et al.* 2012; She *et al.* 2012). In addition, the AL-DS mode gave a higher value of both J_{ini} and J_{ave}. The reason is that the effect of dilutive concentration polarization at the DS side can be reduced when the AL faced the DS. The smoother and denser surface of the AL can improve the fluid shear stress near the membrane surface to reduce accumulation of salt on the membrane and diffusion of salt into the membrane, then reduce both external concentration polarization at the DS side and internal concentration polarization (Mi & Elimelech, 2008; Su *et al.* 2010; Zhao *et al.* 2011).

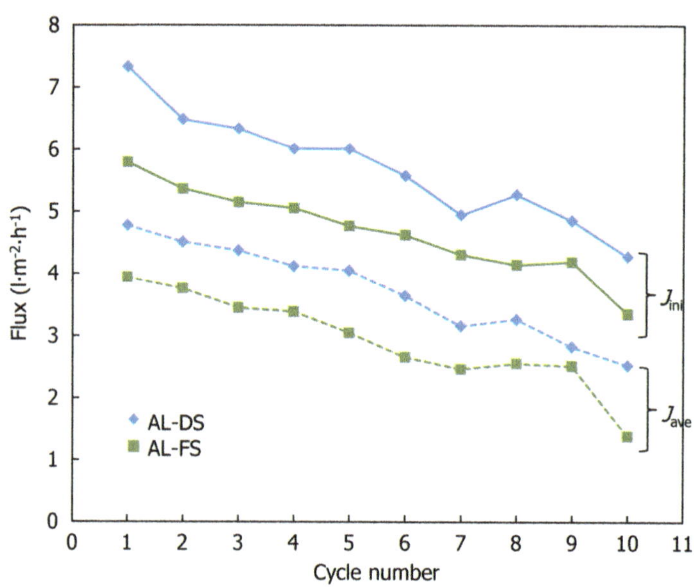

Figure 2.2.1 Initial flux (J_{ini}) and average flux (J_{ave}) of each cycle during the treatment of synthetic municipal wastewater by forward osmosis process.

2.2.3.2 Salt leakage and retention of nutrients and trace metals

Salt leakage from DS to FS was calculated through measurements of the conductivity in the DS along the dilution process. Salt leakage values ranged between 5% to 15% for both modes (Figure 2.2.2). The AL-DS mode showed a slightly lower salt leakage than AL-FS. In most reported studies, the salt rejection of HTI CTA membrane under various membrane orientations and operational modes was around 90% which is consistent with the result observed in this study (Cornelissen *et al.* 2008; Phillip *et al.* 2010; Yip *et al.* 2010; Jin *et al.* 2011). The remaining conductivity in the DS at the end of each cycle of both modes varied between 22.9 and 27.4 mS · cm^{-1}. The reduced conductivity indicates the reliable dilution of DS by the FO process extracting fresh water from the FS, and potential energy saving in further desalination by LPRO.

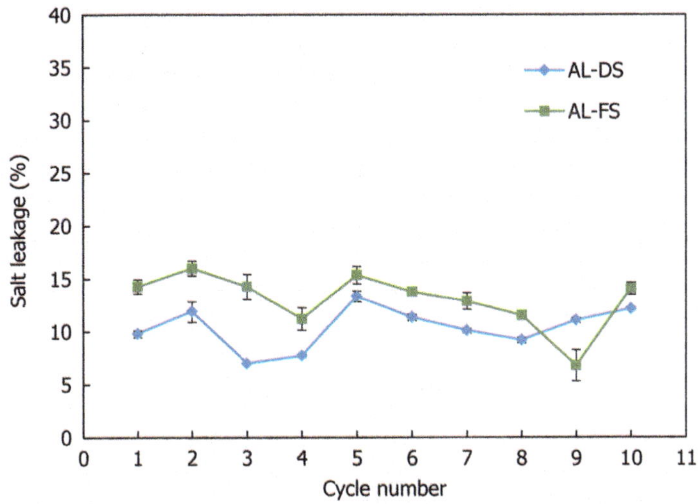

Figure 2.2.2 Salt leakage from the draw solution to feed solution in each cycle during the treatment of synthetic municipal wastewater by forward osmosis process.

The results for the removal of trace metals are presented in Figure 2.2.3. Retention of all trace metals including Cr, Mn, Ni, Cu, Zn and Pb in the FS was high and stable throughout the osmotic concentration process for both modes. Most rejection rates were higher than 99%.

Figure 2.2.4 shows the average retention of nutrients in the FS. Similar to the salt leakage and retention of trace metals, there is no significant difference between the AL-FS and AL-DS mode. The highest rejection of close to 99% was observed for COD and phosphate. However, the rejection was moderate for ammonia and total nitrogen ranging from 67% to 68% and 56% to 59%, respectively, consistent with previous observations in the literature (Cath *et al.* 2006; Holloway *et al.* 2007). The lower rejection of ammonia might be due to its positive charge combined with its low molecular weight.

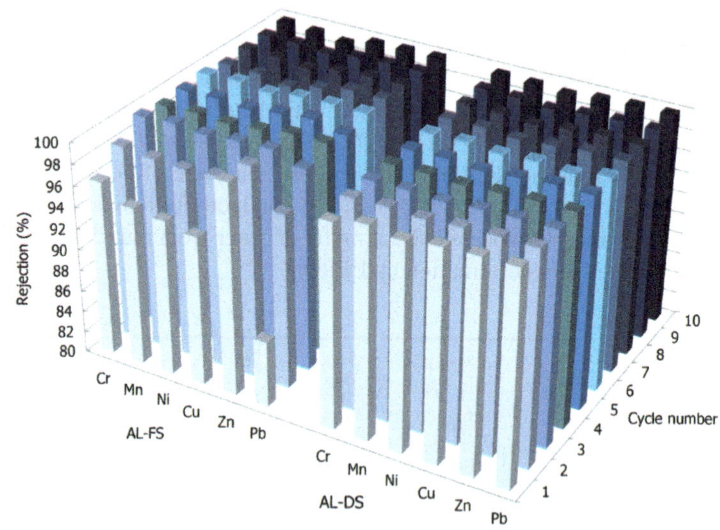

Figure 2.2.3 Rejection of trace metals in synthetic municipal wastewater by forward osmosis process.

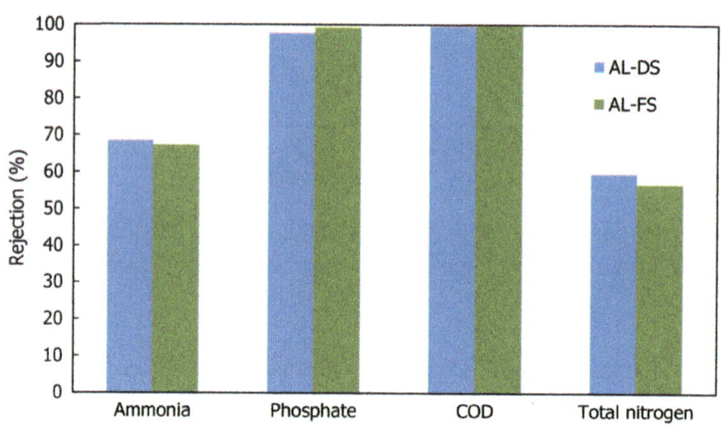

Figure 2.2.4 Rejection of nutrients in synthetic municipal wastewater by forward osmosis process.

The samples of FS and DS were collected from different cycles of the process. By comparing the LC-OCD chromatograms of each component of synthetic municipal wastewater (Figure 2.2.5), and FS and DS (Figure 2.2.6), the inability of the membrane to retain low molecular weight acid compounds (sodium acetate, elution time of about 56 minutes) and low molecular weight neutral compounds (urea, elution time of about 78 minutes) in the FS can be seen.

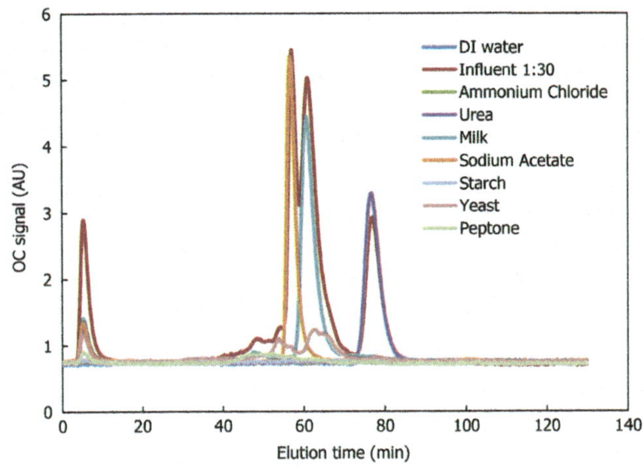

Figure 2.2.5 LC-OCD chromatogram of synthetic municipal wastewater and each component.

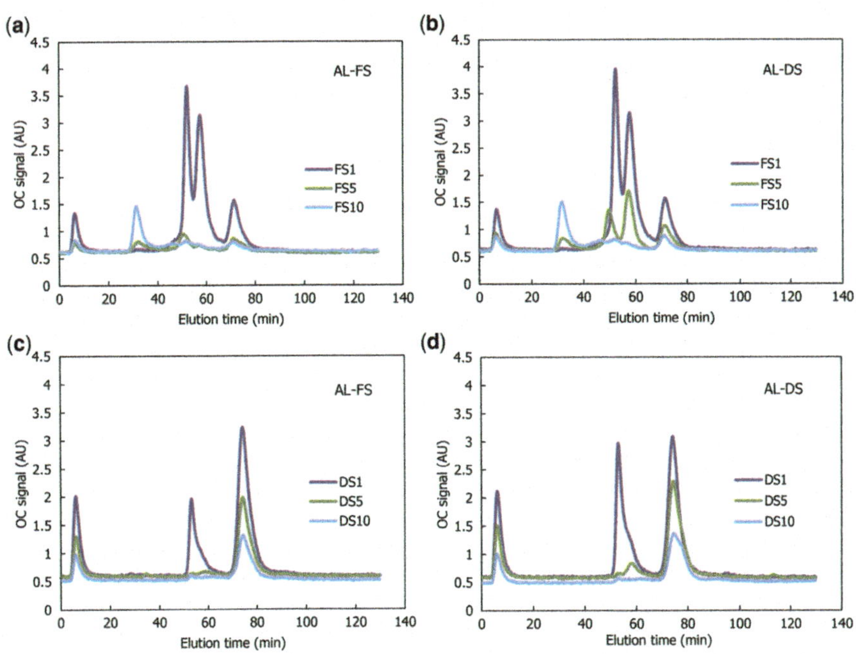

Figure 2.2.6 LC-OCD chromatogram of feed solution (FS) and draw solution (DS) during treatment of synthetic municipal wastewater by forward osmosis process (a and c, AL-FS mode; b and d, AL-DS mode; FS1, FS5 and FS10, FS collected from cycle 1, 5 and 10; DS1, DS5 and DS10, DS collected from cycle 1, 5 and 10).

In addition, the FS underwent some changes during the process (Figure 2.2.6a and 2.2.6b). The reduction of sodium acetate and urea can be explained by the poor rejection by the FO membrane for these compounds. The peak of skimmed milk coupled with yeast extract (elution time of about 60 minutes) disappeared while a biopolymer peak increased after 10 cycles of the process. This is probably due to the hydrolysis and coagulation of some compounds including peptone, skimmed milk, yeast extract and starch in the synthetic municipal wastewater.

2.2.3.3 Fouling characterization and osmotic backwash

After the FO processes for 10 cycles, a thick fouling layer was observed on the membrane surface from both AL-FS and AL-DS modes. LC-OCD analysis shown in Figure 2.2.7a reveals the major composition of the fouling layer. For both AL-FS and AL-DS modes, only a single peak was observed at the elution time of about 35 minutes which indicated the presence of biopolymers. This is consistent with the increase of biopolymers in the feed water indicated in Figure 2.2.6a and 2.2.6b. The actual amount of biopolymer may even higher since the fractions with size larger than 0.45 μm were removed by the pre-filtration in sample preparation before LC-OCD analysis. Due to its hydrophilic and gel-like character, biopolymers tend to attach to the membrane surface while other compounds in the feed water may play a minor role in the formation of the fouling layer without a hydraulic pressure in the FO process.

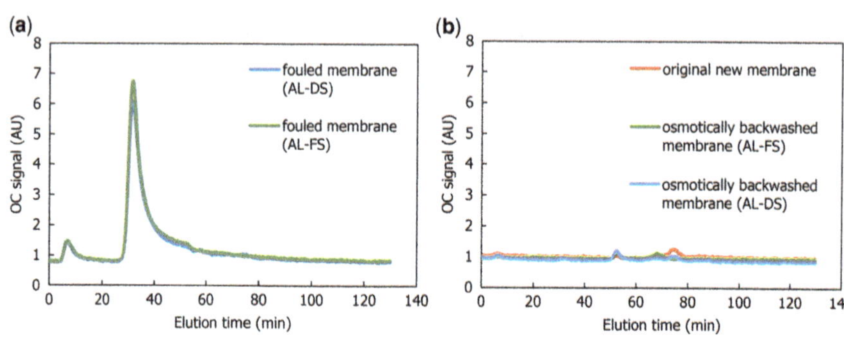

Figure 2.2.7 Foulant on the membrane surface (a, before osmotic backwash; b, after osmotic backwash).

The fouled membranes from both AL-FS and AL-DS modes were cleaned by osmotic backwash. Almost all organic foulants, mainly biopolymers, can be removed from the membrane surface by osmotic backwash. No significant peak can be identified from LC-OCD chromatographs of osmotically backwashed membranes collected from AL-FS and AL-DS modes (Figure 2.2.7b).

In order to identify the performance of osmotic backwash, the fluxes of the fouled membrane before and after osmotic backwash were measured with DI water as FS and 4% NaCl as DS in both AL-FS and AL-DS modes. The AL-DS mode still showed a higher flux than the AL-FS mode regardless of the fouled or osmotically cleaned membrane (Figure 2.2.8). As mentioned above, this is attributed to the denser and smoother AL surface which can reduce dilutive concentration polarization at the DS side (Gray *et al.* 2006; McCutcheon & Elimelech, 2006; Li *et al.* 2012; Cath *et al.* 2013). However, the results also show that osmotic backwash did not result in any improvement in the DI water flux in both modes (Figure 2.2.8), although the membrane surface was cleaned which is proven by the results in Figure 2.2.7. The reason is probably the diffusion of salt into the FO membrane during the osmotic backwash step (Zhao & Zou, 2011). A similar result was also observed in a previous study (Valladares Linares *et al.* 2013). The diffused salt remaining inside the membrane can lead to internal concentration polarization, limiting the effective osmotic pressure in the consequent FO process. This observation suggests that osmotic backwash may not provide an expected performance in the FO process compared to RO since the fouling propensity and driving force are different between FO and RO. An extended soaking of the membrane in DI water may be required to remove residual salt inside the membrane if the osmotic backwash is applied in the FO process for membrane cleaning. It is meaningful to have a systematic investigation to confirm this hypothesis and optimize the osmotic backwash procedure in a further study.

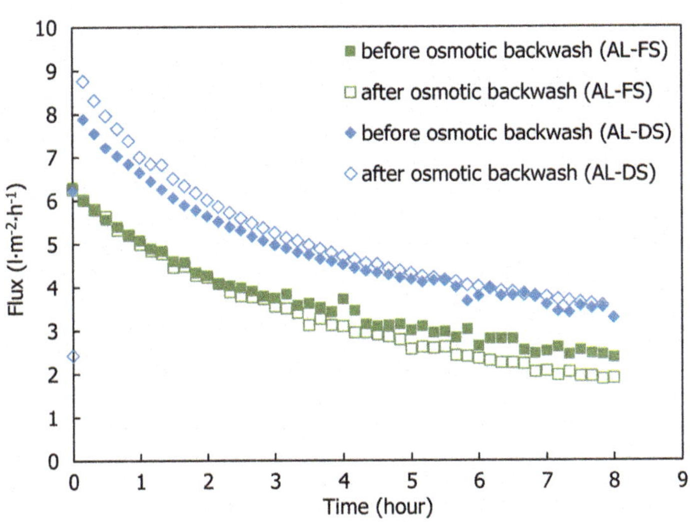

Figure 2.2.8 Deionized water flux from the membrane before and after osmotic backwash.

2.2.4 SUMMARY

Forward osmosis (FO) presents a unique opportunity for integrated wastewater treatment and seawater desalination. This study assesses the efficiency of a submerged FO system to reduce the volume of wastewater that needs to be treated while recovering high quality water that can be further treated for sustainable fresh water production. A semi-batch operation was employed with two membrane orientations in terms of active and support layers. A change of membrane orientation could improve the flux and slightly reduce the salt leakage from the draw solution to the feed solution. The formation of fouling on the membrane resulted in a decrease of the initial flux and average flux with both membrane orientations. The fouling layer on the membrane surface was determined to be caused by biopolymer-like substances. Osmotic backwash removed almost all organic foulants from the membrane surface, but did not improve the flux. There was a moderate to high retention of nutrients (N and P), varying from 56% to 99%, and almost a complete retention for trace metals regardless of membrane orientation. However the membrane showed a limited ability to retain low molecular weight acids and low molecular weight neutral compounds. This study identified a possible role of the FO process to integrate wastewater treatment and seawater desalination for a sustainable solution of the water-energy nexus for coastal cities.

2.2.5 REFERENCES

Achilli A., Cath T. Y., Marchand E. A. and Childress A. E. (2009). The forward osmosis membrane bioreactor: a low fouling alternative to MBR processes. *Desalination*, **239**, 10–21.

Areas C. o. W. M. f. C. U. and N. R. Council. (1993). Managing Wastewater in Coastal Urban Areas. The National Academies Press, Washington, DC.

Bamaga O., Yokochi A., Zabara B. and Babaqi A. (2011). Hybrid FO/RO desalination system: preliminary assessment of osmotic energy recovery and designs of new FO membrane module configurations. *Desalination*, **268**, 163–169.

Bharwada U. (2011). HTI's Forward Osmosis Membrane Bioreactor Process (OsMBR) – A Rugged Versatile and Ecobalanced Process for Industrial Wastewater Plus Reuse: Truly Sustainable Wastewater Treatment Design for a Changing World Scottsdale. Upen Bharwada, Hydration Technology Innovations, LLC 9311 E. Via de Ventura, Scottsdale, Arizona, USA.

Boo C., Lee S., Elimelech M., Meng Z. and Hong S. (2012). Colloidal fouling in forward osmosis: role of reverse salt diffusion. *Journal of Membrane Science*, **390–391**, 277–284.

Boyles W. (1997). The Science of Chemical Oxygen Demand. Technical Information Series Booklet No. 9. HACH.

Cath T. Y., Childress A. E. and Elimelech M. (2006). Forward osmosis: principles, applications, and recent developments. *Journal of Membrane Science*, **281**, 70–87.

Cath T. Y., Drewes J. E. and Lundin C. D. (2009). A Novel Hybrid Forward Osmosis Process for Drinking Water Augmentation using Impaired Water and Saline Water Sources.

Cath T. Y., Hancock N. T., Lundin C. D., Hoppe-Jones C. and Drewes J. E. (2010). A multi-barrier osmotic dilution process for simultaneous desalination and purification of impaired water. *Journal of Membrane Science*, **362**, 417–426.

Cath T. Y., Elimelech M., McCutcheon J. R., McGinnis R. L., Achilli A., Anastasio D., Brady A. R., Childress A. E., Farr I. V., Hancock N. T., Lampi J., Nghiem L. D., Xie M. and Yip N. Y. (2013). Standard methodology for evaluating membrane performance in osmotically driven membrane processes. *Desalination*, **312**, 31–38.

Cornelissen E. R., Harmsen D., de Korte K. F., Ruiken C. J., Qin J.-J., Oo H. and Wessels L. P. (2008). Membrane fouling and process performance of forward osmosis membranes on activated sludge. *Journal of Membrane Science*, **319**, 158–168.

Creel L. (2003). Ripple Effects: Population and Coastal Regions. Page 8 Making the Link: Population Reference Bureau, Washington D.C.

Drewes J. E. and Fox P. (1999). Fate of natural organic matter (NOM) during groundwater recharge using reclaimed water. *Water Science and Technology*, **40**, 241–248.

Drewes J. E., Reinhard M. and Fox P. (2003). Comparing microfiltration-reverse osmosis and soil-aquifer treatment for indirect potable reuse of water. *Water Research*, **37**, 3612–3621.

Fujita Y., Ding W.-H. and Reinhard M. (1996). Identification of wastewater dissolved organic carbon characteristics in reclaimed wastewater and recharged groundwater. *Water Environment Research*, **68**, 867–876.

Gray G. T., McCutcheon J. R. and Elimelech M. (2006). Internal concentration polarization in forward osmosis: role of membrane orientation. *Desalination*, **197**, 1–8.

Günder B. and Krauth K. (1999). Replacement of secondary clarification by membrane separation – results with tubular, plate and hollow fibre modules. *Water Science and Technology*, **40**, 311–320.

HACH (2012). Phosphorus, Reactive (Orthophosphate). DOC 316.53.01113, Method 8178, 7th edn.

Hancock N. T., Xu P., Heil D. M., Bellona C. and Cath T. Y. (2011). Comprehensive bench- and pilot-scale investigation of trace organic compounds rejection by forward osmosis. *Environmental Science & Technology*, **45**, 8483–8490.

Hickenbottom K. L., Hancock N. T., Hutchings N. R., Appleton E. W., Beaudry E. G., Xu P. and Cath T. Y. (2013). Forward osmosis treatment of drilling mud and fracturing wastewater from oil and gas operations. *Desalination*, **312**, 60–66.

Holloway R. W., Childress A. E., Dennett K. E. and Cath T. Y. (2007). Forward osmosis for concentration of anaerobic digester centrate. *Water Research*, **41**, 4005–4014.

Jefferson B., Laine A. L., Stephenson T. and Judd S. J. (2001). Advanced biological unit processes for domestic water recycling. *Water Science & Technology*, **43**, 211–218.

Jin X., Tang C. Y., Gu Y., She Q. and Qi S. (2011). Boric acid permeation in forward osmosis membrane processes: modeling, experiments, and implications. *Environmental Science & Technology*, **45**, 2323–2330.

Johnson P. D., Girinathannair P., Ohlinger K. N., Ritchie S., Teuber L. and Kirby J. (2008). Enhanced removal of heavy metals in primary treatment using coagulation and flocculation. *Water Environment Research*, **80**, 472–479.

Karvelas M., Katsoyiannis A. and Samara C. (2003). Occurrence and fate of heavy metals in the wastewater treatment process. *Chemosphere*, **53**, 1201–1210.

Levine A. D. and Asano T. (2004). Peer reviewed: recovering sustainable water from wastewater. *Environmental Science & Technology*, **38**, 201A–208A.

Li Z.-Y., Yangali-Quintanilla V., Valladares-Linares R., Li Q., Zhan T. and Amy G. (2012). Flux patterns and membrane fouling propensity during desalination of seawater by forward osmosis. *Water Research*, **46**, 195–204.

Lutchmiah K., Cornelissen E. R., Harmsen D. J. H., Post J. W., Lampi K., Ramaekers H., Rietveld L. C. and Roest K. (2011). Water recovery from sewage using forward osmosis. *Water Science and Technology*, **64**, 1443–1449.

Madireddi K., Babcock R. W., Levine B., Huo T. L., Khan E., Ye Q. F., Neethling, Suffet I. H. and Stenstrom M. K. (1997). Wastewater reclamation at Lake Arrowhead, California: an overview. *Water Environment Research*, **69**, 350–362.

McCarty P. L., Bae J. and Kim J. (2011). Domestic wastewater treatment as a net energy producer–can this be achieved? *Environmental Science & Technology*, **45**, 7100–7106.

McCutcheon J. R. and Elimelech M. (2006). Influence of concentrative and dilutive internal concentration polarization on flux behavior in forward osmosis. *Journal of Membrane Science*, **284**, 237–247.

Melin T., Jefferson B., Bixio D., Thoeye C., De Wilde W., De Koning J., van der Graaf J. and Wintgens T. (2006). Membrane bioreactor technology for wastewater treatment and reuse. *Desalination*, **187**, 271–282.

Metcalf and Eddy (2003). Wastewater Engineering Treatment and Reuse, 4th edn., McGraw Hill, Columbus, OH.

Mi B. and Elimelech M. (2008). Chemical and physical aspects of organic fouling of forward osmosis membranes. *Journal of Membrane Science*, **320**, 292–302.

Ng Yan Yan D. and Barron W. (2000). Coastal Pollution Status and the Necessity of Urban Runoff Pollution Controls in Hong Kong. In: 17th International Conference of The Coastal Society, Portland, OR, USA.

Nopens I., Capalozza C. and Vanrolleghem P. A. (2001). Stability Analysis of a Synthetic Municipal Wastewater. Universiteit Gent, Gent.

Phillip W. A., Yong J. S. and Elimelech M. (2010). Reverse draw solute permeation in forward osmosis: modeling and experiments. *Environmental Science & Technology*, **44**, 5170–5176.

Shannon M. A., Bohn P. W., Elimelech M., Georgiadis J. G., Marinas B. J. and Mayes A. M. (2008). Science and technology for water purification in the coming decades. *Nature*, **452**, 301–310.

She Q., Jin X., Li Q. and Tang C. Y. (2012). Relating reverse and forward solute diffusion to membrane fouling in osmotically driven membrane processes. *Water Research*, **46**, 2478–2486.

Su J., Yang Q., Teo J. F. and Chung T.-S. (2010). Cellulose acetate nanofiltration hollow fiber membranes for forward osmosis processes. *Journal of Membrane Science*, **355**, 36–44.

The Barila Group, The Coca-Cola Company, The International Finance Corporation, McKinsey & Company, Nestle S.A., New Holland Agriculture, SABMiller plc, Standard Chartered Bank, and Syngenta AG (2009). Charting Our Water Future, Economic Frameworks to Inform Decision-Making. 2030 Water Resources Group, Washington, DC.

Valladares Linares R., Yangali-Quintanilla V., Li Z. and Amy G. (2012). NOM and TEP fouling of a forward osmosis (FO) membrane: foulant identification and cleaning. *Journal of Membrane Science*, **421–422**, 217–224.

Valladares Linares R., Li Z., Yangali-Quintanilla V., Li Q. and Amy G. (2013). Cleaning protocol for a FO membrane fouled in wastewater reuse. *Desalination and Water Treatment*, **51**, 4821–4824.

Yangali-Quintanilla V., Maeng S. K., Fujioka T., Kennedy M. and Amy G. (2010). Proposing nanofiltration as acceptable barrier for organic contaminants in water reuse. *Journal of Membrane Science*, **362**, 334–345.

Yangali-Quintanilla V., Li Z., Valladares R., Li Q. and Amy G. (2011). Indirect desalination of Red Sea water with forward osmosis and low pressure reverse osmosis for water reuse. *Desalination*, **280**, 160–166.

Yip N. Y., Tiraferri A., Phillip W. A., Schiffman J. D. and Elimelech M. (2010). High performance thin-film composite forward osmosis membrane. *Environmental Science & Technology*, **44**, 3812–3818.

Zhao S. and Zou L. (2011). Relating solution physicochemical properties to internal concentration polarization in forward osmosis. *Journal of Membrane Science*, **379**, 459–467.

Zhao S., Zou L. and Mulcahy D. (2011). Effects of membrane orientation on process performance in forward osmosis applications. *Journal of Membrane Science*, **382**, 308–315.

Chapter 2.3

Indirect desalination of seawater*

2.3.1 INTRODUCTION

The growth of the desalination market in countries with or approaching, physical water scarcity is a fact confirmed by a recent state of the art desalination report (Fritzmann *et al.* 2007). Most of the countries with or approaching water scarcity are located in the Middle East and North Africa (MENA) region (Comprehensive Assessment of Water Management in Agriculture, 2007). In the global scenario, from 2000 to 2005 the installed desalination capacity grew at a compound average rate of 12% (Global Water Intelligence, 2006), and the compound annual growth rate of installed capacity from 1997 to 2007 was 7.9% (Global Water Intelligence, 2008). In the period 2010–2020 the global cumulative contracted capacity of the desalination market will grow at a cumulative average growth rate of 10.5%, reaching 195.8 million m^3 per day in 2020 (GBI Research, 2010). The real price of desalinating water by seawater reverse osmosis (SWRO) is nowadays in the range \$0.59–1.50 per m^3, which is a reduced cost with energy recovery devices, but the cost will not continue decreasing because equipment and energy costs will increase (Fritzmann *et al.* 2007; Ghaffour & Venkat, 2009). The current and forecasted situation means that the price of water will probably increase when subsidies are gradually withdrawn in the Middle East.

Water reuse will play an important role in areas facing water scarcity. Global Water Intelligence predicts a 181% increase of the global water reuse capacity over the years 2005–2010 and, in comparison, the growth of the desalination capacity over the same period was predicted as 102% (Global Water Intelligence, 2005). Clearly, there is a close link between desalination and water reuse, and

---

* This chapter is based, with permission from the copyright holder, on the previously published paper. (Yangali-Quintanilla V., Li Z., Valladares R., Li, Q. and Amy G. (2011). Indirect desalination of Red Sea water with forward osmosis and low pressure reverse osmosis for water reuse. *Desalination*, **280**, 160–166. doi:10.1016/j.desal.2011.06.066)

forward osmosis (FO) membranes can act as bridge between the two processes. Hydration Technology Innovations, LLC (HTI) FO membranes have been used in many FO applications (Cath *et al.* 2005b; McCutcheon *et al.* 2005; Holloway *et al.* 2007; Achilli *et al.* 2009); particularly hybridization of FO with RO has been successfully demonstrated as a water reuse application in studies conducted by Cath *et al.* (2010). Cath and colleagues indicated that the hybrid process of FO and RO is economically favorable for recoveries of water up to 63% (Cath *et al.* 2010). Two companies (HTI and Oasys Water) are presently involved in commercialization of forward osmosis (FO) membranes, but actually only one is commercially available in the market, and the technology has a potential to reduce energy consumption (American Water Intelligence, 2011; Winter, 2011).

This paper presents practical use of FO membranes for demonstrating that a novel FO membrane configuration can achieve indirect desalination of seawater at reduced energy and costs. An innovative plate and frame FO membrane, real seawater as a draw solution, and secondary wastewater effluent as a feed water are used to achieve ultimate RO desalination at low pressure.

2.3.2 MATERIALS, METHODS AND EXPERIMENTAL

2.3.2.1 Membranes and equipment

HTI (Albany, OR) provided flat-sheet membranes (HydroWell, with a support mesh). A schematic of the experimental setup is shown in Figure 2.3.1. A plate and frame FO membrane cell was used for experiments. The cell supports two flat-sheet membranes with a total area of 202 cm^2, with the active layer (thin-film) facing the feed water, and with the support layer facing the draw solution. Two cells were immersed in a tank containing feed water, and were connected to a tank containing the draw solution (DS). A pump (Coleparmer, USA) recirculated the DS inside the cell. FO membrane contactors and spiral-wound FO elements described in previous publications (Cath *et al.* 2005a; Cath *et al.* 2005b; Choi *et al.* 2009; Cath *et al.* 2010; Qin *et al.* 2010) are different from the experimental setup of this study. The conductivity of the draw solution was also monitored with a conductivity meter (WTW, Weilheim, Germany) connected to a computer. A balance (TE6101, Sartorius AG, Göttingen, Germany) was used as flow (and flux) controller when connected to a computer. The low pressure reverse osmosis (LPRO) membrane used was a BW-30 (Dow-Filmtec, Midland, MI). The LPRO setup was comprised of a positive displacement pump (Hydra-Cell, MN), a cross-flow filtration cell accommodating a 139 cm^2 flat-sheet membrane (SEPA CF II, Sterlitech, Kent, WA), needle valves, pressure gauges, a proportional pressure relief valve and stainless steel tubing (Swagelok BV, Netherlands). The LPRO was operated at a net driving pressure of 15 bar, at a flux of 7 L · m^{-2} · h^{-1}, with a recovery of 2%, this limitation of flux and recovery was due to the use of only one SEPA cell. An LC-OCD Model 8

(DOC-Labor, Germany) was used for liquid chromatography organic carbon detection (LC-OCD) analyses of selected water samples.

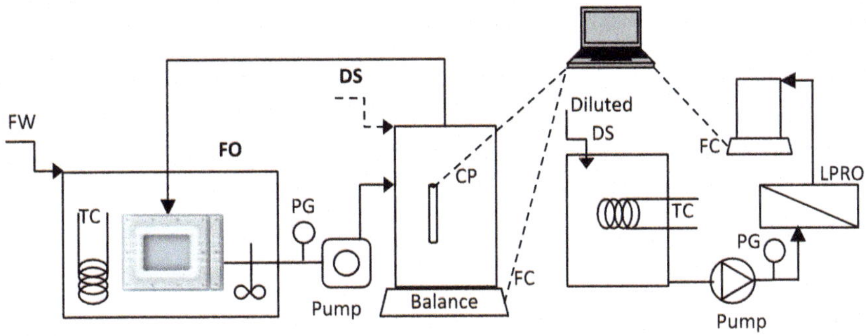

Figure 2.3.1 Schematic of forward osmosis (FO) experimental setup, DS (draw solution), CP (conductivity probe), PG (pressure gauge), FC (flow controller), TC (temperature controller), LPRO (low pressure reverse osmosis).

2.3.2.2 Draw solution and feed water

The draw solution was real Red Sea seawater (pre-filtered with 0.45 μm filters, 40.5 g · L^{-1} as TDS). The dissolved organic carbon (DOC) was approximately 1 mg · L^{-1}. The seawater was collected from the line that provides seawater to the existing reverse osmosis desalination plant at KAUST, located near the town of Thuwal along the Red Sea coast. A secondary wastewater effluent (SWWE) without pre-treatment was collected from the Al Ruwais wastewater treatment plant in Jeddah, Saudi Arabia. The BOD_5 of the wastewater effluent was 20 mg · L^{-1}, and the DOC was 5 mg · L^{-1}. The pH of the feed water was 7.3, the TDS was 2430 mg · L^{-1}, and the temperature was adjusted to 20 ± 0.5 °C. The temperature of the water solutions was controlled at 20 ± 0.5 °C by using chiller/heater devices.

2.3.2.3 Experimental protocol

The experiments were conducted in sequential cycles, as shown in Figure 2.3.2. The figure shows that the experiments started with an initial volume (30 L) of SWWE (named feed water, FW) in the FO tank, with a small volume (1 L) of pre-filtered seawater (named draw solution, DS) in the DS tank. Subsequently, only one pump was used for recirculation of the DS at a flow rate of 100 mL · min^{-1}. The low flow rate in the channel allowed a hydraulic flow of the feed water to inside the channel only driven by osmotic difference. The low flow rate of recirculation allowed a reduced energy consumption of the system, when compared to counter-flow FO membrane system (Cath *et al.* 2005a; Cath *et al.* 2005b; Choi *et al.* 2009; Cath *et al.* 2010; Qin *et al.* 2010; Tan & Ng, 2010). A stirrer operating at 320 rpm was used to provide movement of the feed water inside the tank, with water flowing across the membrane. After 24 hours, the DS increased its volume due to continuous osmosis between the feed water and the draw solution recirculating in the cells. The FW decreased its volume every day, but more FW was introduced to the FW tank after each cycle. The diluted DS was transferred to the feed tank of the LPRO setup. The cycle was repeated every day by replacing the fresh DS, and then filling the LPRO feeding tank.

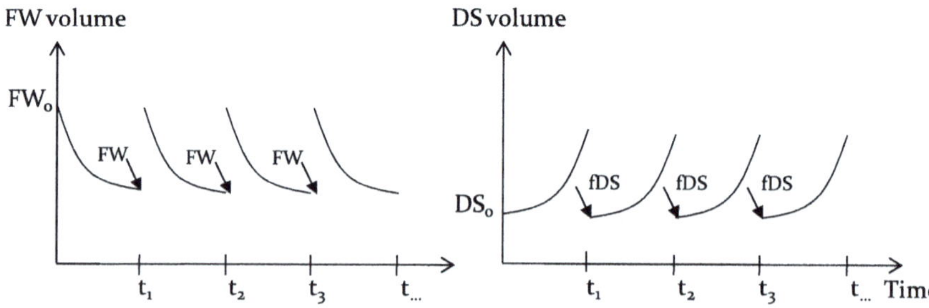

Figure 2.3.2 Cycle of forward osmosis process: FW (feed water), DS (draw solution), fDS (fresh draw solution).

2.3.3 THEORETICAL BACKGROUND

The osmotic flux of the FO membranes was calculated using Equation 2.3.1. Where ΔV is the differential volume change of draw solution (L); A is the membrane area (m^2); and t is the time (h).

$$J = \Delta v / At \tag{2.3.1}$$

The osmotic flux is proportional to the driving osmotic pressure difference, which is demonstrated by the decrease in conductivity. An equation (Equation 2.3.2) for the flux of osmosis membranes when a low concentrated solution is facing the thin-film side of the membrane, and the porous support (mesh) is facing a high concentrated solution was derived by Loeb *et al.* (1997).

$$J_w = \frac{1}{K} \ln\left(\frac{\pi_{\text{Hi}}}{\pi_{\text{Low}}}\right) \tag{2.3.2}$$

Where J_w is the osmotic water flux (L · m^{-2} · h^{-1}), K is the solute resistivity of the membrane (m^2 · h · L^{-1}), π_{Hi} is the osmotic pressure in the high concentrated solution (bar), and π_{Low} is the osmotic pressure in the low concentrated solution (bar). Loeb's equation can be slightly modified and applied to model the flux decline of the dilution experiment. The conductivity can be assumed to be directly proportional to the concentration of the draw solution and hence also proportional to the osmotic pressure, the same can be said for the feed water. In this case $\pi_{SW} = \pi_{Hi}$ and $\pi_{FW} = \pi_{Low}$. Assuming that for the seawater and the feed water, $\ln(\pi_{SW}/\pi_{FW}) = \alpha(\gamma_{SW} - \gamma_{FW}) + \beta$, with γ denoting conductivity ($mS \cdot cm^{-1}$), Equation 2.3.2 can be written as Equation 2.3.3; in this way K' ($mS \cdot cm^{-1} \times m^2 \cdot h \cdot L^{-1}$) can be calculated by fitting the data of conductivity measurements of the feed water and the draw solution. The modelled flux is obtained by using the estimated K' in Equation 2.3.3, and the conductivity data over time.

$$J_w = \frac{1}{K'}(\gamma_{SW} - \gamma_{FW}) \tag{2.3.3}$$

Gray *et al.* (2006), Lay *et al.* (2010) and Tang *et al.* (2010) reported the occurrence of dilutive internal concentration polarization (dilutive ICP) of the FO membrane when the DS is against the support layer (or active layer facing feed water, AL-facing-FW), which is the membrane orientation used during our experiments. This configuration is the most favourable configuration to reduce membrane fouling and to allow a less significant loss of flux when compared to the active layer facing the DS (AL-facing-DS) (Tang *et al.* 2010). Tan and Ng (Tan & Ng, 2010) also reported the occurrence of dilutive ICP in the reverse mode (AL-facing-FW), Tan and Ng concluded that changes in the cross-flow velocities did not affect the water flux across the membrane.

The components of natural organic matter (NOM) present in surface water and secondary wastewater effluents are the most important foulants in water reuse facilities operating with membranes (Kennedy *et al.* 2007; Zheng *et al.* 2010). During FO, interactions between the membrane and the NOM in the feed water cause membrane fouling and therefore a decrease of the membrane flux, besides a decrease of flux due to dilution of the DS. For filtration systems operating in batch cycles, reversible and irreversible fouling can be represented by differences of normalized fluxes (Figure 2.3.3). Reversible fouling means that the fouling can be removed with membrane cleaning such as air scouring or chemical cleaning of the membrane. Reversible fouling involves a relatively medium-term build-up of a foulant layer or the formation of a cake layer at the surface (active layer) of the FO membrane. Irreversible fouling corresponds to that when washing or chemical cleaning does not restore the original flux value, and is caused by more or less permanent deposition of particles on the surface of the membrane, and is characterized by a longer-term decline in flux. After a certain number of cycles (n) and at the end of a filtration period of n cycles, the flux decline is defined as:

$$FD\,(\%) = \frac{(NF_1 - NF_n)}{NF_1} \times 100 \qquad (2.3.4)$$

Where FD is defined as *flux decline*, NF_n is the final normalized flux after n filtration cycles, and NF_1 is the final normalized flux after the first cycle. The apparent irreversible fouling is defined as:

$$Ira(\%) = (NF_1 - NF_{n+1}) \times 100 \qquad (2.3.5)$$

Where Ira is defined as *apparent irreversible fouling*, NF_{n+1} is the final normalized flux after cleaning the membrane after n cycles of operation (air scouring with FW, air scouring with clean water, chemical cleaning) and NF_1 is the final normalized flux after the first cycle. The reversible fouling (Rv) is defined as:

$$Rv(\%) = (1 - Ira) \times 100 \qquad (2.3.6)$$

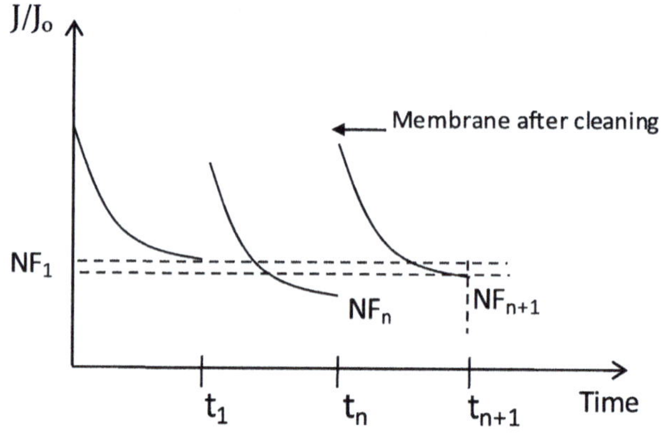

Figure 2.3.3 Scheme for definition of reversible and irreversible fouling: NF (normalized flux).

2.3.4 RESULTS AND DISCUSSION

2.3.4.1 Feed water and draw solution characterization

The characteristics of the SWWE (effluent from Jeddah) are summarized in Table 2.3.1. The pre-filtered seawater (Red Sea water) follows the characterization given in Table 2.3.2.

Table 2.3.1 Wastewater effluent characteristics.

	SWWE Jeddah
Temperature (°C)	20.7
Conductivity (μS · cm^{-1})	4300
pH	7.3
DOC (mg · L^{-1})	5.3
BOD_5 (mg · L^{-1})	20
UVA_{254} (cm^{-1})	0.130
SUVA (L · mg^{-1} · m^{-1})	2.45
Calcium (mg · L^{-1})	108
DO (mg · L^{-1})	6.3
Nitrate (mg · L^{-1})	2.27
Nitrite (mg · L^{-1})	<0.015
Ammonium (mg · L^{-1})	0
Phosphate (mg · L^{-1})	1.05

Table 2.3.2 Seawater (SW) characterization.

	0.45 μm Pre-Filtered SW
Temperature (°C)	20.5
Conductivity (μS · cm^{-1})	57,500
pH	7.8
DOC (mg · L^{-1})	1.12
UVA_{254} (cm^{-1})	0.012
SUVA (L · mg^{-1} · m^{-1})	1.07
TDS (mg · L^{-1})	40,500
SDI	2
Barium (mg · L^{-1})	0.01
Calcium (mg · L^{-1})	571
Magnesium (mg · L^{-1})	1458
Potassium (mg · L^{-1})	488
Sodium (mg · L^{-1})	12,470
Strontium (mg · L^{-1})	7
Bicarbonate (mg · L^{-1})	141
Boron (mg · L^{-1})	2.2
Carbonate (mg · L^{-1})	8.0
Chloride (mg · L^{-1})	23,073
Fluoride (mg · L^{-1})	1.5
Sulfate (mg · L^{-1})	2400

2.3.4.2 Long-term forward osmosis experiments

The forward osmosis experiment was performed during 10 cycles before cleaning the membrane. The FO flux decline during the first 7 cycles is given in Figure 2.3.4. In the same figure, the modelled FO flux is calculated by using Equation 2.3.3. It can be clearly observed that the equation is able to model the measured FO flux decline. The peaks in Figure 2.3.4, occurring at the beginning of each cycle, are more difficult to model due to the mixing of remaining DS in the FO cell and the time of stabilization of the membrane to the fresh DS; this is more evident after the first 3 cycles. The flux fluctuated in the range 1.5–5 $L \cdot m^{-2} \cdot h^{-1}$. Higher fluxes corresponded to fresh draw solutions fillings. The forward osmotic flux decreased due to continuous dilution of the feed water flowing into the osmotic membrane cell.

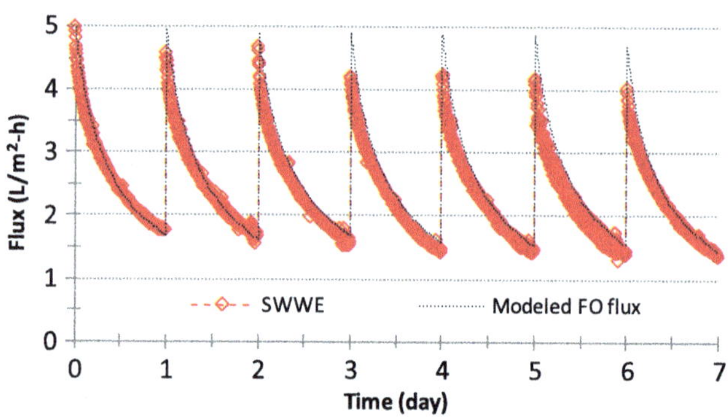

Figure 2.3.4 Forward osmosis flux versus time, and modelled FO flux versus time.

The complete number of cycles (10) before performing the cleaning of the membrane is shown in Figure 2.3.5, showing normalized fluxes. After 10 cycles, the flux decline was 28% due to fouling of the FO membrane. Then, the FO membranes were hydraulically cleaned with "air scouring with clean water" for half an hour. The reversible fouling (*Rv*) was 98.8% and the apparent irreversible fouling (*Ira*) was only 1.2%. The term apparent irreversible fouling is used in order to clarify that less or more flux may be recovered depending on whether concentrated feed water, feed water, or clean water is used during the periodical air scouring. The following 4 cycles after cleaning the membrane show that the flux recovery of the FO membrane was quite acceptable. Replication experiments, with a different batch of feed water and with results not reported in this publication, were performed for a period of 12 days. After this period, the membrane fouling which accumulated was also removed by air scouring, but this time using concentrated FW remaining in the tank, and for a period of 15 minutes. The flux recovered in this case was 90%; therefore, air scouring with clean water for a longer period of time was more effective than using concentrated FW and less time. These results demonstrate that the immersed FO membrane approach using feed tanks has advantages over FO counter-flow membrane contactors, where high cross-flow velocities are necessary to hydraulically control the fouling; alternatively, air scouring in spiral-wound membrane modules can alleviate fouling, and sometimes physical cleaning (scrubbing) is needed to partially restore the initial membrane flux (Cath *et al.* 2009; Qin *et al.* 2010; Tan & Ng, 2010).

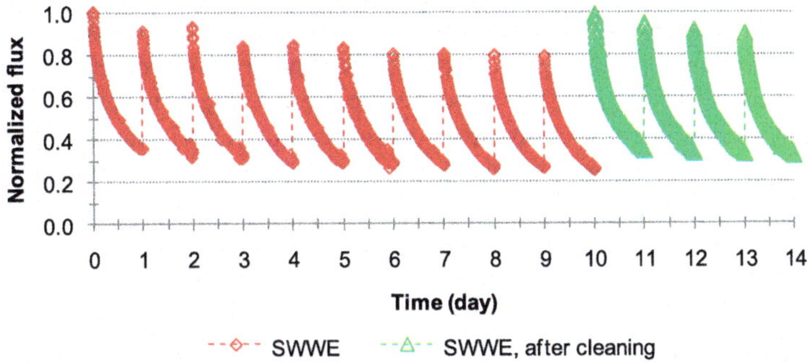

Figure 2.3.5 Normalized forward osmosis flux versus time, SWWE (secondary wastewater effluent).

Desalination of the diluted DS was carried out with a LPRO unit. The operating flux of the LPRO unit was 7 L · m^{-2} · h^{-1} at a pressure of 15 bar, with a recovery of 2%. By relating conductivity to total dissolved solids (TDS), the TDS is shown in Figure 2.3.6; cycling resulted in favorable TDS content for feeding the LPRO system. Calculations with the final quality of the permeate demonstrated that more than 98% of dissolved salts were rejected.

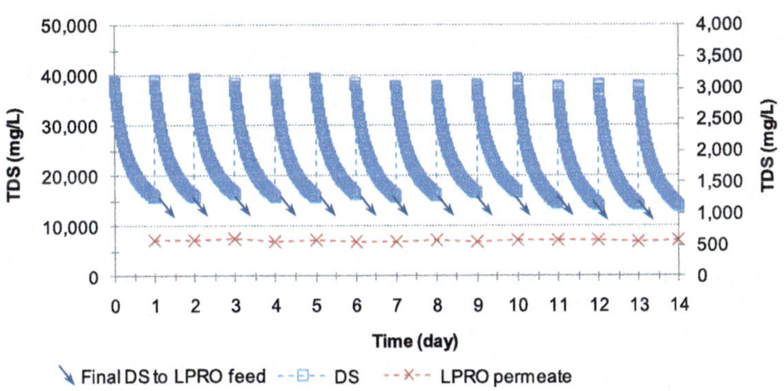

Figure 2.3.6 Concentration of total dissolved solids (TDS) in draw solution (DS) and permeate of LPRO versus time.

The use of seawater is an appropriate draw solution for water reuse applications with FO membranes. Seawater is preferred over concentrate (retentate) from existing desalination plants because: (i) concentrates or brines contain high concentration of salts, and residuals of seawater pretreatment (pH regulators, anti-scalants, coagulants, sodium metabisulfite) can impact FO membrane performance; (ii) shorter-term versus long-term cycles of osmotic operation in order to obtain a suitable dilution of the draw solution; (iii) lower operating costs for desalination of the FO-diluted solution (low-pressure) against high-energy desalination similar to high pressure RO.

The diluted draw solution (diluted seawater) was characterised to determine concentrations of important parameters. This information is presented in Table 2.3.3; as expected, all of the ions and cations were present at a reduced concentration. However, a small increase of dissolved organic carbon (DOC) was identified in the diluted seawater, 1.19 mg · L^{-1} from 1.12 mg · L^{-1}. The increment may possibly be explained by passage of low molecular weight (LMW) neutrals from the feed water (SWWE) to the seawater, which is clearly confirmed in Figure 2.3.7a. Figure 2.3.7a and 2.3.7b are liquid chromatography – organic carbon detection (LC-OCD) chromatograms of the initial seawater and the diluted seawater, and SWWE from Jeddah, respectively. Figure 2.3.7b shows that biopolymers, humic substances, building blocks and low molecular weight acids did not pass through the FO membrane.

Table 2.3.3 FO-diluted seawater characterization.

	0.45 μm Pre-Filtered SW after Dilution
Conductivity (μS · cm^{-1})	21,900
Temperature (°C)	20.5
pH	7.6
DOC (mg · L^{-1})	1.19
TDS (mg · L^{-1})	15,640
UVA$_{254}$ (cm^{-1})	0.025
SUVA (L · mg^{-1} · m^{-1})	2.10
Calcium (mg · L^{-1})	257
Magnesium (mg · L^{-1})	639
Potassium (mg · L^{-1})	188
Sodium (mg · L^{-1})	5620
Strontium (mg · L^{-1})	3.9
Bicarbonate (mg · L^{-1})	65
Boron (mg · L^{-1})	1
Chloride (mg · L^{-1})	10,288
Fluoride (mg · L^{-1})	0.6
Sulfate (mg · L^{-1})	1122
Nitrate (mg · L^{-1})	0.0
Nitrite (mg · L^{-1})	$<$0.015
Ammonium (mg · L^{-1})	0.0
Phosphate (mg · L^{-1})	0.52

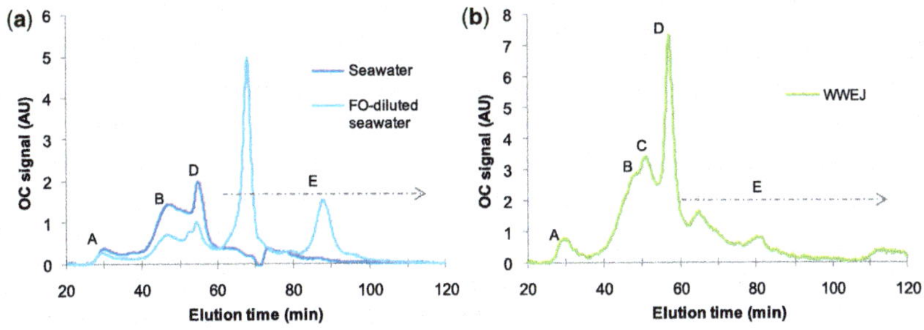

Figure 2.3.7 (a) LC-OCD chromatograms of seawater and FO-diluted seawater; (b) LC-OCD chromatogram of wastewater effluent from Jeddah (WWEJ); A = biopolymers; B = Humic substances; C = building blocks; D = low molecular weight acids; E = low molecular weight neutrals.

2.3.5 ENERGY, COST AND WATER REUSE CONSIDERATIONS

2.3.5.1 Comparison of energy use

The energy consumption for desalinating water with RO membranes lies between 2.5–4 kWh · m^{-3} (Fritzmann *et al.* 2007; Ghaffour & Venkat, 2009), this as a result of the development of new efficient membranes and the use of energy recovery devices over the last decade or so. The total energy consumption associated with the proposed technology (FO membrane cells immersed in tanks) of FO-LPRO revealed a conservative estimated range of 1.3–1.5 kWh · m^{-3} for desalinating diluted seawater with water recovery from a SWWE. A breakdown of the energy consumption calculation for FO-LPRO, and a comparison to RO (Fritzmann *et al.* 2007) is presented in Figure 2.3.8.

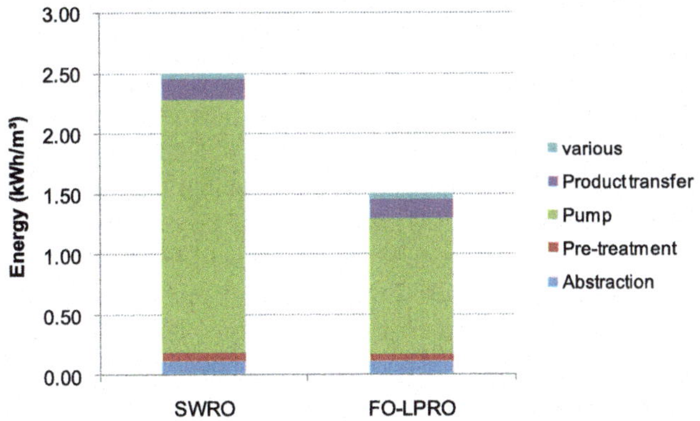

Figure 2.3.8 Comparison of energy consumption between desalination with RO and desalination with immersed FO-LPRO.

A comparison with existing SWWE water reclamation facilities (de Koning *et al.* 2008; Van Houtte & Verbauwhede, 2008) makes FO-LPRO competitive; existing water reuse installations using membrane filtration (microfiltration or ultrafiltration) and RO have an overall energy demand of 1.5–1.7 kWh · m^{-3}. Therefore, indirect desalination with "immersed" FO membranes and LPRO is an attractive consideration at almost half of the energy demand of high pressure RO desalination.

2.3.5.2 Cost analysis

In this analysis, four configurations were considered. The first and second configurations were already defined: FO-LPRO, and FO only. The third and fourth technologies correspond to typical configurations for water reuse and desalination, UF-LPRO and high pressure SWRO, respectively. The plant capacity for all configurations was assumed as 100 $m^3 \cdot h^{-1}$. The cost analysis assumed an interest rate of 6% amortized in a period of 20 years. The cost of energy was assumed as 0.08 USD · kWh^{-1}. The assumption for the FO plate and frame membrane module cost was 100 USD · m^{-2}, the FO membrane cost is 30 USD · m^{-2}. Capital expenditure (CAPEX) and operational expenditure (OPEX) assumptions corresponded to market prices for constructed ultrafiltration (UF), LPRO in water reuse, and RO membrane desalination projects (Elarde & Bergman, 2001, 2003; Yangali-Quintanilla *et al.* 2010). CAPEX and OPEX for FO was based on own estimations. The benefit-cost ratio (BCR) was calculated by dividing the extra CAPEX to the present value of savings in operation and maintenance costs. BCRs greater or equal than 1 are defined as optimal, and BCRs less than 1 are not economically attractive (NEA). When comparing FO-LPRO vs. SWRO (Figure 2.3.9), optimal BCRs are achieved at average FO fluxes greater or equal to 5.45 L · $m^{-2} \cdot h^{-1}$. However, a minimum average FO flux of 10.52 L · $m^{-2} \cdot h^{-1}$ is required to reach an optimal BCR to replace UF-LPRO with FO-LPRO. Nevertheless, this calculation is based only on production costs. For water reuse considerations, FO-LPRO may present more advantages if compared to UF-LPRO: better quality of product water, less energy consumption, less consumption of chemicals, which translate into less environmental impact.

Figure 2.3.10 presents a comparison of production costs per m^3 of water. FO-LPRO is still not economically viable if compared to SWRO; average FO fluxes lower than 5.5 L · $m^{-2} \cdot h^{-1}$ represent a cost of ~0.91 USD · m^{-3}. This average FO flux barrier is expected to be changed with FO membranes of second generation (American Water Intelligence, 2011; Winter, 2011; Yu *et al.* 2011). Fluxes greater than 10.5 L · $m^{-2} \cdot h^{-1}$ will be needed to lower costs of FO-LPRO when compared to UF-LPRO (~0.77 USD · m^{-3}). However, as already mentioned, decisions to move from UF-LPRO may be beyond cost considerations, being those considerations related to environmental aspects. The winner in production costs (Figure 2.3.10) is FO *per se*, the technology can produce water with good enough quality to be used in applications mentioned in the following section, which presents alternatives of water reuse for direct use of diluted draw solutions; in this way, even lower energy use than the values previously mentioned can be achieved.

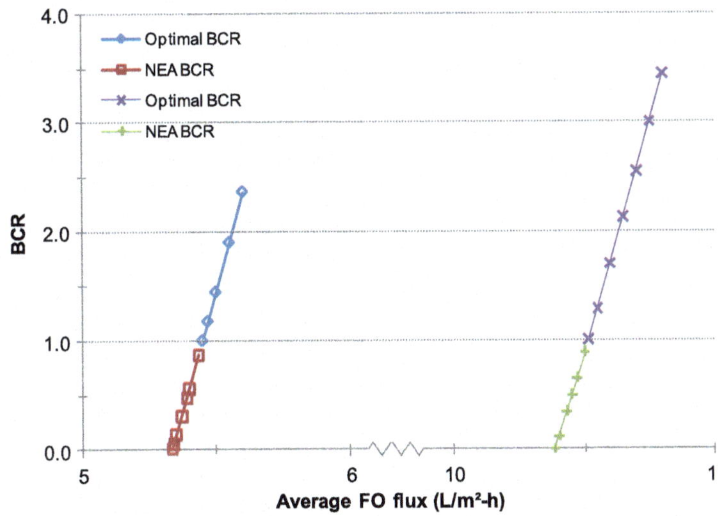

Figure 2.3.9 Cost analysis for benefit-cost ratio (BCR) between forward osmosis – low pressure reverse osmosis FO-LPRO vs seawater reverse osmosis SWRO and FO-LPRO vs UF-LPRO (100 $m^3 \cdot h^{-1}$); FO-LPRO vs SWRO (◊, □); FO-LPRO vs UF-LPRO (×, +); not economically attractive (NEA).

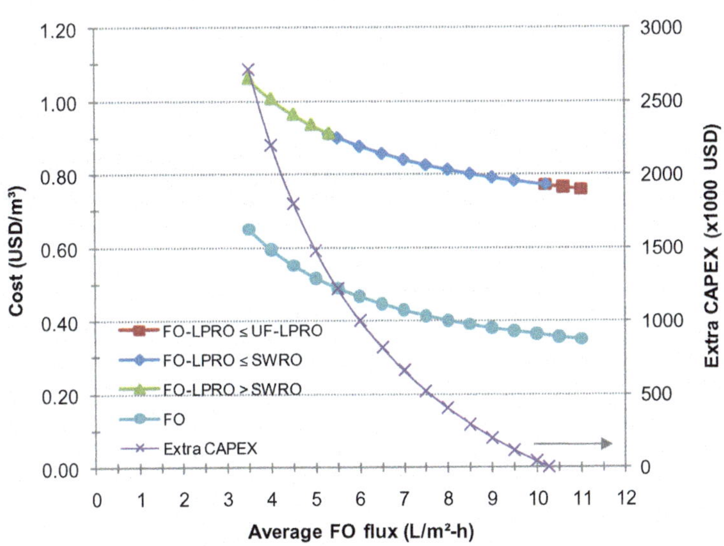

Figure 2.3.10 Cost analysis for water production of FO-LPRO, UF-LPRO, SWRO and FO (100 $m^3 \cdot h^{-1}$); extra capital expenditure (CAPEX) for FO-LPRO vs. SWRO is shown in left vertical axis.

2.3.5.3 Alternative water reuse of diluted draw solutions

It was mentioned that low salinities can be reached by the FO system described in the present publication (~15 g · L^{-1} as TDS). It is important to mention that this salinity can be even lowered to 6–10 g · L^{-1}, when: 1) using a reduced volume of DS at the beginning of each FO cycle, 2) using a less concentrated DS (normal seawater has a TDS of 35 g · L^{-1}), and 3) using more FO membrane area. Thus, the final TDS after the FO process can be controlled. This condition opens possibilities for direct use of a diluted draw solution. One option can be the use of the low salinity water as water for aquaculture. Low salinity (4–10 g · L^{-1}) shrimp farming has been widely used in Thailand (Szuster & Flaherty, 2001) and there is interest in Saudi Arabia to move from seawater aquaculture to brackish water aquaculture (shrimps) employing *partial* desalination. A major aquaculture (prawn) company in Saudi Arabia is looking into available alternatives to increase provision of clean brackish water, and one possibility could be the aforementioned condition of diluted seawater with FO. Irrigation of crops with saline waters has been investigated in Saudi Arabia by Hussain and Al-Saati (Hussain & Al-Saati, 1999). They indicated that mixing saline waters with normal irrigation water is an option; therefore, a better hypothesized option may be the direct use or mixing of diluted seawater (less saline water) with normal irrigation water or with treated wastewaters. The trade-offs between using a plain secondary wastewater effluent versus a mixed water should be further investigated, but definitely one advantage of the latter is the lower presence of toxic heavy metals and other micropollutants, therefore a minimized presence or absence of toxic heavy metals in crops and soils is expected.

2.3.6 SUMMARY

The use of energy still remains the main component of the costs of desalting water. Forward osmosis (FO) can help to reduce the costs of desalination, and extracting water from impaired sources can be beneficial in this regard. Experiments with FO membranes using a secondary wastewater effluent as a feed water and Red Sea water as a draw solution demonstrated that the technology is promising. FO coupled with low pressure reverse osmosis (LPRO) was implemented for indirect desalination. The system consumes only 50% (~1.5 kWh · m^{-3}) of the energy used for high pressure seawater RO (SWRO) desalination (2.5–4 kWh · m^{-3}), and produces a good quality water extracted from the impaired feed water. Fouling of the FO membranes was not a major issue during long-term experiments over 14 days. After 10 days of continuous FO operation, the initial flux declined by 28%. Cleaning the FO membranes with air scouring and clean water recovered the initial flux by 98.8%. A cost analysis revealed FO *per se* as viable technology. However, a minimum average FO flux of 10.5 L · m^{-2} · h^{-1} is needed to compete with water reuse using UF-LPRO, and 5.5 L · m^{-2} · h^{-1} is needed to recover and desalinate water at less cost than SWRO.

2.3.7 REFERENCES

Achilli A., Cath T. Y., Marchand E. A. and Childress A. E. (2009). The forward osmosis membrane bioreactor: a low fouling alternative to MBR processes. *Desalination*, **239**, 10–21.

American Water Intelligence (2011). The Race to Commercialize Forward Osmosis. American Water Intelligence, Media Analytics Ltd., Oxford, UK.

Cath T. Y., Adams D. and Childress A. E. (2005a). Membrane contactor processes for wastewater reclamation in space: II. Combined direct osmosis, osmotic distillation, and membrane distillation for treatment of metabolic wastewater. *Journal of Membrane Science*, **257**, 111–119.

Cath T. Y., Gormly S., Beaudry E. G., Flynn M. T., Adams V. D. and Childress A. E. (2005b). Membrane contactor processes for wastewater reclamation in space: Part I. Direct osmotic concentration as pretreatment for reverse osmosis. *Journal of Membrane Science*, **257**, 85–98.

Cath T. Y., Drewes J. E. and Lundin C. D. (2009). A Novel Hybrid Forward Osmosis Process for Drinking Water Augmentation using Impaired Water and Saline Water Sources.

Cath T. Y., Hancock N. T., Lundin C. D., Hoppe-Jones C. and Drewes J. E. (2010). A multi-barrier osmotic dilution process for simultaneous desalination and purification of impaired water. *Journal of Membrane Science*, **362**, 417–426.

Choi Y.-J., Choi J.-S., Oh H.-J., Lee S., Yang D. R. and Kim J. H. (2009). Toward a combined system of forward osmosis and reverse osmosis for seawater desalination. *Desalination*, **247**, 239–246.

Comprehensive Assessment of Water Management in Agriculture (2007). Water for Food, Water for Life: A Comprehensive Assessment of Water Management in Agriculture. Earthscan, London; International Water Management Institute, Colombo.

de Koning J., Bixio D., Karabelas A., Salgot M. and Schäfer A. (2008). Characterisation and assessment of water treatment technologies for reuse. *Desalination*, **218**, 92–104.

Elarde J. and Bergman R. (2001). The cost of membrane filtration for municipal water supplies. In: Membrane Technology Conference. American Water Works Association, San Antonio, Texas.

Elarde J. and Bergman R. (2003). The cost of membrane softening and desalting for municipal water supplies. In: Membrane Technology Conference. American Water Works Association, Atlanta, Georgia.

Fritzmann C., Löwenberg J., Wintgens T. and Melin T. (2007). State-of-the-art of reverse osmosis desalination. *Desalination*, **216**, 1–76.

GBI Research (2010). Desalination Market to 2020 – Technology Driven Cost Reduction in Membrane Based Processes Set to Drive Sustainability Investments into the Market. GBI Research, Dublin, Ireland.

Ghaffour N. and Venkat R. (2009). The True Cost of Water Desalination: Review and Evaluation. In IDA World Congress Atlantis, The Palm, Dubai, UAE.

Global Water Intelligence (2005). Water Reuse Markets 2005–2015: A Global Assessment and Forecast. Global Water Intelligence, Media Analytics Ltd., Oxford, UK.

Global Water Intelligence (2006). Desal's Double Digit Future. Global Water Intelligence, Media Analytics Ltd., Oxford, UK.

Global Water Intelligence (2008). Desalination in 2008 Global Market Snapshot. Global Water Intelligence DesalData/IDA, Media Analytics Ltd., Oxford, UK.

Gray G. T., McCutcheon J. R. and Elimelech M. (2006). Internal concentration polarization in forward osmosis: role of membrane orientation. *Desalination*, **197**, 1–8.

Holloway R. W., Childress A. E., Dennett K. E. and Cath T. Y. (2007). Forward osmosis for concentration of anaerobic digester centrate. *Water Research*, **41**, 4005–4014.

Hussain G. and Al-Saati A. J. (1999). Wastewater quality and its reuse in agriculture in Saudi Arabia. *Desalination*, **123**, 241–251.

Kennedy M., Yangali-Quintanilla V., Heijman B. G. J., Amy G. and Schippers J. C. (2007). Colloidal and Non-colloidal NOM Fouling of UF Membranes. In: Membrane Technology Conference and Exposition. American Water Works Association, Tampa, Florida.

Lay W. C. L., Chong T. H., Tang C. Y., Fane A. G., Zhang J. and Liu Y. (2010). Fouling propensity of forward osmosis: investigation of the slower flux decline phenomenon. *Water Science and Technology*, **61**, 927–935.

Loeb S., Titelman L., Korngold E. and Freiman J. (1997). Effect of porous support fabric on osmosis through a Loeb-Sourirajan type asymmetric membrane. *Journal of Membrane Science*, **129**, 243–249.

McCutcheon J. R., McGinnis R. L. and Elimelech M. (2005). A novel ammonia – carbon dioxide forward (direct) osmosis desalination process. *Desalination*, **174**, 1–11.

Qin J.-J., Kekre K. A., Oo M. H., Tao G., Lay C. L., Lew C. H., Cornelissen E. R. and Ruiken C. J. (2010). Preliminary study of osmotic membrane bioreactor: effects of draw solution on water flux and air scouring on fouling. *Water Science & Technology*, **62**, 1353–1360.

Szuster B. W. and Flaherty M. (2001). Inland low-salinity shrimp farming in the central plains region of Thailand.

Tan C. H. and Ng H. Y. (2010). Modeling of external and internal concentration polarization effect on flux behaviour of forward osmosis. *Water Science & Technology*, **8**, 533–539.

Tang C. Y., She Q., Lay W. C. L., Wang R. and Fane A. G. (2010). Coupled effects of internal concentration polarization and fouling on flux behavior of forward osmosis membranes during humic acid filtration. *Journal of Membrane Science*, **354**, 123–133.

Van Houtte E. and Verbauwhede J. (2008). Operational experience with indirect potable reuse at the Flemish Coast. *Desalination*, **218**, 198–207.

Winter C. (2011). Innovator: Robert McGinnis of Oasys Water. BussinessWeek.

Yangali-Quintanilla V., Maeng S. K., Fujioka T., Kennedy M. and Amy G. (2010). Proposing nanofiltration as acceptable barrier for organic contaminants in water reuse. *Journal of Membrane Science*, **362**, 334–345.

Yu Y., Seo S., Kim I.-C. and Lee S. (2011). Nanoporous polyethersulfone (PES) membrane with enhanced flux applied in forward osmosis process. *Journal of Membrane Science*, **375**, 63–68.

Zheng X., Ernst M., Huck P. M. and Jekel M. (2010). Biopolymer fouling in dead-end ultrafiltration of treated domestic wastewater. *Water Research*, **44**, 5212–5221.

Part 3
Fouling

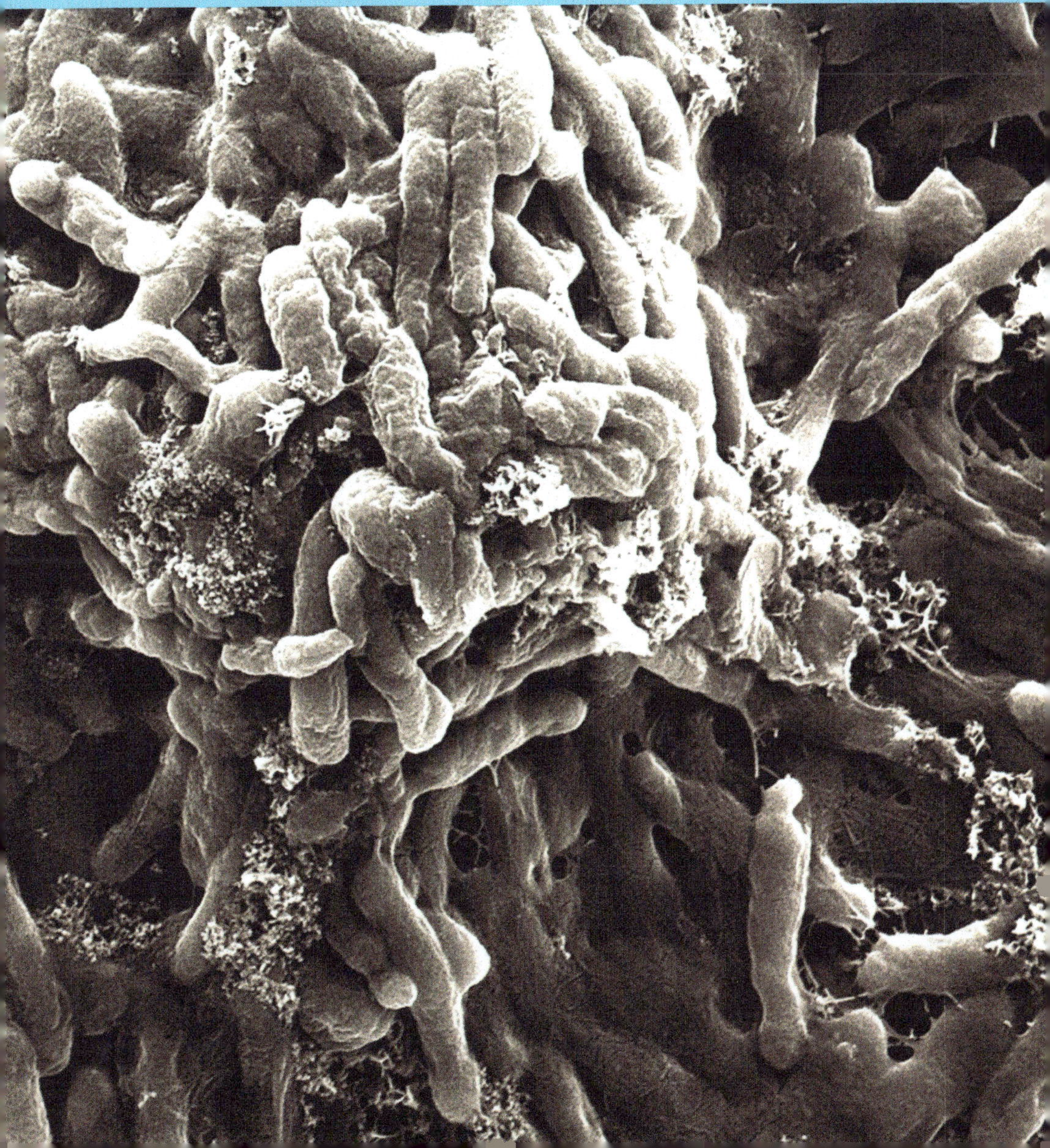

Chapter 3.1

Fouling propensity during desalination of seawater*

3.1.1 INTRODUCTION

Due to increasing population, industry, agriculture, economic growth and urbanization, water supply will undoubtedly become scarcer in the years to come. New technologies are being explored to recycle and reuse treated wastewater and/or desalinate seawater or brackish water. Membranes have been largely used for specialized applications for several decades, largely for water treatment to supply the growing demand for fresh water. In addition to thermal technologies, such as Multi Stage Flash, seawater (SW) desalination with membrane technologies including reverse osmosis (RO) and nanofiltration (NF) have been successfully practiced and commercialized in areas proximate to the ocean. However, both RO and NF are processes driven by a high hydraulic pressure. These processes require an intensive energy input and the water recovery is often limited. Membrane fouling, brine disposal and environmental impacts are also critical constraints to RO and NF systems.

Forward osmosis (FO) is a novel membrane process that can potentially be used as an energy-saving alternative to conventional membrane processes, such as RO in water treatment. The driving force in the FO process is an osmotic pressure created by the osmotic gradient when a high concentration of a saline solution flows along one side of the membrane versus a low concentration of feed along the other side. Since the FO process utilizes natural osmosis as the pressure source, the energy consumption can be significantly reduced compared to other pressure driven membrane processes (e.g., RO). FO has been considered as a possible technology

---

* This chapter is based, with permission from the copyright holder, on the previously published paper. (Li Z.-Y., Yangali-Quintanilla V., Valladares-Linares R., Li Q., Zhan T. and Amy G. (2012). Flux patterns and membrane fouling propensity during desalination of seawater by forward osmosis. *Water Research*, **46**, 195–204. doi:10.1016/j.watres.2011.10.051)

for water treatment and related processes. Lab and pilot scale research of FO have been conducted for many applications, such as desalination (McCutcheon *et al.* 2005, 2006), concentration of brine (Tang & Ng, 2008), hybrid process for water reuse (Cath *et al.* 2010), power generation (Loeb, 2002; Achilli *et al.* 2009a), osmotic bioreactor (Achilli *et al.* 2009b), liquid food concentration (Petrotos & Lazarides, 2001), and medical and pharmaceutical applications (Wright *et al.* 2003; Yang *et al.* 2009).

Although FO has been increasingly studied for many applications, there are several research areas needed to move the FO process forward including innovation of favorable FO membranes and modules, understanding of FO membrane fouling, and identification of an appropriate draw solution (DS). Most work has been focused on developing new FO membrane (such as well-constructed cellulose acetate FO membrane (Zhang *et al.* 2010), high performance thin film membrane (Yip *et al.* 2010) and thin film composite FO hollow fiber membrane (Wang *et al.* 2010a) and draw solutes (such as ammonia-carbon dioxide (McCutcheon *et al.* 2005)) and water soluble magnetic nanoparticles (Ling *et al.* 2010)). However, very few studies on FO membrane fouling under natural water conditions have been reported. In one study, Mi and Elimelech (2008) researched organic fouling on the FO membrane by using some model foulants including humic acid, alginate and protein, all foulants caused flux decline during the FO process. Another study from the same team (Mi & Elimelech, 2010b) showed that alginate fouling on an FO membrane was almost fully reversible. Wang *et al.* (2010b) used a latex microparticle as a model foulant and observed the surface coverage of the FO membrane by latex particles. Lay *et al.* (2010) employed a silicon dioxide nanoparticle as a model foulant and explained a slow flux decline phenomenon in FO. Gypsum scaling was also able to form on the membrane during the FO process, but was fully reversible by water rinsing (Mi & Elimelech, 2010a). However scaling was not observed on the FO membrane in an osmotic membrane bioreactor with synthetic wastewater as the feed (Lay *et al.* 2011). When activated sludge was used as the feed in an osmotic membrane bioreactor, both reversible and irreversible fouling were not found on the FO membrane under selected operational condition (Cornelissen *et al.* 2008).

So far, most studies of the FO process have only been based on some simple model foulants for fouling analysis in short-term tests. Although FO is considered to be a promising technology for SW desalination, there have been no published studies on fouling propensity of FO with real SW, particularly in long-term tests. In the present study, we investigated flux patterns and membrane fouling propensity of FO with natural SW as the feed (SWFO). SWFO experiments were performed for 18 to 30 day periods. To the best of our knowledge, these are the first results about fouling on an FO membrane surface with real SW as the feed. The results reported in this study can be valuable for developing a successful protocol of pretreatment and membrane cleaning during SWFO.

3.1.2 MATERIALS AND METHODS

3.1.2.1 Feed and draw solution

Natural SW from the Red Sea was used as the feed in this study. SW was collected from the line that provides natural SW to a commercial RO desalination plant at the King Abdullah University of Science and Technology (KAUST). Before the SWFO process, natural SW was pre-filtered by a capsule filter (pore size 10 μm, Whatman, US) to remove large particles. The quality of SW pre-filtered by a 10 μm capsule filter is given in Table 3.1.1. In order to prevent the growth of bacteria and algae which may change the properties of the feed, 0.02% NaN_3 was added to the pre-filtered SW and SWFO was operated under dark condition to inhibit bioactivity. In addition, 4% NaCl solution (with 0.02% NaN_3) was used as a simulated seawater (SSW) for a baseline measurement. The conductivity of the SSW was adjusted to 54 ± 1 mS · cm^{-1} which was equal to SW used as the feed in this study. NaCl is also the most commonly used draw solute in FO because it has high solubility, low cost and is easily to reconcentrate to high concentration without scaling problems by conventional methods downstream of FO, such as low pressure RO, or possibly, membrane distillation (MD). The draw solution (DS) was prepared by dissolving NaCl in deionized water obtained from a Milli-Q ultrapure water purification system (Millipore, MA) at a concentration of 2 M.

Table 3.1.1 Characterization of SW pre-filtered by a 10 μm capsule filter.

Analyte Conductivity (mS · cm^{-1})	SW 54 ± 1
pH	8.2 ± 0.1
DOC (mg · L^{-1})	0.98
UVA_{254} (cm^{-1})	0.016
SUVA (L · mg^{-1} · m^{-1})	1.63
TDS (mg · L^{-1})	37,400
Sodium (mg · L^{-1})	11,340
Potassium (mg · L^{-1})	477
Calcium (mg · L^{-1})	557
Magnesium (mg · L^{-1})	1427
Silicon (mg · L^{-1})	0.24
Iron (mg · L^{-1})	0.12
Strontium (mg · L^{-1})	7.0
Boron (mg · L^{-1})	2.2
Bicarbonate (mg · L^{-1})	137
Carbonate (mg · L^{-1})	7.2
Chloride (mg · L^{-1})	21,473
Fluoride (mg · L^{-1})	1.7
Sulfate (mg · L^{-1})	2312

3.1.2.2 Forward osmosis set-up and fouling tests

A commercial flat sheet FO membrane was used in this study. The membrane is made of cellulose triacetate (CTA) embedded about a polyester (PES) screen mesh (HTI, LLC, Albany, OR). The active layer (AL) is the shiny side of the membrane (Figure 3.1.1). Before using, the membrane was soaked in deionized (DI) water for 24 hours to remove the glycerin which was used to replace the water during shipment.

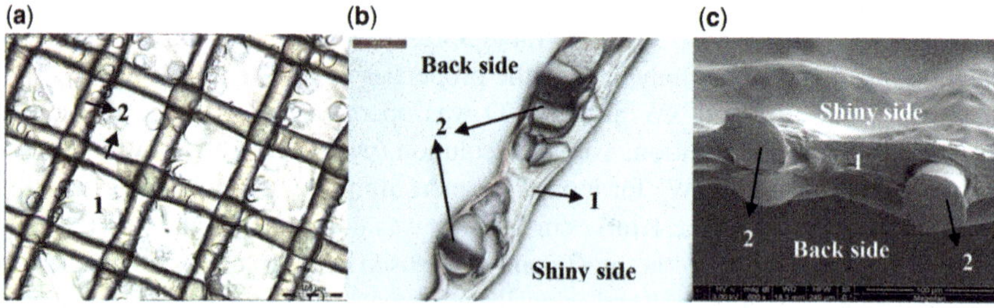

Figure 3.1.1 HTI CTA FO membrane (a, top view of the membrane under optical microscopy; b, cross-section of the membrane under optical microscopy; c, cross-section of the membrane under scanning electron microscopy; 1, CTA layer for filtration and salt rejection; 2, embedded PES mesh screen for mechanical support; scale bar 100 μm in a and c, 50 μm in b).

A bench-scale FO membrane system was employed in this study. The membrane cell was structured with equal channels on both sides providing an effective membrane area of 2×10 cm^2 and a channel height of 0.2 cm for filtration. Two peristaltic pumps (Core-Parmer) were applied for the circulation of feed and draw solutions, respectively, to generate equal fluid flows with the same crossflow velocity in closed loops on each side of membrane. The AL of the membrane was set against the feed. It is smoother than the back side (support layer) and can avoid the effect of surface roughness on the formation of fouling layer. A digital scale (Mettler Toledo) connected to a computer was used to monitor the weight loss of the feed tank at intervals of 20 minutes during FO, and the weight change of the feed was converted to the permeate flux. All experiments were performed at a temperature of 21 ± 1 °C under dark conditions.

During FO process, a low crossflow rate is preferred to reduce energy consumption. In order to study the effect of crossflow on the performance of the FO process, the crossflow was varied from a low velocity (4.2 cm · s^{-1}, i.e. crossflow rate of 100 ml · min^{-1}) to a high velocity (16.7 cm · s^{-1}, i.e. crossflow rate of 400 ml · min^{-1}). The FO process was conducted in a semi-batch mode. Each batch was started with 1 L of feed and 1 L of DS and interrupted when 300 ml of permeate was extracted from the feed to the DS. Then concentrated feed and diluted DS were replaced by fresh aliquots and the SWFO process was continued to the next batch.

Before SW was used as the feed, a baseline measurement with SSW was performed. During SWFO, the decline of flux is related to two factors, i.e., formation of fouling and loss of osmotic pressure caused by external concentration polarization (ECP) and internal concentration polarization (ICP). Since SSW contained only NaCl and NaN_3 in DI water, there was not an effect from other foulants such as natural organic matter (NOM). The decline of flux was only due to the combined effects from the concentrative ECP (CECP) on the feed side, dilutive ICP (DICP), and dilutive ECP (DECP) on the DS side (Figure 3.1.2a). After the baseline measurement, SW was employed as the feed. The fouling layer would form on the AL of the membrane. A fouling resistance and a concentrative ICP (CICP) in the fouling layer (Figure 3.1.2b) were created. The fouling related CICP could cause an enhanced osmotic pressure in the fouling layer on the feed side. This enhanced osmotic pressure is due to not only the accumulation of salt from the feed bulk into the fouling layer on the membrane surface but also the revise diffusion of the salt from DS to the feed side (Lee *et al.* 2010). The salt diffused from DS to the feed was trapped by the fouling layer and also contributed to the enhanced osmosis pressure on the feed side. The enhanced osmotic pressure within fouling layer reduced the net driving force, consequently caused the significant decline of flux.

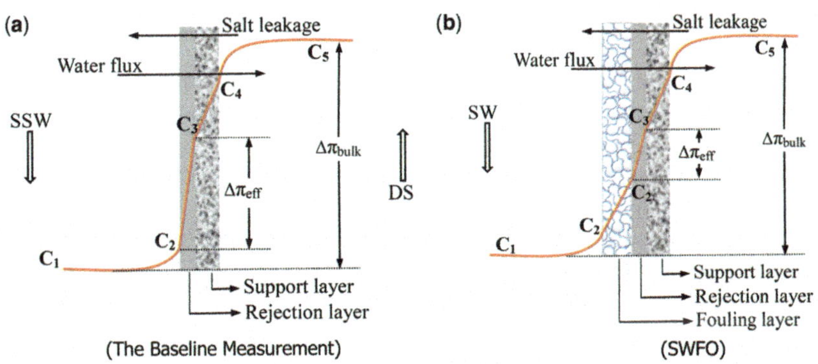

Figure 3.1.2 Illustrations of driving force profiles, expressed as the gradient of salt concentration during the baseline measurement and SWFO ($C_1 < C_2$ due to CECP; $C_2 < C_{2f}$ due to fouling related CICP; $C_3 < C_4$ due to DICP; $C_4 < C_5$ due to DECP; $\Delta\pi_{bulk}$ is the theoretical osmotic pressure; $\Delta\pi_{eff}$ is the effective osmotic pressure).

When SW was used as the feed (i.e. SWFO), a total of 10 batches (B_1 to B_{10}) were operated with each crossflow velocity before cleaning or replacing the membrane. In other words, 3 L of permeate passed through the membrane under each crossflow velocity. In each batch, the first 40 minutes was applied to stabilize the filtration system. The initial flux of each batch was measured from the filtration time of 40 minutes. The average flux of each batch was calculated by Equation 1:

$$J_{ave} = \frac{\Delta V}{At} \tag{3.1.1}$$

where J_{ave} is the average flux of each batch of FO; ΔV is the volume of permeate extracted from the feed to DS in each batch (300 mL in this study); A is the effective area of the membrane (20 cm^2 in this study); t is the filtration time of each batch.

In order to explore the reversibility of the fouling, the fouled membrane was collected after B_{10} of SWFO with the crossflow velocity of 4.2 cm · s^{-1}. The fouled membrane was flushed *in situ* by DI water at a crossflow velocity of 16.7 cm · s^{-1} for 10 minutes. Then, the feed was switched to SW again and one more batch of SWFO (i.e. B_{11}) was processed with a crossflow velocity of 4.2 cm · s^{-1}.

3.1.2.3 Analytical methods

The cations and anions in SW were analyzed by Inductively Coupled Plasma/Mass Spectrometry (ICP-MS 7500; Agilent, US) and Ion Chromatography (ICS, 1600; Dionex, US). Total organic carbon (TOC) and dissolved organic carbon (DOC) were analyzed by an Organic Carbon Analyzer (TOC-V_{CPH}, Shimadzu, Japan). Total dissolved solid (TDS) was measured by the standard methods from Eaton *et al.* (2005).

NOM in SW and DS was characterized by Liquid Chromatography-Organic Carbon Detection (LC-OCD, Model 8, DOC-Labor, Germany).

The conductivity of the feed and DS was monitored by a digital conductivity meter (Lab960, Schott, Germany) and converted to the salt concentration for the calculation of salt rejection.

The osmotic pressure of feed and DS was measured by a Cryoscopic Osmometer (Osmomat 030, Gonotec, Germany) and expressed as Osmolality (Osmol · kg^{-1}).

The fouling on the membrane surface was analyzed by Scanning Electron Microscopy-Energy Dispersive X-ray Apparatus (SEM-EDX, Magellan, FEI).

The biopolymer (extracellular polymeric substances, EPS) on the AL of the membrane was stained by Alcian Blue 8GX and 4′,6-diamidine-2-phenylindole (DAPI) solution (Bar-Zeev *et al.* 2009; Villacorte *et al.* 2009), then observed and photographed by an optical microscope (DP72, Olympus, Japan) with bright illumination and UV epifluorescence, respectively.

3.1.3 RESULTS AND DISCUSSION

3.1.3.1 Flux patterns during FO

The osmotic pressure of SSW and SW was 1.17 ± 0.03 Osmol · kg^{-1}. When the same DS (i.e. 2 M NaCl solution) was used for all FO processes in this study, the approximate osmotic pressures of SSW and SW made the flux comparable between the baseline measurement and SWFO. The operational time for collecting 300 mL of permeate in each batch varied from 39 to 41 hours with a crossflow velocity of 16.7 cm · s^{-1} and from 48 to 64 hours with a crossflow velocity of 4.2 cm · s^{-1}. The flux patterns of the baseline measurement and SWFO were presented in Figures 3.1.3 and 3.1.4.

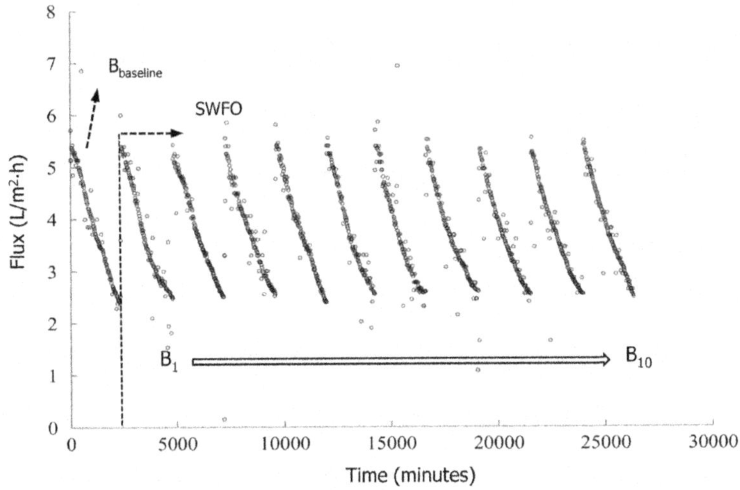

Figure 3.1.3 Flux patterns of the baseline measurement and SWFO with the crossflow velocity of 16.7 cm · s^{-1}.

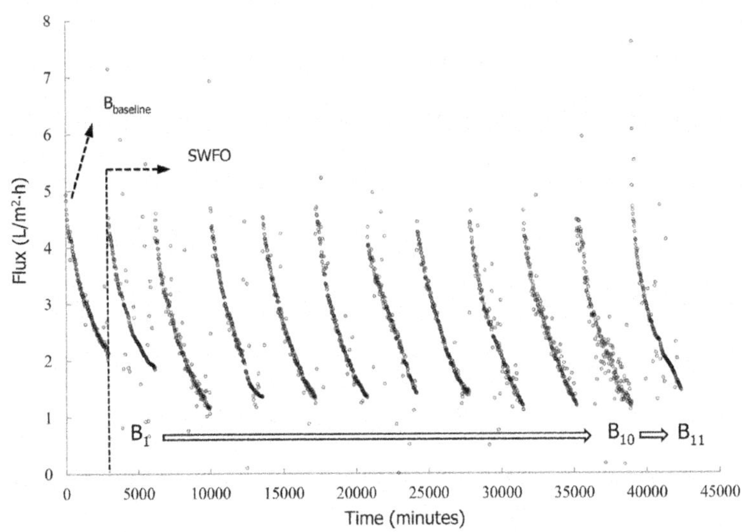

Figure 3.1.4 Flux patterns of the baseline measurement and SWFO with the crossflow velocity of 4.2 cm · s^{-1}.

The reproducibility of the process was demonstrated by the data collected from the different batches of FO process. The flux decreased significantly in each batch of FO. The rate of flux decline was about 53% with a crossflow velocity of 16.7 cm · s^{-1} while it reached more than 70% with crossflow velocity of 4.2 cm · s^{-1}. In addition, the flux patterns were similar from $B_{baseline}$ (the batch of the baseline measurement) to B_{10} (the tenth batch of SWFO) with a crossflow velocity of 16.7 cm · s^{-1} while there were some changes during the initial several batches with a crossflow velocity of 4.2 cm · s^{-1}. In this study, the AL of the membrane was set against the feed. According to the structure and separation capacity of the FO membrane, fouling on the feed side of the membrane could be classified as external types (such as cake layer or gel layer) which is similar to RO membrane fouling, rather than internal types (such as pore blocking) in other membrane processes like ultrafiltration or microfiltration. The flux decline during SWFO was attributed to a number of factors including diluting the feed, CECP, fouling related CICP and the fouling related resistance on the feed side, and concentrating the DS, DICP and DECP on the DS side. The combined effects from these factors caused the significant decrease of flux during SWFO. Higher crossflow velocity provided greater shear stress and vortex, that not only inhibiting the formation of a fouling layer on the membrane surface but also reduced ECP on both sides of the membrane by enhancing the back diffusion and convection of salts from the surface of the membrane or fouling layer to the bulk feed or DS. Therefore, a higher flux was obtained when a crossflow velocity of 16.7 cm · s^{-1} was applied.

The variation of J_{ave} of the baseline measurement and different stages of SWFO is represented in Figure 3.1.5. The J_{ave} of $B_{baseline}$ decreased about 19% when crossflow velocity was varied from 16.7 to 4.2 cm · s^{-1}. During the baseline measurement, NaCl solutions at different concentrations were employed as the feed (SSW) and DS, respectively. Thus, there was not fouling related CICP and resistance in the baseline measurement. In addition, the HTI FO membrane uses an embedded PES mesh screen to provide mechanical strength of the membrane. The thickness and porosity of the support layer in this membrane is not as distinct as other thin film membranes. The membrane related DICP could be minimized. The decrease of J_{ave} during the baseline measurement with low crossflow velocity mainly arose from promoted CECP on the feed side and DECP on the DS side. When SWFO was run, a minor decrease of J_{ave} with a crossflow velocity of 16.7 cm · s^{-1} was also observed between $B_{baseline}$ and B_1. However, the decrease of J_{ave} was much more significant and continued to B_2 with the crossflow velocity of 4.2 cm · s^{-1}. The decrease of J_{ave} between baseline measurement and SWFO was probably due to the formation of fouling on the membrane surface when SW was used as the feed. Once SW contacted AL of the membrane, the foulants would start to adsorb onto the membrane surface by the interaction between foulants and the membrane. For example, the biopolymeric fouling layer was observed on the surface of glass slides after soaking the slides in SW (Bar-Zeev *et al.* 2009). The adhesion of biopolymer on the membrane surface during SWFO could be enhanced by the permeate flow which can promote the transport of solutes and foulants from the feed bulk to the membrane surface. On the other hand, crossflow velocity had a greater impact on ECP rather than ICP. The formation of fouling layer on the membrane was more severe with low crossflow velocity while it could be reduced, but not eliminated, by high crossflow velocity. Therefore fouling related resistance and CICP negatively affected the flux with both high and low crossflow velocities, and this negative effect was more significant with low crossflow velocity.

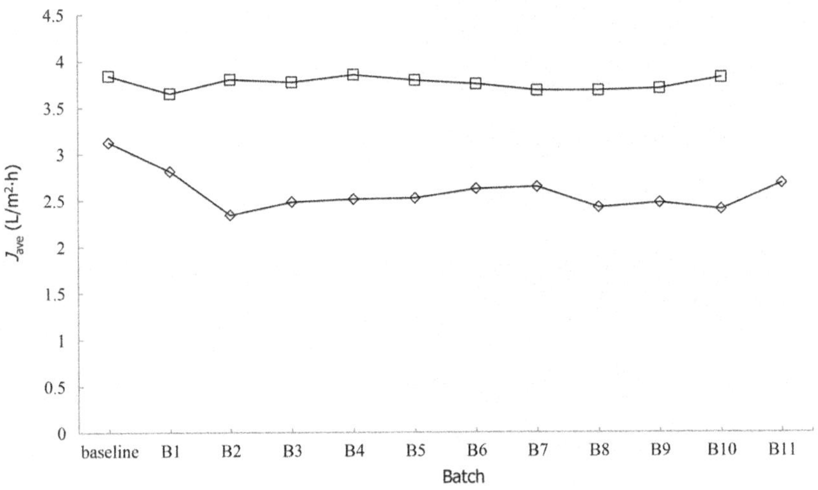

Figure 3.1.5 Average flux of the baseline measurement and SWFO (□, the crossflow velocity of 16.7 cm · s^{-1}; ◇, the crossflow velocity of 4.2 cm · s^{-1}).

After initial batches of SWFO (B_1 for the crossflow velocity of 16.7 cm · s^{-1}, B_1 and B_2 for the crossflow velocity of 4.2 cm · s^{-1}), J_{ave} was nearly constant afterward to B_{10}. Similar to other membrane processes, it can be hypothesized that the initial stage of SWFO was the main period for the build-up of fouling on the membrane surface, particularly for the semi-batch operation adopted for SWFO in this study. The fouling layer on the membrane surface formed in the initial batches of SWFO and became stable afterward. The fresh feed used in following batches, especially with high crossflow velocity, partially flushed the fouling layer rather than making it thicker or compacter. Consequently, the J_{ave} of SWFO was stable in the later batches of SWFO.

After B_{10} of SWFO with a cross flow velocity of 4.2 cm · s^{-1}, the membrane was flushed *in situ* by DI water. The result (B_{11} with the crossflow velocity of 4.2 cm · s^{-1} in Figure 3.1.5) indicates that J_{ave} could be partially recovered and almost equal to that of B_1 with the same crossflow velocity (i.e., 4.2 cm · s^{-1}) after flushing by DI water. The recovery of the flux by DI water flushing may be attributed to the removal of the fouling on the membrane surface.

3.1.3.2 Identification of major foulants

The surfaces of the original virgin membrane and fouled membranes collected from the different batches of SWFO were characterized by SEM images combined with EDX spectrometry graphs (Figures 3.1.6 and 3.1.7).

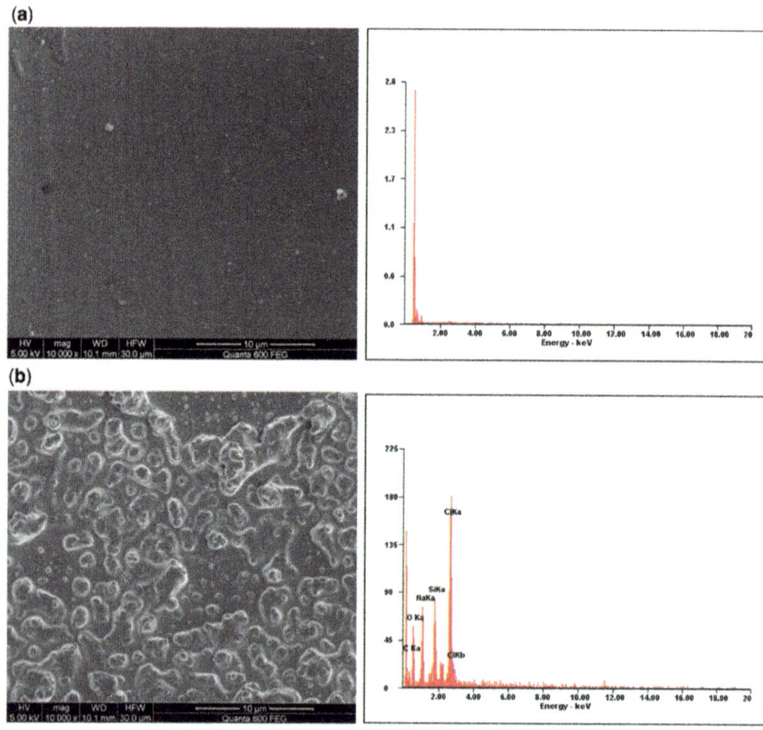

Figure 3.1.6 SEM-EDX graphs of the membrane surface (a, the original virgin membrane; b, the fouled membrane after B_{10} of SWFO with a crossflow velocity of 16.7 cm · s^{-1}).

Only the weak peaks of carbon and oxygen were detected on the surface of the original virgin membrane (Figure 3.1.6a). They were from the raw materials (CTA and PES) of the membrane and probably the residual glycerin from manufacture process of the membrane. After 10 batches of SWFO with a crossflow velocity of 16.7 cm · s^{-1}, the membrane surface was completely covered by a fouling layer (Figure 3.1.6b). The fouling layer was made up of the scaling-like foulants surrounded by biopolymeric substances. EDX graphs reveal some strong peaks from the fouled membrane surface. The strong peaks of sodium and chlorine were definitely due to the salinity of SW. The enhanced peaks of oxygen and carbon were due to the NOM in the fouling layer. Although the amount of silica in SW is much lower than other inorganic elements (e.g. calcium and magnesium), a strong peak of silicon was observed while other inorganic elements were negligible on the surface of the fouled membrane. Thus, the scaling-like foulants could be due to the deposition of silica.

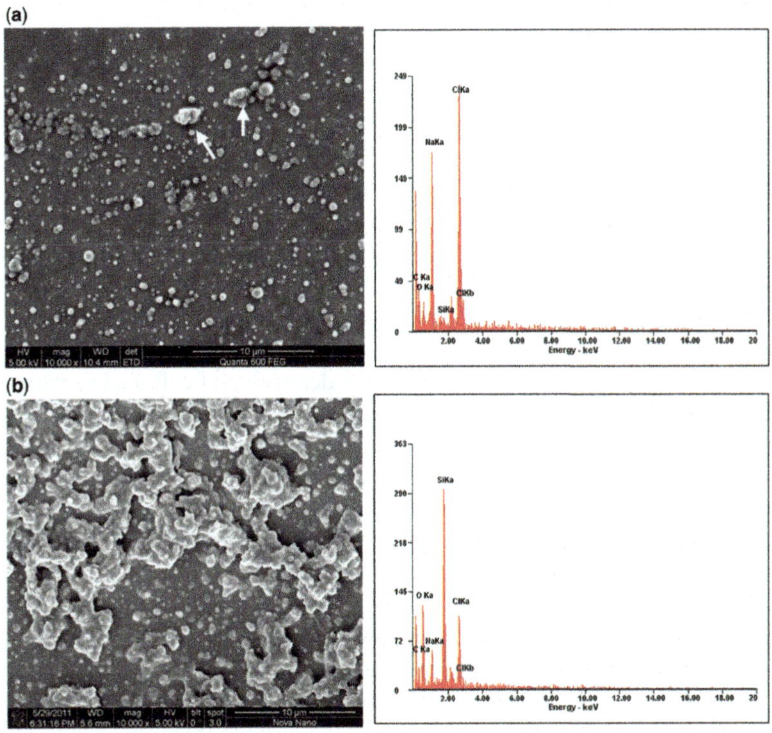

Figure 3.1.7 SEM-EDX graphs of the membrane surface (a, the fouled membrane after B_1 of SWFO with crossflow velocity of 4.2 cm · s^{-1}; b, the fouled membrane after B_{11} of SWFO with crossflow velocity of 4.2 cm · s^{-1}).

Silica is generally found in water supplies in three different forms, namely, dissolved silica (also known as 'reactive silica', or monosilicic acid), colloidal silica and particulate silica (Comb, 1996). In order to identify the major form of silica in SW used in this study, a parallel analysis was conducted by measuring silicon concentration in SW pre-filtered by 10 and 0.45 μm capsule filters, respectively. There was not a significant difference in silicon concentration in the 10 μm filtered SW versus 0.45 μm filtered SW. It could be concluded that the dominant form of silica in SW used as the feed in this study was dissolved silica as monosilicic acid

(H_4SiO_4). Thus, the formation of silica scaling was likely due to the polymerization of dissolved silica on the membrane surface rather than the deposition of colloidal or particulate silica.

According to the flux results, the fouling should be more severe with the low crossflow velocity. Figure 3.1.7a and 3.1.7b were obtained from the membranes collected from B_1 and B_{11} of SWFO with the crossflow velocity of 4.2 cm · s^{-1}. After the B_1 of SWFO, some particle-like foulants were deposited on the membrane (Figure 3.1.7a). Except carbon and oxygen, the EDX graph shows the strong peaks of sodium and chlorine and the weak peak of silica. At some sites, the polymerization of silica (marked by the arrows in Figure 3.1.7a) on the membrane surface was initiated. Figure 3.1.7a also reveals that the biopolymeric fouling was not severe during B_1 of SWFO with the crossflow velocity of 4.2 cm · s^{-1}. A distinct polymerized silica cluster was observed in Figure 3.1.7b. B_{11} could be considered to be the SWFO process with a pre-fouled membrane. The distinct polymerized silica in Figure 3.1.7b was formed in previous batches of SWFO (i.e. B_2–B_{10}) before B_{11}. It was found that the biopolymeric foulants were flushed out by DI water after B_{10} of SWFO. After water flushing, the biopolymeric substances disappeared and the residual silica fouling showed a typical pattern of polymerized silica (Sahachaiyunta *et al.* 2002) rather than deposition of colloidal or particulate silica (Figure 3.1.7b). This indicates the reversibility of the organic fouling layer and explains the increase of permeate flux in B_{11} (Figures 3.1.4 and 3.1.5) as well. The biopolymeric fouling related CICP and resistance could be the major reasons for the decrease of flux in SWFO with a low crossflow velocity. The results reveal that the organic fouling could be reduced by maintaining the crossflow velocity at a certain level during SWFO or water flushing without any chemical treatment after SWFO.

The polymerization of silica during SWFO could be initiated by associating the dissolved silica onto the membrane surface when silica was concentrated by CECP during the initial period of SWFO. The hydroxyl group in CTA could play a role in attracting the silica to the membrane surface (Sheikholeslami & Tan, 1999) while metals such as magnesium, calcium and iron in SW can facilitate the polymerization of silica (Sahachaiyunta *et al.* 2002; Sheikholeslami *et al.* 2002). During SWFO, CECP led to an increase of salts (such as $CaCl_2$, $MgCl_2$) and silica concentration near the membrane surface. Polymerization of silica would be significantly accelerated by increased concentration of silica with the presence of calcium and magnesium (Sheikholeslami & Tan, 1999). In addition, in the form of monosilicic acid, silica is relatively un-ionized at most natural pH levels, with about 10% of monosilicic acid ionized at a pH of 8.5 which is very close to the pH of SW. The polymerization of monosilicic acid is favored by partial ionization (Sheikholeslami *et al.* 2002). The monosilicic acid could initially and rapidly form polysilicic acid of low molecular weight and then large polymeric species (Comb, 1996). Once the polymerized silica fouling forms on the membrane, it is difficult to remove while deposition of colloidal silica and particulate silica can be washed away by physical flushing (Koo *et al.* 2001). The hypothesized process for the polymerization of silica during SWFO is demonstrated in Figure 3.1.8.

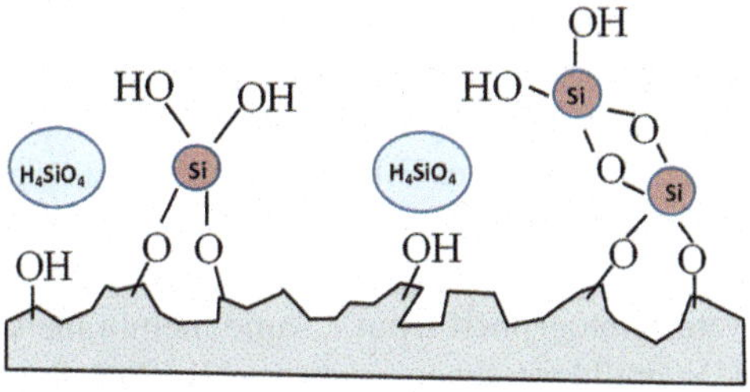

Figure 3.1.8 Polymerization of dissolved silica on the membrane surface during SWFO.

During the polymerization process, monosilicic acid and polysilicic acid also reacts with organic compound to form anhydrides (Ning, 2003) that promotes NOM fouling. Furthermore, the polymerized silica clusters change the morphology of the membrane surface and work as a scaffold in the fouling layer. The space inside the polymerized silica clusters could provide favorable sites without strong hydrodynamic vortex and mixing for the accumulation and adhesion of NOM, especially biopolymers. Therefore, a fouling matrix comprising polymerized silica and adhered NOM can form on the membrane surface during the SWFO process.

Silica fouling is one of the major unresolved problems during brackish water RO due to the high concentration of silica in brackish water. Open ocean seawater contains much less silica than brackish water. Therefore this compound is not considered as a typical cause of mineral fouling during seawater RO (SWRO). Calcium carbonate and magnesium hydroxide are the most common causes for SWRO membrane scaling. However, polymerized silica was found to be dominant inorganic foulant while scaling caused by calcium and magnesium was not found during SWFO in this study. The combined effects from the CTA membrane properties (containing hydroxyl group), SW properties (pH, salt content) and hydrodynamics of the FO process (CECP and fouling related CICP) led to a rapid and prevailing formation of polymerized silica, creating a fouling layer complemented with NOM on the surface of the FO membrane. The silica-NOM fouling layer and its related CICP may hinder the accumulation and deposition of other mineral scalants on the membrane surface. In addition, the driving force during SWFO was only osmotic pressure. The lack of high applied hydraulic pressure during SWFO also prevented, or retarded, the formation of other mineral scaling since calcium or magnesium scaling are favored at high operating pressure during RO and NF processes. The results suggest that a pretreatment such as coagulation

(Sheikholeslami *et al.* 2002), chemical precipitation (Al-Rehaili 2003), silica gel seeding (Bremere *et al.* 2000), or hybrid process involving ultrafiltration (Cheng *et al.* 2009) to remove silica from SW could improve the performance during the SWFO process.

The presence of the biopolymers surrounding the polymerized silica clusters is visually demonstrated in Figure 3.1.9. The membrane sample was the same as that in Figure 3.1.6b. EPS was stained by Alcian Blue and visualized by microscopy. EPS (the blue substances in Figures 3.1.9a and 3.1.9b) showed a sticky and gel-like form, and could play a critical role in the fouling related CICP. In addition, bacteria tended to accumulate at sites where EPS adhered to the membrane. Figures 3.1.9b and 3.1.9c were photogrphed from a same membrane sample under bright illumination and UV epifluorescence, respectively. The fluorescence in Figure 3.1.9c indicates the presence of bacteria. The position of bacteria in Figure 3.1.9c is in accordance with the position of the EPS mass in Figure 3.1.9b. EPS are mainly polysaccharides, proteins, nucleic acids and lipids. They provide the mechanical stability of biofilms, immobilize bacteria, and also serve as a nutrient source for bacteria (Flemming & Wingender, 2010).

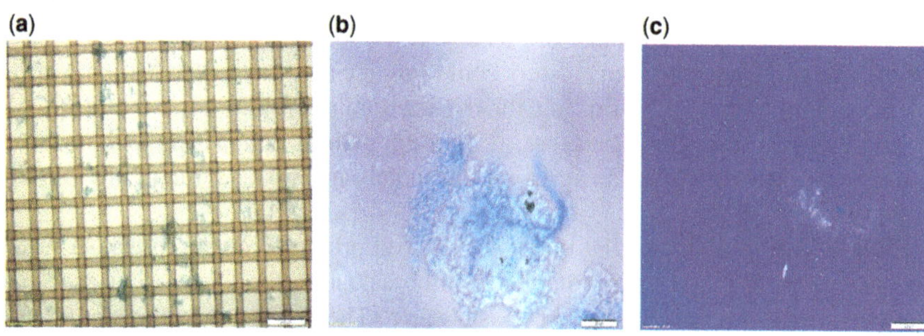

Figure 3.1.9 Visualization of the biopolymers on the membrane surface after B_{10} of SWFO with the crossflow velocity of 16.7 cm · s^{-1} (a and b, EPS viewed with Alcian Blue stain under bright illumination; c, bacteria accumulated in EPS mass viewed with DAPI stain under UV epifluorescence; Scale bar 200 μm in a, 20 μm in b and c).

Considering the energy-saving and water recovery rate, it is desirable to maintain a FO process for long-term operation with a minimum, or even without crossflow. Therefore, biofouling may be a big challenge during SWFO since it is easy to form and adhere to the membrane surface. In addition to negative effects including reduction of membrane lifetime and increase of fouling resistance, biofouling layer is a favorable structure to hinder the back diffusion of salt from the membrane surface to the feed bulk, thus cause a significant drop of the effective osmotic pressure which is the only driving force during SWFO.

3.1.3.3 Salt and foulant rejection

The salt rejection was about 98% and almost constant for both the baseline measurement and all batches of SWFO (Figure 3.1.10). Crossflow velocity did not show a significant effect on the salt rejection. It has been reported that the HTI CTA membrane can provide a salt rejection up to about 95% (McCutcheon *et al.* 2006; Cornelissen *et al.* 2008; Yip *et al.* 2010).

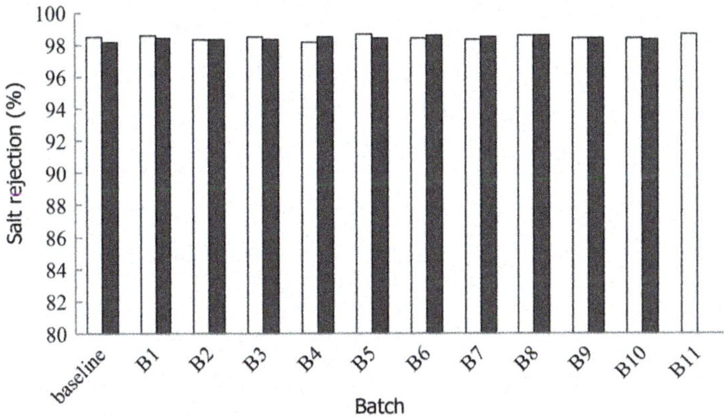

Figure 3.1.10 Salt rejection during SWFO (■, crossflow velocity of 16.7 cm · s^{-1}; □, crossflow velocity of 4.2 cm · s^{-1}).

During the RO process, the fouling layer, especially a biofilm on the membrane surface, leads to a reduction in salt rejection. The reason is known as biofilm enhanced osmotic pressure or cake enhanced concentration polarization (Hoek & Elimelech, 2003; Herzberg *et al.* 2009; Herzberg, 2010). In the case of FO, it has been proven that reverse NaCl flux (i.e., salt leakage) from the DS to the feed side may increase with increasing NaCl concentration in the DS solution when DI water is used as the feed (Phillip *et al.* 2010). It can be hypothesized that the principle of salt transport through an FO membrane is similar to an RO membrane. The driving force leading to the transport of NaCl through the membrane is the chemical potential gradient across the membrane and relates to the difference in concentration (chemical activity) across the membrane (Ho & Sirkar, 1992). The salt transport comprises the diffusional salt flux across the dense (or active) layer and convective salt flux in the porous layer. The salt passage is enhanced by increasing concentration gradient of salt. In addition, convective salt flux is entrained by water flux in RO. During SWFO, the convective salt flux is opposed by the permeate flux. CECP, CICP, DECP and DICP on the feed and DS side, respectively, reduced the concentration gradient between the feed and DS. The effective concentration gradient of salt from the DS side to the feed side was restricted by salt concentration at the interfaces of the fouling layer-AL and

AL-support layer (C_{2f} and C_3 in Figure 3.1.2b). Both C_{2f} and C_3 were dominated by ICP which could not be reduced by the crossflow of the feed and DS. Reduced chemical potential gradient combined with a water flux opposed to the salt transport led to the high and stable salt rejection during SWFO in this study.

NOM in the feed and DS was characterized by LC-OCD (Figure 3.1.11). The DS sample was collected from B_1 of SWFO with the crossflow velocity of 4.2 cm · s^{-1}. LC-OCD is able to fractionate NOM into about 10 classes of components that are detected by an organic carbon detector. A nearly complete rejection of NOM was attained during SWFO. A weak peak was detected from the DS at the elution time corresponding to humic-like substances while the biopolymer peak is completely absent. The absence of NOM, especially in the form of biopolymers in the DS indicates their retention and possible role in fouling.

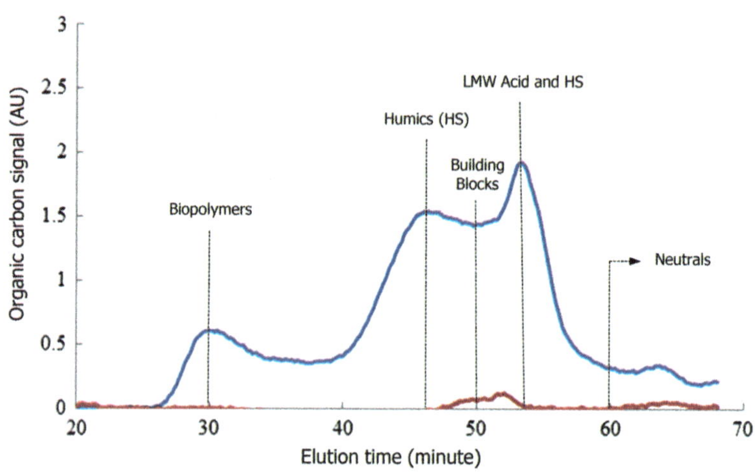

Figure 3.1.11 NOM rejection during SWFO (blue line is SW; brown line is the DS collected from B_1 of SWFO with the crossflow velocity of 4.2 cm · s^{-1}).

3.1.4 SUMMARY

The membrane fouling propensity of natural seawater during forward osmosis was studied. Seawater from the Red Sea was used as the feed in a forward osmosis process while a 2 M sodium chloride solution was used as the draw solution. The process was conducted in a semi-batch mode under two crossflow velocities, 16.7 cm · s^{-1} and 4.2 cm · s^{-1}. For the first time reported, silica scaling was found to be the dominant inorganic fouling (scaling) on the surface of membrane active layer during seawater forward osmosis. Polymerization of dissolved silica was the major mechanism for the formation of silica scaling. After ten batches of seawater forward osmosis, the membrane surface was covered by a fouling layer of assorted polymerized silica clusters and natural organic matter, especially biopolymers. Moreover, the absorbed biopolymers also provided bacterial attachment sites. The accumulated organic fouling could be partially removed by water flushing while the polymerized silica was difficult to remove. The rate of flux decline was about 53% with a crossflow velocity of 16.7 cm · s^{-1} while reaching more than 70% with a crossflow velocity of 4.2 cm · s^{-1}. Both concentration polarization and fouling played roles in the decrease of flux. The salt rejection was stable at about 98% during seawater forward osmosis. In addition, an almost complete rejection of natural organic matter was attained. The results from this study are valuable for the design and development of a successful protocol for a pretreatment process before seawater forward osmosis and a cleaning method for fouled membranes.

3.1.5 REFERENCES

Achilli A., Cath T. Y. and Childress A. E. (2009a). Power generation with pressure retarded osmosis: an experimental and theoretical investigation. *Journal of Membrane Science*, **343**, 42–52.

Achilli A., Cath T. Y., Marchand E. A. and Childress A. E. (2009b). The forward osmosis membrane bioreactor: a low fouling alternative to MBR processes. *Desalination*, **239**, 10–21.

Al-Rehaili A. M. (2003). Comparative chemical clarification for silica removal from RO groundwater feed. *Desalination*, **159**, 21–31.

Bar-Zeev E., Berman-Frank I., Liberman B., Rahav E., Passow U. and Berman T. (2009). Transparent exopolymer particles: potential agents for organic fouling and biofilm formation in desalination and water treatment plants. *Desalination and Water Treatment*, **3**, 7.

Bremere I., Kennedy M., Mhyio S., Jaljuli A., Witkamp G.-J. and Schippers J. (2000). Prevention of silica scale in membrane systems: removal of monomer and polymer silica. *Desalination*, **132**, 89–100.

Cath T. Y., Hancock N. T., Lundin C. D., Hoppe-Jones C. and Drewes J. E. (2010). A multi-barrier osmotic dilution process for simultaneous desalination and purification of impaired water. *Journal of Membrane Science*, **362**, 417–426.

Cheng H.-H., Chen S.-S. and Yang S.-R. (2009). In-line coagulation/ultrafiltration for silica removal from brackish water as RO membrane pretreatment. *Separation and Purification Technology*, **70**, 112–117.

Comb L. F. (1996). Silica chemistry and reverse osmosis. *Ultrapure Water*, **13**, 41–43.

Cornelissen E. R., Harmsen D., de Korte K. F., Ruiken C. J., Qin J.-J., Oo H. and Wessels L. P. (2008). Membrane fouling and process performance of forward osmosis membranes on activated sludge. *Journal of Membrane Science*, **319**, 158–168.

Eaton A. D., Clesceri L. S., Rice E. W. and Greenberg A. E. (2005). Standard Methods for the Examination of Water and Wastewater, 21st edn. Joint publication of the American Public Health Association (APHA), the American Water Works Association (AWWA), and the Water Environment Federation (WEF), New York.

Flemming H.-C. and Wingender J. (2010). The biofilm matrix. *Nature Reviews Microbiology*, **8**, 623–633.

Herzberg M. (2010). Osmotic effects of biofouling in reverse osmosis (RO) processes: physical and physiological measurements and mechanisms. *Desalination and Water Treatment*, **15**, 287–291.

Herzberg M., Kang S. and Elimelech M. (2009). Role of Extracellular Polymeric Substances (EPS) in biofouling of reverse osmosis membranes. *Environmental Science & Technology*, **43**, 4393–4398.

Ho W. W. and Sirkar K. K. (1992). Membrane Handbook. Springer Science & Business Media, New York.

Hoek E. M. V. and Elimelech M. (2003). Cake-enhanced concentration polarization: a new fouling mechanism for salt-rejecting membranes. *Environmental Science & Technology*, **37**, 5581–5588.

Koo T., Lee Y. J. and Sheikholeslami R. (2001). Silica fouling and cleaning of reverse osmosis membranes. *Desalination*, **139**, 43–56.

Lay W. C. L., Chong T. H., Tang C. Y., Fane A. G., Zhang J. and Liu Y. (2010). Fouling propensity of forward osmosis: investigation of the slower flux decline phenomenon. *Water Science and Technology*, **61**, 10.

Lay W. C. L., Zhang Q., Zhang J., McDougald D., Tang C., Wang R., Liu Y. and Fane A. G. (2011). Study of integration of forward osmosis and biological process: membrane performance under elevated salt environment. *Desalination*, **283**, 123–130.

Lee S., Boo C., Elimelech M. and Hong S. (2010). Comparison of fouling behavior in forward osmosis (FO) and reverse osmosis (RO). *Journal of Membrane Science*, **365**, 34–39.

Ling M. M., Wang K. Y. and Chung T.-S. (2010). Highly water-soluble magnetic nanoparticles as novel draw solutes in forward osmosis for water reuse. *Industrial & Engineering Chemistry Research*, **49**, 5869–5876.

Loeb S. (2002). Large-scale power production by pressure-retarded osmosis, using river water and sea water passing through spiral modules. *Desalination*, **143**, 115–122.

McCutcheon J. R., McGinnis R. L. and Elimelech M. (2005). A novel ammonia–carbon dioxide forward (direct) osmosis desalination process. *Desalination*, **174**, 1–11.

McCutcheon J. R., McGinnis R. L. and Elimelech M. (2006). Desalination by ammonia–carbon dioxide forward osmosis: influence of draw and feed solution concentrations on process performance. *Journal of Membrane Science*, **278**, 114–123.

Mi B. and Elimelech M. (2008). Chemical and physical aspects of organic fouling of forward osmosis membranes. *Journal of Membrane Science*, **320**, 292–302.

Mi B. and Elimelech M. (2010a). Gypsum scaling and cleaning in forward osmosis: measurements and mechanisms. *Environmental Science & Technology*, **44**, 2022–2028.

Mi B. and Elimelech M. (2010b). Organic fouling of forward osmosis membranes: fouling reversibility and cleaning without chemical reagents. *Journal of Membrane Science*, **348**, 337–345.

Ning R. Y. (2003). Discussion of silica speciation, fouling, control and maximum reduction. *Desalination*, **151**, 67–73.

Petrotos K. B. and Lazarides H. N. (2001). Osmotic concentration of liquid foods. *Journal of Food Engineering*, **49**, 201–206.

Phillip W. A., Yong J. S. and Elimelech M. (2010). Reverse draw solute permeation in forward osmosis: modeling and experiments. *Environmental Science & Technology*, **44**, 5170–5176.

Sahachaiyunta P., Koo T. and Sheikholeslami R. (2002). Effect of several inorganic species on silica fouling in RO membranes. *Desalination*, **144**, 373–378.

Sheikholeslami R. and Tan S. (1999). Effects of water quality on silica fouling of desalination plants. *Desalination*, **126**, 267–280.

Sheikholeslami R., Al-Mutaz I. S., Tan S. and Tan S. D. (2002). Some aspects of silica polymerization and fouling and its pretreatment by sodium aluminate, lime and soda ash. *Desalination*, **150**, 85–92.

Tang W. and Ng H. Y. (2008). Concentration of brine by forward osmosis: performance and influence of membrane structure. *Desalination*, **224**, 143–153.

Villacorte L. O., Kennedy M. D., Amy G. L. and Schippers J. C. (2009). The fate of Transparent Exopolymer Particles (TEP) in integrated membrane systems: removal through pre-treatment processes and deposition on reverse osmosis membranes. *Water Research*, **43**, 5039–5052.

Wang R., Shi L., Tang C. Y., Chou S., Qiu C. and Fane A. G. (2010a). Characterization of novel forward osmosis hollow fiber membranes. *Journal of Membrane Science*, **355**, 158 167.

Wang Y., Wicaksana F., Tang C. Y. and Fane A. G. (2010b). Direct microscopic observation of forward osmosis membrane fouling. *Environmental Science & Technology*, **44**, 7102–7109.

Wright J., Johnson R. and Yum S. (2003). DUROS® osmotic pharmaceutical systems for parenteral & site-directed therapy. *Drug Delivery Technology*, **3**, 64–73.

Yang Q., Wang K. Y. and Chung T.-S. (2009). A novel dual-layer forward osmosis membrane for protein enrichment and concentration. *Separation and Purification Technology*, **69**, 269–274.

Yip N. Y., Tiraferri A., Phillip W. A., Schiffman J. D. and Elimelech M. (2010). High performance thin-film composite forward osmosis membrane. *Environmental Science & Technology*, **44**, 3812–3818.

Zhang S., Wang K. Y., Chung T.-S., Chen H., Jean Y. C. and Amy G. (2010). Well-constructed cellulose acetate membranes for forward osmosis: minimized internal concentration polarization with an ultra-thin selective layer. *Journal of Membrane Science*, **360**, 522–535.

Chapter 3.2

NOM and TEP fouling*

3.2.1 INTRODUCTION

Forward osmosis (FO), an osmotic-driven membrane process, has been gaining attention for being a low-energy alternative for several applications usually performed by pressure-driven membranes, such as reverse osmosis (RO) and nanofiltration (NF). Recent studies have focused on the use of forward osmosis (FO) membranes for oil/gas drilling wastewater reclamation, landfill leachate treatment (Lambert & Lampi, 2010) and water desalination (Cath *et al.* 2010).

Natural organic matter (NOM) found in water has been shown to be a serious fouling agent for RO and NF membranes (Hong & Elimelech, 1997; Cho *et al.* 2000; Lee *et al.* 2004; Lee *et al.* 2010). Holloway *et al.* (2007) compared the fouling behavior of FO and RO for wastewater concentrate treatment, and demonstrated a slower flux decline rate in FO than in RO (Holloway *et al.* 2007). Mi *et al.* (2010) studied organic fouling of FO membranes caused by sodium alginate (simulating polysaccharides in wastewater). They showed that for osmotic pressure processes, where there is no hydraulic pressure, fouling is highly reversible using a cleaning solution of deionized (DI) water with 50 mM of NaCl as feed water and 4 M NaCl as draw solution (DS); it was even suggested that chemical cleaning is not needed (Mi & Elimelech, 2010). High recovery of permeate fluxes was also demonstrated in Lee's *et al.* (2010) research, using mainly an increased crossflow velocity in the feed stream to clean the organic fouling (composed of alginate, humic acid and bovine serum albumin) of a FO membrane, without employing chemical cleaning (Lee *et al.* 2010).

---

* This chapter is based, with permission from the copyright holder, on the previously published paper. (Valladares Linares R., Yangali-Quintanilla V., Li Z. and Amy G. (2012). NOM and TEP fouling of a forward osmosis (FO) membrane: Foulant identification and cleaning. *Journal of Membrane Science*, **421–422**, 217–224. doi:10.1016/j.memsci.2012.07.019)

Transparent exopolymer particles (TEP) are deformable, gel-like, transparent particles of acid mucopolysaccharides suspended in the water mass that appear in many forms and are ubiquitous to freshwater and marine environments. TEP can be divided according to its size as particulate TEP (>0.4 μm) and colloidal TEP, usually in a range of 0.05 to 0.4 μm (Villacorte *et al.* 2009). TEP are among the biopolymers that might be associated with fouling on the support layer (SL) of an FO membrane in a configuration of active layer facing the feed solution (AL-FS) when using seawater as a DS; several studies suggest that these particles are responsible for biofouling of membranes (Bar-Zeev *et al.* 2009). Significantly larger concentrations of TEP are found in seawater compared to wastewater effluents and diverse raw fresh waters (Villacorte *et al.* 2009; Berman *et al.* 2011). Therefore, we may hypothesize that TEP fouling of the FO membrane can occur when seawater is involved.

There has been a significant amount of research regarding chemical cleaning of RO and NF membranes. Experiments performed by Hong and Elimelech (1997) identified that ethylenediaminetetraacetic acid (EDTA) treatment of NOM-fouled NF membranes improves the cleaning efficiency significantly (Hong & Elimelech, 1997). In another study, bovine serum albumin (BSA) fouled membranes were treated with as solution of NaOCl (100 ppm), effectively removing the protein (Kuzmenko *et al.* 2005). Unfortunately, there are no studies that refer to such processes applied to an FO fouled membrane; this is particularly important when referring to TEP fouling, which cannot be easily controlled by feed water disinfection or microfiltration (Bar-Zeev *et al.* 2009; Villacorte *et al.* 2009; Berman *et al.* 2011).

This research is focused on identification of the NOM fractions that might cause fouling of the active layer of a FO membrane used in an AL-FS configuration, fed with a secondary wastewater effluent (SWWE), using seawater as DS. Additionally, the study proved the presence of TEP fouling on the support layer of a FO membrane. A comparison of cleaning techniques was performed for both sides of the membrane, determining the reversibility of the fouling on the active layer (AL) of the membrane, and cleaning performance for the SL.

3.2.2 EXPERIMENTAL

3.2.2.1 FO membranes and cell configuration

The FO membranes used in this study were provided by Hydration Technology Innovations (HTI, Scottsdale, AZ). The HTI membrane was originally received as flat sheet coupons, each one measuring 4" × 6", and then cut into the proper size for the membrane cell. The FO membrane is made of cellulose triacetate (CTA) embedded in a polyester mesh support, with a thickness of 30 to 50 μm (Tang *et al.* 2010), making it thinner than a thin-film composite (TFC) with porous support layer membrane (NF/RO) (approximate thickness of 125 μm (Yip *et al.* 2010)), which provides the membrane with a higher flux in FO mode. The structural parameter S, defined as the relation between the support layer thickness (t_s) and the tortuosity (τ) over the porosity (ε) of the membrane ($S = t_s\tau/\varepsilon$), is significantly higher for a TFC-NF/RO membrane ($S = 2155$ μm) compared to the HTI FO membrane ($S = 595$ μm) (Yip *et al.* 2010); this difference translates into a significantly higher resistance to mass transfer for the former one, reducing the flux of water through the membrane, leading to a higher performance of the HTI FO membrane.

The membrane cell was a custom-made plate and frame assembly made of poly (methyl methacrylate) (PMMA). The cell can hold two flat-sheet membranes, each with an area of 202 cm^2, forming a channel with a cross section of 1.5 cm width × 9 cm height. No spacer was used in the experiments. The recovery of the FO process for these experiments was 7%, however, recovery can be further increased (~50%) by changing the configuration of the feed water tank or increasing the number of cells immersed.

The AL of the membrane was facing the FO feed, leaving the support side in contact with the high osmotic pressure solution (seawater, DS). This configuration (AL-FS) has been proven to be the most effective in preventing fouling of an FO membrane (Cornelissen *et al.* 2008; Mi & Elimelech, 2008; Wang *et al.* 2010).

3.2.2.2 Water samples

The feed for the FO process was a SWWE from the Al-Ruwais wastewater treatment plant in the city of Jeddah, Saudi Arabia, which uses activated sludge aeration tanks after primary treatment. The biological oxygen demand (BOD_5) of the wastewater effluent was 20 mg · L^{-1}, the chemical oxygen demand (COD) was 20 mg · L^{-1}, the total organic carbon (TOC) was 4.60 mg · L^{-1}, the pH was 7.3, the conductivity was 2087 μS · cm^{-1} and the osmotic pressure was 0.35 bar. The total suspended solids (TSS) and volatile suspended solids (VSS) of the effluent were 2.4 and 2.1 mg · L^{-1}, respectively.

For the high osmotic pressure solution or DS, pre-filtered (0.45 μm pore size) Red Sea water was used as filtered seawater (FSW); seawater was obtained from the seawater pipeline of the RO plant operating at KAUST. The value for total dissolved solids (TDS) was 40,500 mg · L^{-1}, the conductivity was 54,300 μS · cm^{-1} and the osmotic pressure was 30.3 bar. The pH of the water was 7.8 and the dissolved organic carbon (DOC) was 1.12 mg · L^{-1}. Both the SWWE and the FSW, as well as the diluted draw solution (DDS) characterization can be found in Table 3.2.1.

3.2.2.3 FO membrane fouling procedure

The FO membrane cell was immersed in a feed tank, connected with tubing to the DS container; this DS was recirculated by a gear pump (GP) (Coleparmer) inside the cell formed by the membranes and frame. Figure 3.2.1 shows a detailed scheme of the setup.

Table 3.2.1 Wastewater effluent, filtered seawater and diluted seawater characteristics.

	SWWE	FSW	DDS
Temperature (°C)	20.7	20.5	20.5
Conductivity ($\mu S \cdot cm^{-1}$)	4300	57,500	21,900
Osmotic pressure (bar)	0.35	30.30	10.25
pH	7.3	7.8	7.6
DOC ($mg \cdot L^{-1}$)	5.3	1.12	1.19
UVA_{254} (cm^{-1})	0.130	0.012	0.025
SUVA ($L \cdot mg^{-1} \cdot m^{-1}$)	2.45	1.07	2.10
Nitrate ($mg \cdot L^{-1}$)	2.27	1.56	0.62
Nitrite ($mg \cdot L^{-1}$)	<0.015	<0.015	<0.015
Ammonium ($mg \cdot L^{-1}$)	0.85	0.4	0.12
Phosphate ($mg \cdot L^{-1}$)	2.18	0	0
Calcium ($mg \cdot L^{-1}$)	108	571	257
Magnesium ($mg \cdot L^{-1}$)	64	1458	639
Potassium ($mg \cdot L^{-1}$)	29	488	188
Sodium ($mg \cdot L^{-1}$)	768	12,470	5620
Strontium ($mg \cdot L^{-1}$)	n.a.	7	3.9
Bicarbonate ($mg \cdot L^{-1}$)	45	141	65
Boron ($mg \cdot L^{-1}$)	1.00	2.2	1.8
Carbonate ($mg \cdot L^{-1}$)	0.3	8.0	1.5
Chloride ($mg \cdot L^{-1}$)	1505	23,073	10,289
Fluoride ($mg \cdot L^{-1}$)	0	1.5	0.7
Sulfate ($mg \cdot L^{-1}$)	71	2400	1122
Barium ($mg \cdot L^{-1}$)	0	0.01	0

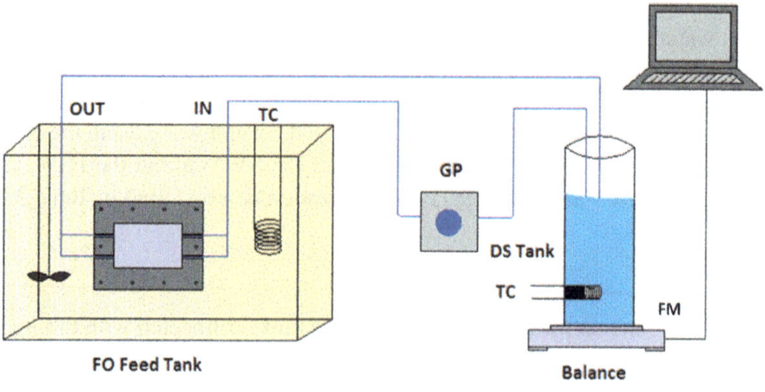

Figure 3.2.1 FO system layout.

At the beginning, 500 mL of seawater were poured in the DS tank; at the same time, the FO feed tank was filled with 18 L of SWWE. The two flat sheet FO membranes were set in the membrane cell, which was then immersed in the FO tank. The recirculation pump was started at a flow rate of 50 mL · min^{-1} (transverse velocity for the DS of 0.06 cm · s^{-1}), along with the recording devices for flow and conductivity. This transverse flow inside the channel allowed the DS water to extract water from the feed, a process driven only by osmotic difference; no negative pressure inside the cells was observed during the time of recirculation of the DS. The dilution process took 24 hours, increasing the level in the DS tank, while the SWWE was concentrated (volume decreased). A stirring device was used to create turbulence in the feed water and achieve proper mixing with a velocity gradient of about 50 s^{-1}. The temperature of both FO feed and DS was maintained at 20 ± 0.5 °C using a temperature controller (TC) with submergible stainless steel coils.

A balance (TE6101, Sartorius AG, Göttingen, Germany) was used as flow monitor (FM) when connected to a computer, for the purpose of comparing the flux recovery obtained with the different cleaning techniques implemented. This was done by measuring the difference in weight for the DS tank over an interval of time, resulting in the flow of water through the membrane. As the total area of the membrane used is known, flux can readily be calculated.

An evaporator (Rotavapor R-210, Buchi, Switzerland) was used to concentrate SWWE at 30 °C without involving a membrane process, in order to identify and analyze the difference in concentrated feed water (CFW) resulting from the FO process (sample taken on the last day of FO operation) and the evaporation process, and further determine which fraction of the NOM contributed to the fouling. This sample was referred to as evaporated wastewater effluent (EWWE).

3.2.2.4 NOM characterization

A Liquid Chromatograph coupled with Organic Carbon Detection analyser (LC-OCD) (Model 8, DOC LABOR, Germany) was used to analyse the organic carbon content of the water samples, following the procedure established by Huber *et al.* (2011). The process consists of size-exclusion chromatography columns which separates the organic carbon into several fractions based on size and hydrophobic and ionogenic characteristics. 1000 µL of wastewater effluent samples are used with a dilution of 4:1. For seawater samples, the amount of water is usually 2000 µL, with no dilution factor (1:1). Results were calculated using peak area integration.

The recommended DOC concentration for any sample injected into the LC-OCD analyzer should not be higher than 5 mg · L^{-1}. In this case, a target of 2 mg · L^{-1} of DOC was used for all of the samples, based on previous analyses of TOC and DOC.

An ultrasonic (US) bath was used for 30 minutes to extract the NOM fraction that was attached to the surface of a 2 cm × 2 cm sample of fouled membrane. The same US procedure was followed for a new FO membrane that was used as control, as well as for the SWWE and DS (filtered seawater).

3-Dimensional Fluorescence Excitation Emission Matrixes (3-D FEEM) were obtained with a Fluoromax-4 Spectrofluorometer (Horiba, USA) for the different water samples. The preparation of the samples consisted only in the filtration of the water with a 0.45 μm filter. The results are interpreted based on the identification of DOC fractions in previous spectrofluorescence studies (Amy *et al.* 2010). No pH adjustment was necessary because all of the samples tested were in a narrow pH range (7.3 to 7.8).

The fouling on both sides of the FO membrane surface was visually analyzed by Scanning Electron Microscopy (SEM) (Nova NanoSEM, FEI, USA).

In order to analyze the presence of TEP on the FO membrane, Alcian Blue dye was used to stain the particles and make them visible by microscopy. The Villacorte *et al.* (2009) protocol for preparing the cationic dye Alcian Blue was employed (Villacorte *et al.* 2009). An Olympus BX61 Motorized Transmitted and Reflected Light Research Microscope (Japan) was used with different objectives (60× and 100×) to take images of a fouled membrane after the 10-day cycle FO process.

The cut pieces of membrane were submerged in a petri dish with DI water to rinse for 2 minutes; afterwards, the pieces were immersed in a solution of Alcian Blue for 10 minutes to allow the dye to be absorbed by the TEP; finally, a second rinse was made with DI water to remove excess stain before putting the samples under the microscope.

A qualitative analysis of the images was realized; a quantitative analysis based on grayscale images was not possible because of the interference of the support mesh (dark color) with the stained TEP cluster.

3.2.2.5 FO membrane cleaning

3.2.2.5.1 Air scouring cleaning

Cleaning with air scouring was performed for 15 minutes with two air-stone diffusers (one on each side of the cell), at a cross flow rate of 4166 mg · L^{-1} in the feed water tank, resulting in 3.13 liters of air per liter of feed water. This flow rate was considered to prevent damage of the membrane and, at the same time from previous trials, to remove macroscopic fouling deposited on the surface of the membrane. Two different conditions were tested to determine the reversible fouling of the FO membrane with air scouring. First, the membrane was cleaned inside the CFW; afterwards, a second cleaning was done with DI water as feed and the flux was measured again. Both were *in situ* cleaning techniques.

3.2.2.5.2 Chemical cleaning

For the active layer of the membrane, which was in contact with the SWWE, the washing solution contained 0.8% sodium ethylenediaminetetraacetic acid (EDTA) and 1% Alconox, a detergent composed of sodium dodecylbenzenesulfonate (10–30%), sodium carbonate (7–13%), tetrasodium pyrophosphate (10–30%) and sodium phosphate (10–30%), adapted from cleaning protocols for ROGA RO spiral wound elements with CTA membranes. The cleaning was performed *in situ*, leaving the membrane cell immersed for 5 minutes on each side in the solution and rinsing both the tank and the cell with tap water; meanwhile, DI was recirculated instead of DS to prevent membrane drying. The pH of the solution was set to 8.0 to prevent membrane damage (hydrolization of CTA) based on manufacturer specifications.

The cleaning potential of the two solutions was tested to analyze the best option for TEP removal on the support side of the FO membrane; this cleaning was not performed *in situ*. The first solution (Solution 1) had the same composition as the one used for the chemical cleaning of the membrane fouling on the active layer. The second one contained 1% aqueous solution of NaOCl (pH 7.6) (Solution 2). The membrane was submerged in a petri dish with cleaning solution before staining the TEP with the dye, using two pieces of the fouled membrane, one for each solution. After being submerged for 10 minutes, a quick 1-minute rinse with DI water was done before the dyeing process.

3.2.3 RESULTS AND DISCUSSION

3.2.3.1 FO membrane process

Figure 3.2.2 shows representative results for the normalized flux of the FO membrane process for each cycle (day) of an experiment. The membrane was tested with DI as the feed and pre-filtered seawater as DS for the first day. After this, SWWE was used as feed, running the experiment for 11 days (day 2 to 12 in Figure 3.2.2) with a recovery of 7% per cycle, before the first cleaning (membrane cell submerged in the CFW) with air scouring for 15 minutes (day 13). The next day (day 14), the feed water was changed to DI water, and air scouring was performed again for the same period of time. Chemical cleaning was made with the cleaning solution described in Section 4.2.6.2 (day 15), after which a test with DI water as feed was made to determine the flux recovery.

After 11 days of experiment with the SWWE as feed (day 2 to 12 in Figure 3.2.2), the fluxes of the FO process varied between 7.2 and 2.0 $L \cdot m^{-2} \cdot h^{-1}$. The draw solution was diluted, on average, from 51,500 $\mu S \cdot cm^{-1}$ to 22,500 $\mu S \cdot cm^{-1}$ in one cycle (1 day). The average flux was 3.42 $L \cdot m^{-2} \cdot h^{-1}$. The flux decline of the FO process after 11 days was 25%, sufficient to start the cleaning process discussed in Section 3.2.3.4.

The driving force for the FO process was determined for the first day of this 11-cycle experiment (day 2 in Figure 3.2.2), considering the osmotic difference of the feed water and the DS, which was 29.95 bar. The driving force at the end of day 2 was only 9.05 bar, which explains part of the flux decline observed. The values for the ideal flux were calculated considering the internal concentration polarization (ICP) model used by Gray *et al.* (2006), assuming a resistance of solute diffusion for the seawater (K) of 10.26 d · m^{-1} obtained from a previous work using the same DS (Yangali-Quintanilla *et al.* 2011), and water and solute permeability coefficients of the active layer of the membrane A and B of 0.027 m · atm^{-1} · d^{-1} and 0.011 m · d^{-1}, respectively. The results for the modelled and the observed flux vary significantly (up to 22%), showing that fouling is responsible for a fraction of the reduced flux. Therefore, upon further analysis, calculations for recovery can be related to the flux decline resulting from fouling, not taking into consideration the loss of osmotic pressure difference, which is present in all of the FO cycles.

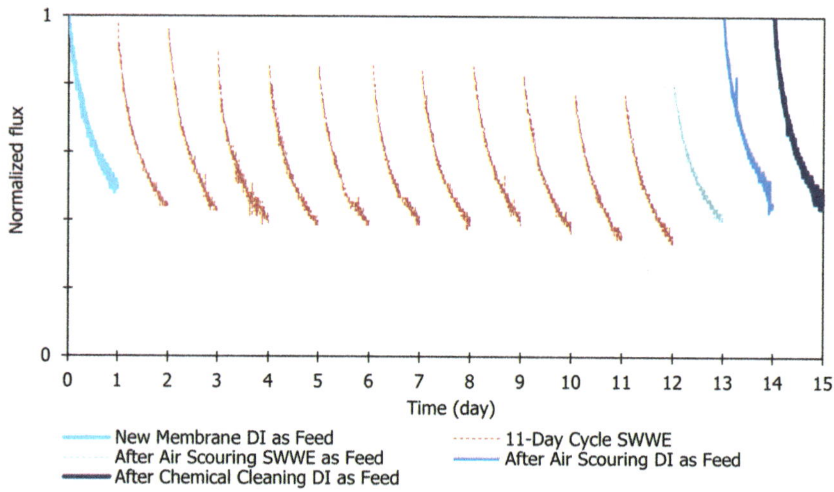

Figure 3.2.2 Flux decline of the FO process.

3.2.3.2 Fouling of the active layer of FO membrane

Figure 3.2.3a shows the chromatograms for a sample of CFW and EWWE, and Figure 3.2.3b for a sample of DS and DDS. The peak in the organic carbon curve representing biopolymers (polysaccharides and proteins) is larger in the EWWE than in the CFW, suggesting that these compounds are not present in the feed water remaining after the FO process. The chromatograms for the DS and DDS samples show that there is no transport of biopolymers through the membrane from the feed to the DS. There is a significant difference in concentration for the biopolymers comparing the CFW and the EWWE, but not for the remainder of the CDOC fractions. For the CFW, the biopolymers were present at 1.44 mg · L^{-1}, while the corresponding peak for the EWWE showed 2.52 mg · L^{-1}. The humic substances were present at concentrations of 5.40 and 5.25 mg · L^{-1}, respectively; building blocks were also similar in concentration for both samples, with 2.64 and 2.51 mg · L^{-1}, accordingly; and so are the low molecular weight (LMW) neutrals

(2.29 and 2.37 mg · L^{-1}, respectively). These fractions show a slight increase in the CFW because of the volume of effluent concentrated by the FO process in relation to the evaporated volume (EWWE). The DDS showed a low concentration of biopolymers compared to the DS, 0.05 mg · L^{-1} and 0.08 mg · L^{-1}, respectively, supporting the hypothesis that biopolymers were not transported through the membrane, and thus only the original concentration of biopolymers in the seawater was being diluted when the FO process took place. However, for the LMW neutrals, the concentration in the DDS is considerably higher than the concentration in the DS (0.74 to 0.11 mg · L^{-1}), because this CDOC fraction is transported from the feed water through the membrane due to the size of the compounds, which are smaller than the molecular weight cutoff (MWCO) of the membrane, established in previous studies as 200 Da (Valladares Linares *et al.* 2011).

Figure 3.2.3c shows the organic carbon curve after the extraction of the foulants from the membrane with an ultrasonic bath, comparing it with the control samples (new piece of membrane, SWWE and DS). A single biopolymers peak can be observed for the curve of the fouled membrane extraction sample, supporting the hypothesis that these substances were forming a fouling layer which attached to the surface membrane. The rest of the chromatogram is very similar to the new membrane control sample, which confirms that the remainders of the CDOC fractions have little or no impact on the fouling layer formation. Moreover, the new FO membrane releases LMW neutrals as seen in the control curve, even after being submerged and washed with DI water; these are presumably preservatives which are released to the DDS and thus are not present in the fouled membrane extraction sample.

Figure 3.2.4 shows the 3-D FEEM resulting from the analysis of the SWWE, CFW and EWWE on the same intensity scale, which corroborates the results previously obtained with the LC-OCD analysis. Likewise, the 3-D FEEM show a peak in the protein-like substances (emission range 320–350 nm, excitation range 270–280 nm (Amy *et al.* 2010)), which is likely associated with biopolymers, in the EWWE that is not detectable in the CFW, suggesting that this NOM fraction is not present in the feed water, and it is not being transported through the membrane to the DS (based on LC-OCD results). Therefore, biopolymers (protein-like substances) are likely responsible for the FO membrane fouling.

The results from the LC-OCD analysis combined with the 3-D FEEM evidence suggest that biopolymers and protein-like substances are accumulating on the active layer of the membrane, thus being partially responsible of the fouling of the FO membrane when SWWE is being fed to the process. Figure 3.2.5a, b show a SEM image of the active layer of the fouled membrane, where a uniform fouling layer is covering the surface, composed of bacteria as well as biopolymers and protein-like substances. These DOC fractions have been previously reported to be good predictors of fouling in low and high pressure membranes (Lee *et al.* 2004), and have also been identified as important foulant agents in RO (Li *et al.* 2007) and membrane bioreactor (MBR) systems (Wang & Li, 2008).

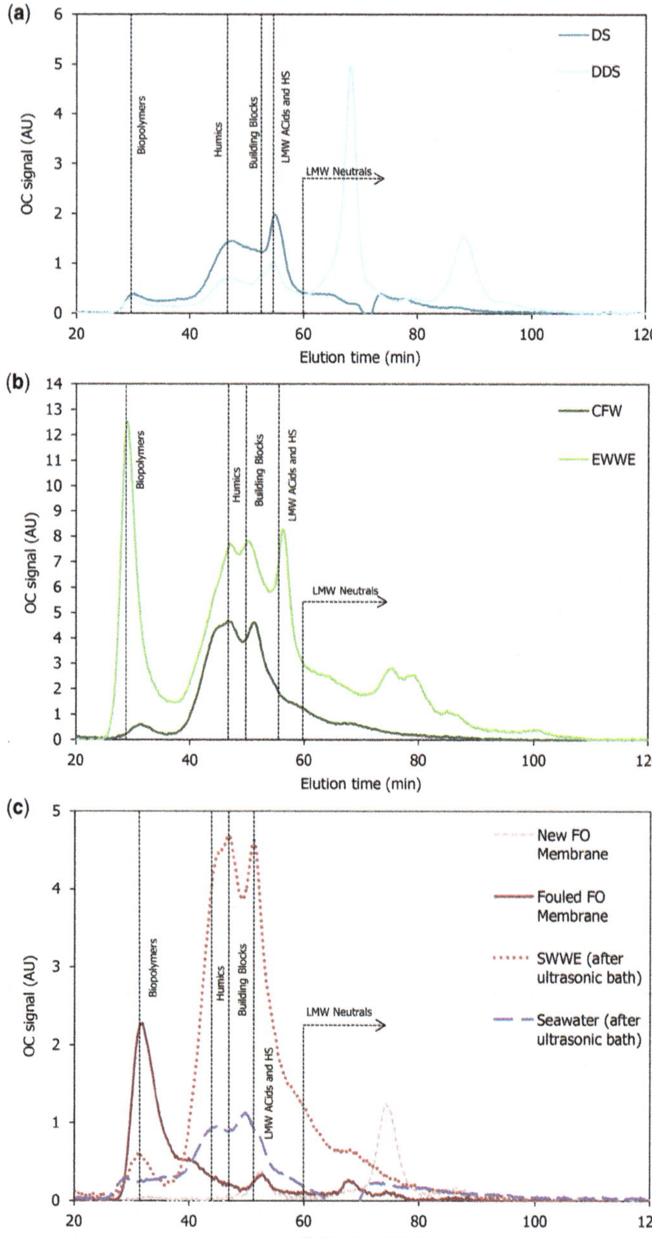

Figure 3.2.3 LC-OCD chromatograms for (a) CFW and EWWE; (b) DS and DDS; (c) fouling extraction of the used FO membrane compared to a virgin FO membrane, feed water and draw solution.

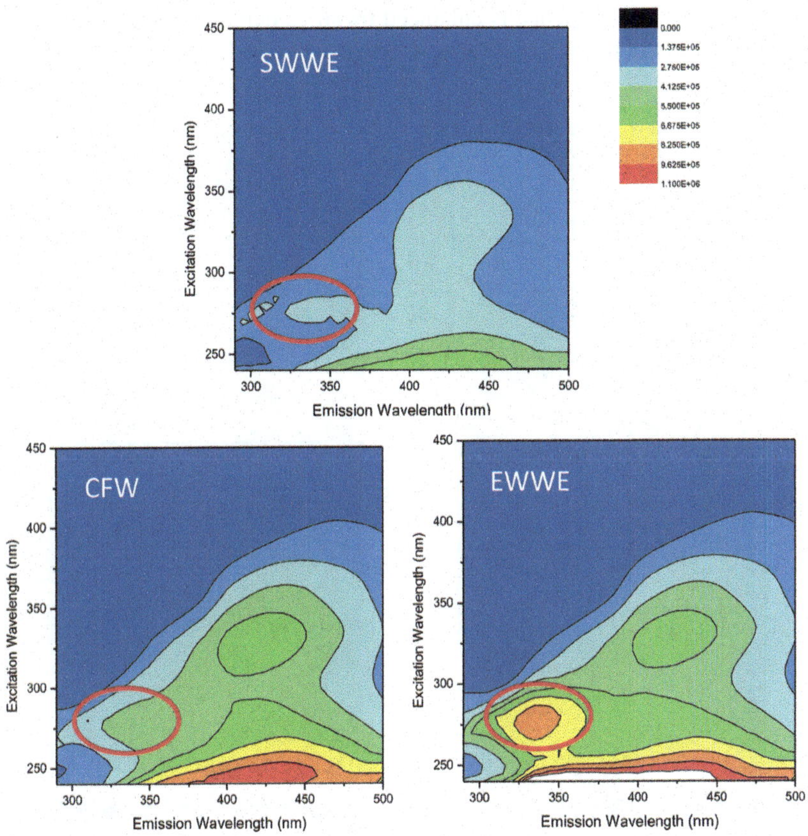

Figure 3.2.4 3-D FEEM of SWWE, CFW and EWWE.

3.2.3.3 Fouling of the support layer of FO membrane

TEP analysis was made on both the active and support layer of the membrane with microscopy and proper dyes. Figure 3.2.6a shows a picture of the support layer of a used FO membrane after the dyeing process (protocol in Section 3.2.2.4). Clusters of TEP were identified as foulants on this side of the membrane; there were large concentrations of these particles accumulating on the support side. This observation is consistent with the SEM images in Figure 3.2.5c, d, which show the support layer of the fouled membrane covered with bacteria and gel-like substances. The mechanism of TEP fouling may involve the adhesion of these particles to the membrane surface due to the stickiness of the substances forming TEP: polysaccharides, proteins, and nucleic acids; this mechanism may be coupled with bridging of TEP and calcium ions found in the seawater. Previous studies have confirmed that TEP presence is enhanced by silica deposits on the surface when the membranes are in contact with seawater (Li *et al.* 2012).

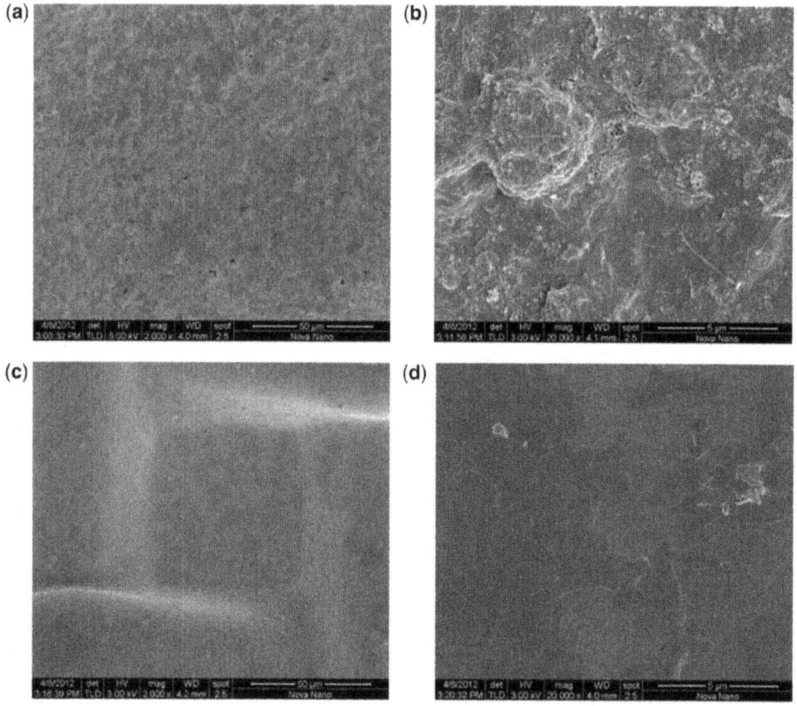

Figure 3.2.5 SEM images of a fouled FO membrane. (a) and (b) NOM fouling on the active layer; (c) and (d) TEP fouling on the support layer.

For the active layer, there was no attachment of TEP to the surface (Figure 3.2.6b), suggesting that the SWWE does not contain a significant amount of these particles or that they might not be depositing on the membrane surface. Based on this observations and the fact that the support layer is facing the DS (AL-FS orientation), the identified TEP was derived from the seawater, similar to other TEP studies mentioned in the literature (Villacorte *et al.* 2009).

Problems associated with TEP fouling of either filtration or RO desalination membranes are attributable to these particles, rather than from bacteria in the water, meaning that disinfection of the feed water would not be useful to prevent TEP from causing fouling (Bar-Zeev *et al.* 2009; Berman *et al.* 2011), and microfiltration would not significantly remove colloidal TEP (Villacorte *et al.* 2009). Consequently, Section 3.2.3.4 describes the results obtained for the two cleaning solutions proposed for TEP removal.

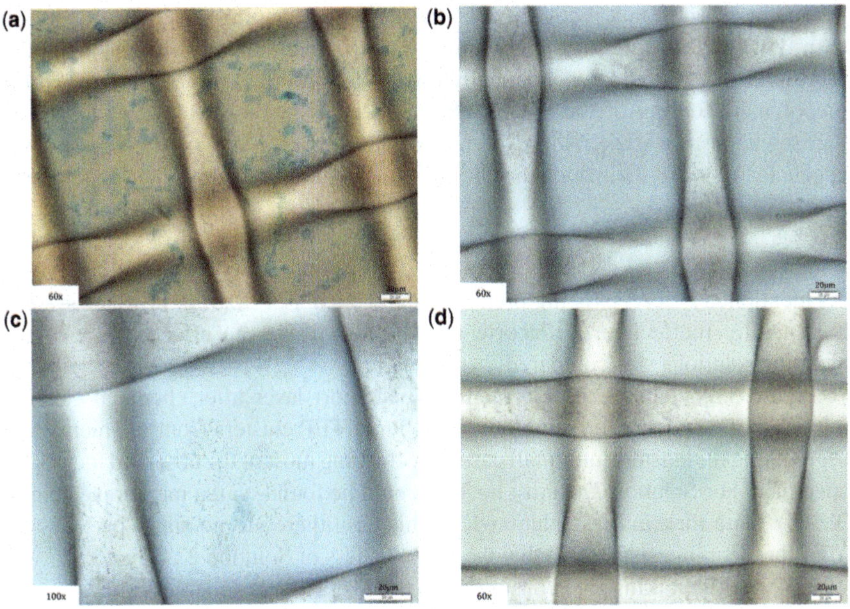

Figure 3.2.6 Bright light microscopy images of a fouled FO membrane stained with Alcian Blue dye. (a) TEP fouling on the support layer (60×); (b) active layer (60×); (c) support layer of a cleaned FO membrane (100×) with Solution 1; (d) support layer of a cleaned FO membrane (60×) with Solution 2.

3.2.3.4 Cleaning of the FO membrane – active layer

Equation 3.2.1 was used to determine flux recovery:

$$\text{Flux recovery}\,(\%) = \frac{J_i - J_f}{J_i} \times 100 \tag{3.2.1}$$

where J_i is the average flux recorded for the initial or referential cycle and J_f is the average flux recorded for a cycle run after a certain cleaning technique was used. The flux recovery for the air scouring realized within the CFW (day 13 in Figure 3.2.2) was 90.8%. For the air scouring cleaning with DI water, the recovery was 88.9% (day 14 in Figure 3.2.2). It is important to note that the initial flux (J_i) for the CFW cleaning was obtained with SWWE as feed (day 2 in Figure 3.2.2); on the other hand, the initial flux (J_i) for the DI cleaning was obtained with DI as feed (day 1 in Figure 3.2.2), hence, the difference in recovery. Nevertheless, both results indicate a high reversibility of the NOM fouling. This high reversibility may due to the lack of hydraulic pressure on the FO process which results in the formation of a loose foulant layer on the surface of the membrane, which was also described by Mi *et al.* (2010) in a previous study (2010).

The increase in flux recovery after the chemical cleaning (DI as feed) was less than 3% of the original flux, revealing that the hydraulically reversible fouling represented 90.8% and the chemically reversible fouling 91.8% of the total NOM-biopolymers fouling of the FO membrane. Therefore, the flux that could not be recovered, or the irreversible fouling considering cleaning of the active layer, accounts for only 8.2%, which may be attributed to TEP fouling of the support layer.

3.2.3.5 Cleaning of the FO membrane – support layer

As mentioned before, existing pretreatment against TEP fouling in water systems is generally ineffective (Villacorte *et al.* 2009), thus, the effectiveness of two cleaning solutions was compared.

Figure 3.2.6c shows the FO membrane support layer after chemical cleaning with Solution 1. The solution removed the large TEP clusters from the membrane, segregating the clusters into small particles, cleaning most of the observed membrane specimen. For Solution 2, almost no TEP could be found on the membrane (Figure 3.2.6d). Note that, unlike Solution 1, in this case there are no small particles left after the cleaning, suggesting that the effectiveness of Solution 2 is greater.

TEP fouling, taking into consideration the results for reversible and irreversible NOM-biopolymers fouling, was equal to or less than 8.2% of the total fouling of the FO membrane. After the membrane was cut for TEP identification and cleaning, flux recovery tests could not be performed. Future studies should determine the effectiveness of the chemical cleaning supported with flux decline and recovery experiments.

3.2.4 SUMMARY

The study of forward osmosis (FO) membranes has increased due to the already demonstrated advantages compared to high-energy membrane processes such as reverse osmosis (RO). This research focuses on characterization of the natural organic matter (NOM) fraction causing fouling on the active layer (AL) of a FO membrane in a novel plate and frame module configuration, facing secondary wastewater effluent as a feed solution (FS) and seawater used as a draw solution (DS). In addition, transparent exopolymer particles (TEP) were observed on the support layer (SL) of the membrane in contact with the DS. The NOM fouling layer, after characterizing the water samples and membranes used, was found to be composed of biopolymers and protein-like substances, which adversely affect the flux of water through the FO membrane. However, NOM fouling showed high reversibility, up to 90% when air scouring for 15 minutes is used as a cleaning technique. The irreversible fouling in this work was found to be 8.2% after chemical cleaning. On the support layer of the membrane, TEP formed clusters clearly identifiable with an optical microscope and a TEP-specific dye. Chemical cleaning with 1% NaOCl for 10 minutes proved to be the most effective method to remove TEP.

3.2.5 REFERENCES

Amy G., Croue J.-P., Salinas S. and Kennedy M. (2010). Pretreatment Options for Controlling Organic Foulants in Seawater Reverse Osmosis (SWRO). In ARWADEX 2010, Riyadh.

Bar-Zeev E., Berman-Frank I., Liberman B., Rahav E., Passow U. and Berman T. (2009). Transparent exopolymer particles: potential agents for organic fouling and biofilm formation in desalination and water treatment plants. *Desalination and Water Treatment*, **3**, 7.

Berman T., Mizrahi R. and Dosoretz C. G. (2011). Transparent exopolymer particles (TEP): a critical factor in aquatic biofilm initiation and fouling on filtration membranes. *Desalination*, **276**, 184–190.

Cath T. Y., Hancock N. T., Lundin C. D., Hoppe-Jones C. and Drewes J. E. (2010). A multi-barrier osmotic dilution process for simultaneous desalination and purification of impaired water. *Journal of Membrane Science*, **362**, 417–426.

Cho J., Amy G. and Pellegrino J. (2000). Membrane filtration of natural organic matter: factors and mechanisms affecting rejection and flux decline with charged ultrafiltration (UF) membrane. *Journal of Membrane Science*, **164**, 89–110.

Cornelissen E. R., Harmsen D., de Korte K. F., Ruiken C. J., Qin J.-J., Oo H. and Wessels L. P. (2008). Membrane fouling and process performance of forward osmosis membranes on activated sludge. *Journal of Membrane Science*, **319**, 158–168.

Gray G. T., McCutcheon J. R. and Elimelech M. (2006). Internal concentration polarization in forward osmosis: role of membrane orientation. *Desalination*, **197**, 1–8.

Holloway R. W., Childress A. E., Dennett K. E. and Cath T. Y. (2007). Forward osmosis for concentration of anaerobic digester centrate. *Water Research*, **41**, 4005–4014.

Hong S. and Elimelech M. (1997). Chemical and physical aspects of natural organic matter (NOM) fouling of nanofiltration membranes. *Journal of Membrane Science*, **132**, 159–181.

Huber S. A., Balz A., Abert M. and Pronk W. (2011). Characterisation of aquatic humic and non-humic matter with size-exclusion chromatography – organic carbon detection – organic nitrogen detection (LC-OCD-OND). *Water Research*, **45**, 879–885.

Kuzmenko D., Arkhangelsky E., Belfer S., Freger V. and Gitis V. (2005). Chemical cleaning of UF membranes fouled by BSA. *Desalination*, **179**, 323–333.

Lambert M. and Lampi K. (2010). Forward Osmosis, Reality vs. Perception, "an HTI Perspective". In: AMTA/Statkraft AS – 2nd Osmosis Membrane Summit, San Diego, CA.

Lee N., Amy G., Croué J.-P. and Buisson H. (2004). Identification and understanding of fouling in low-pressure membrane (MF/UF) filtration by natural organic matter (NOM). *Water Research*, **38**, 4511–4523.

Lee S., Boo C., Elimelech M. and Hong S. (2010). Comparison of fouling behavior in forward osmosis (FO) and reverse osmosis (RO). *Journal of Membrane Science*, **365**, 34–39.

Li Q., Xu Z. and Pinnau I. (2007). Fouling of reverse osmosis membranes by biopolymers in wastewater secondary effluent: role of membrane surface properties and initial permeate flux. *Journal of Membrane Science*, **290**, 173–181.

Li Z.-Y., Yangali-Quintanilla V., Valladares-Linares R., Li Q., Zhan T. and Amy G. (2012). Flux patterns and membrane fouling propensity during desalination of seawater by forward osmosis. *Water Research*, **46**, 195–204.

Mi B. and Elimelech M. (2008). Chemical and physical aspects of organic fouling of forward osmosis membranes. *Journal of Membrane Science*, **320**, 292–302.

Mi B. and Elimelech M. (2010). Organic fouling of forward osmosis membranes: fouling reversibility and cleaning without chemical reagents. *Journal of Membrane Science*, **348**, 337–345.

Tang C. Y., She Q., Lay W. C. L., Wang R. and Fane A. G. (2010). Coupled effects of internal concentration polarization and fouling on flux behavior of forward osmosis membranes during humic acid filtration. *Journal of Membrane Science*, **354**, 123–133.

Valladares Linares R., Yangali-Quintanilla V., Li Z. and Amy G. (2011). Rejection of micropollutants by clean and fouled forward osmosis membrane. *Water Research*, **45**, 6737–6744.

Villacorte L. O., Kennedy M. D., Amy G. L. and Schippers J. C. (2009). The fate of Transparent Exopolymer Particles (TEP) in integrated membrane systems: removal through pre-treatment processes and deposition on reverse osmosis membranes. *Water Research*, **43**, 5039–5052.

Wang X.-M. and Li X.-Y. (2008). Accumulation of biopolymer clusters in a submerged membrane bioreactor and its effect on membrane fouling. *Water Research*, **42**, 855–862.

Wang Y., Wicaksana F., Tang C. Y. and Fane A. G. (2010). Direct microscopic observation of forward osmosis membrane fouling. *Environmental Science & Technology*, **44**, 7102–7109.

Yangali-Quintanilla V., Li Z., Valladares R., Li Q. and Amy G. (2011). Indirect desalination of Red Sea water with forward osmosis and low pressure reverse osmosis for water reuse. *Desalination*, **280**, 160–166.

Yip N. Y., Tiraferri A., Phillip W. A., Schiffman J. D. and Elimelech M. (2010). High performance thin-film composite forward osmosis membrane. *Environmental Science & Technology*, **44**, 3812–3818.

Chapter 3.3

Draw solute induced calcium carbonate scaling*

3.3.1 INTRODUCTION

Scaling formed by mineral salts, such as calcium salts and silica, is an important aspect in high pressure membrane processes, i.e. reverse osmosis (RO) and nanofiltration (NF) (Rahardianto *et al.* 2008; Al-Amoudi, 2010; Antony *et al.* 2011). Scaling is an impediment for the application of high pressure membrane process since it causes a significant decline of permeate flux and irreversible damage to the membrane. Generally, scaling in RO and NF process is due to the supersaturation of salts at the interface of the membrane and feed solution, followed by crystallization and growth (Antony *et al.* 2011). Various factors, such as temperature, hydrodynamic conditions, pH, and interaction of solutes, can affect the formation of scaling (Sheikholeslami, 2003; Lee & Lee, 2005; Rahardianto *et al.* 2008; Jawor & Hoek, 2009).

In recent years, forward osmosis (FO) has been studied and applied as a novel membrane process to achieve a more sustainable solution of the water-energy nexus. FO is an osmotically driven membrane process. The driving force of the FO process is the osmotic pressure created by a salinity difference (gradient) between a feed water (FS) and draw solution (DS) which are separated by a FO membrane (Chung *et al.* 2012). Given that researchers have identified the potential of FO or FO hybrid systems for desalination (Cath *et al.* 2006; Coday & Cath, 2014), scaling in the FO process is gaining more attention recently. So far, the amount of research about scaling in the FO process is still limited and most researchers have been

*This chapter is based, with permission from the copyright holder, on the previously published paper. (Li Z., Valladares Linares R., Bucs Sz. S., Aubry C., Ghaffour N., Vrouwenvelder J.S. and Amy G. (2015). Calcium carbonate scaling in seawater desalination by ammonia–carbon dioxide forward osmosis: mechanism and implications. *Journal of Membrane Science*, **481**, 36–43. doi:10.1016/j.memsci.2014.12.055) focusing on gypsum scaling (Mi & Elimelech, 2010a; Arkhangelsky *et al.* 2012; Chen *et al.* 2013; Zhang *et al.* 2014).

The similar feed conditions as the RO process were simulated in the FO process, with a near-supersaturated gypsum concentration in the FS. The mechanism of scaling formation in the FO process was found to be similar to RO process. The scaling in FO process can be induced by both concentration polarization and reversely diffused ions from DS, which lead to supersaturation of salts on the membrane surface. Furthermore, under the unique operational condition (driven by the osmotic pressure) of the FO process, the gypsum scaling on the membrane active layer (AL) was almost fully reversible when the AL faces the FS (AL-FS orientation) (Mi & Elimelech, 2010a; Arkhangelsky *et al.* 2012). However, when the active layer faced DS (AL-DS orientation) and the porous support layer faced FS, the scaling became partially irreversible, which was due to the formation of crystals of gypsum inside the porous structure of support layer (Arkhangelsky *et al.* 2012). In addition, membrane material also affects the formation of gypsum scaling and its reversibility. The gypsum scaling caused less water flux decline and the scaling was more reversible on a cellulose acetate (CA) membrane than a polyamide (PA) membrane in the FO process (Mi & Elimelech, 2010a). In addition to gypsum, silica scaling has also been observed in FO processes (Li *et al.* 2012; Mi & Elimelech, 2013). The formation of silica scaling is dominated by silica polymerization and facilitated by the presence of hydroxyl groups on the membrane material or organic compounds in the FS. Therefore, silica scaling was more irreversible than gypsum.

On the other hand, ammonia–carbon dioxide (NH_3/CO_2) and ammonium bicarbonate (NH_4HCO_3) have been considered as a promising DS in the FO desalination process (McCutcheon *et al.* 2005, 2006; McGinnis *et al.* 2013). The attractiveness of this DS is that the fresh water can be recovered from the diluted DS after FO process by moderate heating (near 60 °C). The solute is decomposed into NH_3 and CO_2, which can easily be removed from water and used to regenerate the DS. In order to obtain a high concentration of ammonium salt, the DS can also be prepared by mixing NH_4HCO_3 and ammonium hydroxide (NH_4OH) in water at proper proportions (McCutcheon *et al.* 2006).

In the present study, we reported a new scaling phenomenon in seawater desalination by the FO process using NH_4HCO_3 as the DS. For the first time reported, this calcium carbonate ($CaCO_3$) scaling is different from the scaling caused by supersaturation of salts in RO and FO processes mentioned above. The mechanism of the scaling formation was investigated.

3.3.2 MATERIALS AND METHODS

3.3.2.1 FS, DS and FO set-up

All salts used in this study were purchased from Sigma-Aldrich (St. Louis, MO). The properties of natural seawater used in this study were reported in our previous publication (Li *et al.* 2012). This study was initially conducted as the post-treatment method for our previous FO study in which seawater could be diluted before desalination (Valladares Linares *et al.* 2013). Therefore, the 50% diluted natural seawater (DSW, 50% diluted by deionized (DI) water) was employed as FS in this study. Natural seawater was pre-filtered by a capsule filter (pore size 0.45 μm, Polycap™ AS 36, Whatman®) before the dilution. 2% NaCl solution (synthetic diluted seawater, SDSW) were used as FS in baseline testing to compare with natural seawater while DI water was used as the FS to investigate the effect of scaling on the performance of membrane. The DS in this study was 2M NH_4HCO_3. The experimental FO set-up was described in our previous publication (Li *et al.* 2012). In this study, the FO cell was installed vertically. The pH of FS and DS in this study was weakly alkaline (pH 8.2 ± 0.2). The variation of pH after the FO process was observed to be minor in both FS and DS (pH 8.5 ± 0.2 after the FO process).

3.3.2.2 FO membrane, and the measurement of intrinsic permeability and separation properties

The FO membrane was provided by Hydration Technology Innovations (HTI, Scottsdale, AZ, USA) and received as a flat coupon. The water permeability (*A* value) and salt rejection (*R* value) of the membrane was tested in a RO set-up (SEPA CF II, Sterlitech, Kent, WA, USA). The transmembrane pressure varied from 5 to 9 bar and the crossflow velocity was 10 $cm \cdot s^{-1}$. DI water was used as feed for the *A* value measurement. A cocktail of salts in solution (containing NaCl, KCl, $CaCl_2$, $FeCl_3$, $MgCl_2 \cdot 6H_2O$, $SrCl_2 \cdot 6H_2O$, $BaCl_2$, $Na_2O_3Si \cdot 9H_2O$ and H_3BO_3 with a concentration of 10 $mg \cdot L^{-1}$ of each target ion at pH 6.8 ± 0.5), was used as the feed to measure *R* for target ions listed in Table 3.3.1. The procedures to detect *A* and *R* value were discussed elsewhere (Wei *et al.* 2011).

3.3.2.3 Experimental protocol for FO testing

The initial volumes of FS and DS were 1 L each, and testing was terminated when 100 mL permeate was collected to the DS vessel. The bottom-up co-current fluid flow was operated at both the FS and DS side of the cell under the same crossflow velocity. All filtration processes were conducted at an ambient temperature of 21 ± 1 °C. The variation of weight of DS solution was measured by a digital balance (Sartorius AG, Germany) and converted to flux. The reverse draw solute (NH_4HCO_3) from DS to FS was detected by measuring total nitrogen (TN) in FS and DS before and after the FO process, and subsequently converted to the amount of NH_4HCO_3. Specific reverse salt flux, $J_{specific}$ is defined as the ratio of reverse NH_4HCO_3 flux ($mmol \cdot m^{-2} \cdot h^{-1}$) and forward water flux ($L \cdot m^{-2} \cdot h^{-1}$) to eliminate the effect of the draw solute concentration (Hancock & Cath, 2009; Phillip *et al.* 2010).

DI water was used as the blank sample for TN measurement. In a parallel testing, after the FO process with DSW as the FS (i.e. 100 mL permeate was collected), both the FS and DS sides of the membrane were simultaneously flushed by DI water *in situ* at a crossflow velocity of 16.7 $cm \cdot s^{-1}$ for 20 minutes. The recovery of DI water flux was measured with 2M NH_4HCO_3 as the DS at a crossflow velocity of 8.3 $cm \cdot s^{-1}$ to investigate the reversibility of the scaling. In order to determine the composition of the scaling layer and the impact of water flushing on removal of the scalants, the membrane samples (2 × 2 cm) before and after hydraulic cleaning were collected and soaked in 10 mL 2% HNO_3 solution overnight with shaking to dissolve the scalants for further analysis.

3.3.2.4 Analytical methods

The top surface (active layer, AL) and back surface (support layer, SL) of the original membrane were characterized by Fourier Transform Infrared Spectroscopy in the attenuated total reflectance mode (ATR-FTIR, Perkin Elmer Spectrum 100, USA). The cations were analyzed in 2% HNO_3 solution by Inductively Coupled Plasma/Mass Spectrometry (ICP-MS 7500; Agilent, USA). TN was measured by an Organic Carbon Analyzer (TOC-VCPH, Shimadzu, Japan). Adhesive force measurements were performed with an Atomic Force Microscope (Nano-Observer AFM microscope, CSI, France) using a 10 μm calcite probe (Novascan, USA). The roughness of AL and SL was measured by AFM. The membrane surfaces after scaling testing and water flushing were visualized by a Scanning Electron Microscopy-Energy Dispersive X-ray Apparatus (SEM-EDX, Magellan, FEI, USA) and an Optical Microscope (DP72, Olympus, Japan). The particle sizes in the feed DSW were analyzed by a particle size analyzer (Zetasizer, Nano Series, Malvern, UK).

3.3.3 RESULTS AND DISCUSSION

3.3.3.1 Characterization of FO membrane

The commercial HTI FO membrane is made of cellulose triacetate (CTA) embedded about a polyester (PES) screen. There is no distinct porous support layer in this membrane (Wang *et al.* 2012). Both AL and SL are relatively dense. The water permeability of CTA FO membrane and its rejection of the ions presenting in seawater are shown in Table 3.3.1. The membrane showed the rejection rates to cations following the order: trivalent (Fe^{3+}) > divalent (Ca^{2+}, Mg^{2+}, Sr^{2+} and Ba^{2+}) > monovalent (K^+ and Na^+). The rejection of silica (in the form of monosilicic acid, $Na_2O_3Si \cdot 9H_2O$) was up to about 90%. The membrane showed a poor rejection of boron. This is due to diffusion of the non-ionic form of boron which passes through at the near neutral pH used in the study (Koseoglu *et al.* 2008). Figure 3.3.1 reveals that there is no distinct difference between

Table 3.3.1 Intrinsic water permeability and salt rejection of HTI CTA FO membrane.

Transmembrane Pressure (bar)	Water Permeability ($\times 10^{-12}$ m · s^{-1} · Pa^{-1})	Rejection (%)								
		K	**Na**	**B**	**Ca**	**Mg**	**Fe**	**Sr**	**Si**	**Ba**
5	1.95	81.88	89.43	19.14	92.90	92.58	99.18	92.61	89.90	92.75
6		83.38	85.33	19.53	91.54	92.63	98.82	92.66	84.71	92.48
7		86.89	87.08	14.01	91.67	92.75	97.81	92.57	86.88	92.10
8		87.23	88.22	21.51	92.81	93.12	98.25	92.76	90.04	92.50
9		88.25	89.66	24.61	93.21	93.70	98.21	93.58	86.83	93.36

Note: All measurements were conducted under RO mode with a constant feed flow linear velocity of 0.1 m · s^{-1}. Membrane active layer faced the feed water.

AL and SL, except for the surface roughness. The AL is smoother than the SL (Figure 3.3.1).

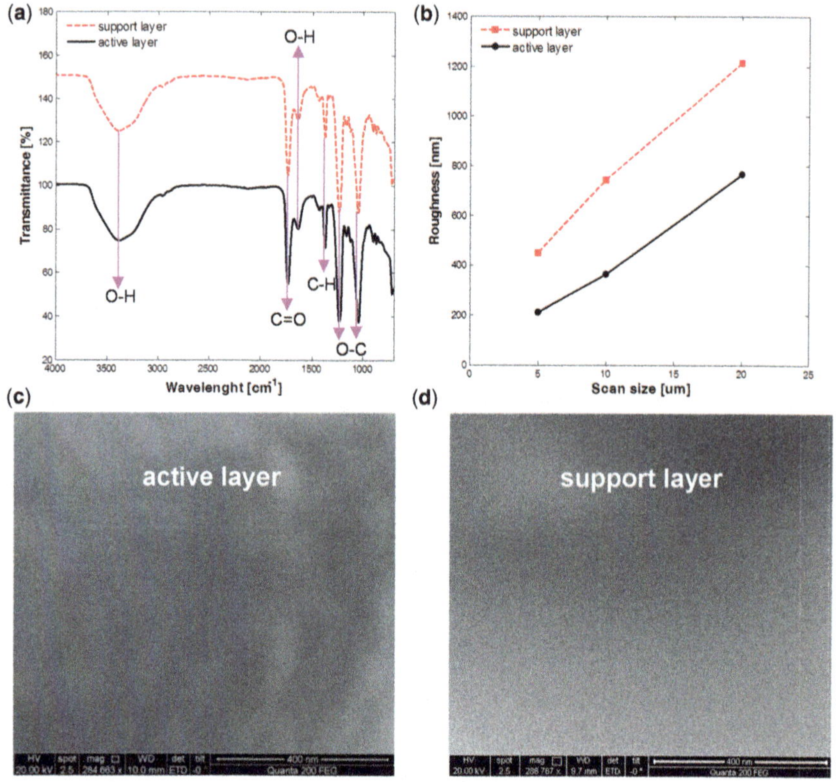

Figure 3.3.1 The properties of virgin HTI CTA FO membrane. (a) The FTIR-ATR spectra of CTA FO membrane surface; (b) The roughness of CTA FO membrane surface; (c) SEM image of the membrane activate layer; (d) SEM image of the membrane support layer.

3.3.3.2 Water and reverse solute flux

The reverse NH_4HCO_3 flux indicated by $J_{specific}$ was measured by analyzing TN, then converted to the amount of NH_4HCO_3 in different FSs, using different membrane orientations and crossflow velocities (Figure 3.3.2). TN can indicate the flux of $NH4^+$. According to Donnan exclusion mechanism and ion-pairing effect, the rejection of $NH4^+$ approximately represents the rejection of NH_4HCO_3 (Rice *et al.* 2011). The crossflow velocities varying from 2.1 to 16.7 $cm \cdot s^{-1}$ under AL-FS did not show a significant effect on initial water flux as well as $J_{specific}$. Membrane orientation played a more important role on $J_{specific}$. Under the AL-FS mode, $J_{specific}$ was 152.0 $mmol \cdot L^{-1}$

while it was 76.4 mmol·L^{-1} in the AL-DS mode when DI was used as the FS (Figure 3.3.2). This is attributed to the higher water flux and external concentration polarization at DS side in the AL-DS mode. Higher water flux induced more severe external dilutive concentration polarization at the interface between membrane and DS at DS side (Hancock & Cath, 2009; Zhao *et al.* 2011). The external dilutive concentration polarization at DS side reduced not only the effective osmotic pressure but also the amount of draw solute accessing to the membrane. Accordingly, a higher $J_{specific}$ (i.e. reverse diffusion of draw solute) was observed with DSW compared to SDSW as FS under AL-FS mode.

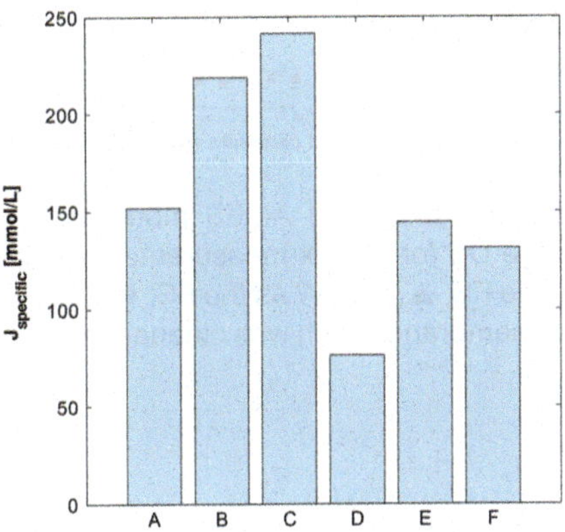

	Feed water	Crossflow velocity (cm/s)	Membrane orientation	Initial flux (L/m²h)
A	DI	8.3	AL-FS	10.6
B	SDSW	8.3	AL-FS	7.24
C	DSW	8.3	AL-FS	6.75
D	DI	8.3	AL-DS	16.4
E	DI	16.7	AL-FS	11.4
F	DI	2.1	AL-FS	10.2

Figure 3.3.2 Effects of feed water, membrane orientation and crossflow velocity on the specific reverse salt (NH_4HCO_3) flux.

The AL-DS mode gave higher permeate water flux than the AL-FS mode when DI water was the FS (Figure 3.3.3). This can be explained by the reduction or elimination of dilutive internal concentration polarization (ICP) in the AL-DS mode (Zhao & Zou, 2011). When DI water is the FS, the dilutive ICP gives less impact on the drop of osmotic pressure in AL-DS mode than AL-FS mode since less salt (draw solute) can diffuse into the membrane when DS faces the AL of membrane. Water flux is not significantly affected by the membrane orientation when SDSW was used as the FS. When a saline water is used as the FS, the concentrative concentration polarization at FS side can be promoted by a rough or porous SL in the AL-DS mode, which may counteract the positive effect from the reduction or elimination of dilutive ICP mentioned above (McCutcheon *et al.* 2006; Zhao & Zou, 2011). Furthermore, a dramatic decline in the permeate flux was observed when DSW was employed as the FS in the AL-DS mode compared to the AL-FS mode. After 100 mL permeate was collected, the flux decline is about 47.1% under the AL-FS mode and 93.6% under the AL-DS mode. The membrane surface was fully covered by a scale-like layer under both modes.

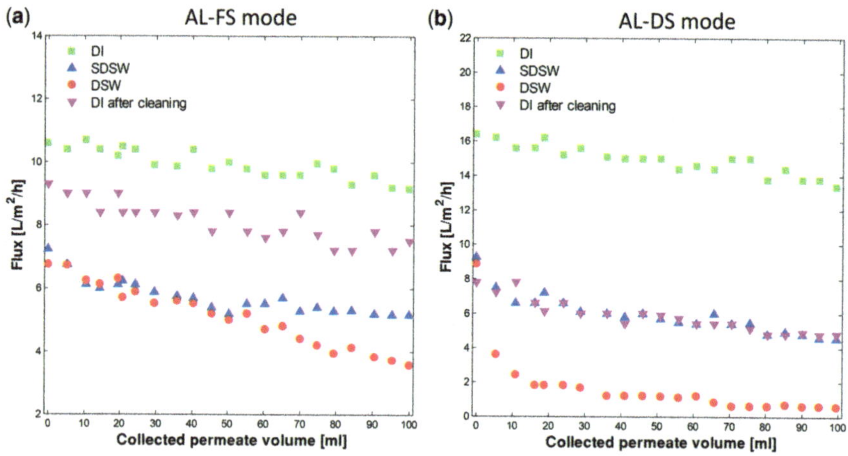

Figure 3.3.3 Flux patterns during FO process under (a) AL-FS mode and (b) AL-DS mode. 2M NH_4HCO_3 was used as the DS for all flux measurements at the flow velocity of 8.3 cm · s^{-1}. (■, DI water as the FS; ▲, SDSW as the FS; ●, DSW as the FS; ▼, DI water flux measured from the membrane which was cleaned by water flushing after the scaling testing).

3.3.3.3 Characterization of scaling in seawater desalination using NH_3/CO_2 FO process

A scaling layer was formed on the membrane surface at the FS side after the FO process with DSW as the FS under both AL-FS and AL-DS modes (Figure 3.3.4). The scalant showed acicular crystals which accumulated mainly in clusters. EDX spectra indicated that the scalant consisted predominantly of $CaCO_3$. Generally, there are three polymorphs of anhydrous $CaCO_3$, which include aragonite, calcite and vaterite. The scaling observed in this study showed a typical aragonite cluster which has a radial pattern. This is attributed to the high ratio of Mg/Ca in seawater (Falini *et al.* 1994). Furthermore, the solubility product constant (K_{sp}) of $CaCO_3$ ($K_{sp} = 8.7 \times 10^{-9}$ at 25 °C) is much lower than $MgCO_3$ ($K_{sp} = 1.0 \times 10^{-5}$ at 25 °C) (Atkins & Jones, 2010), thus Ca^{2+} in the FS has a greater tendency to pair with CO_3^{2-} to form the crystal. Meanwhile, the amount of Ca^{2+} in feed seawater is much greater than other elements (e.g., Fe, Sr, Ba) which have low K_{sp} value as well in the form of carbonate salts. Therefore $CaCO_3$ was the predominant compound of the scaling layer in this study.

In order to analyze the detailed composition of the scalant, the membrane sample with the scaling layer was soaked in HNO_3 solution and the dissolved salts were analyzed by ICP-MS (Table 3.3.2). A similar composition of the scaling layer was detected from the membrane samples under AL-FS and AL-DS modes. The large amount of Na and *K* is due to the residual seawater since the membrane was

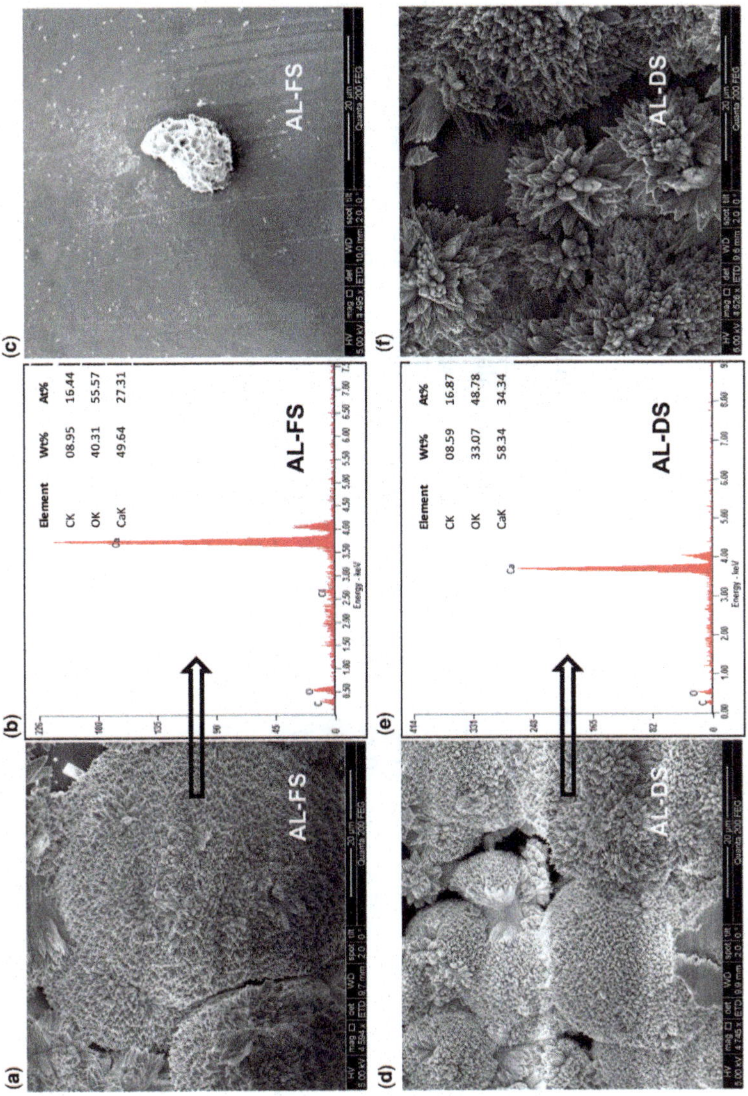

Figure 3.3.4 SEM image and EDX spectra of scaling on the FO membrane surface at the FS side. (a and d), SEM image of the membrane after scaling testing; (b and e), EDX spectra; (c and f), SEM image of the membrane after hydraulic cleaning.

Table 3.3.2 The composition of scaling on the membrane surface at the FS side before and after hydraulic cleaning under different membrane orientations.

Membrane Orientation	Sample No.	Composition of the Scalant (×10^{-3} mg·cm^{-2})								
		K	**Na**	**B**	**Ca**	**Mg**	**Fe**	**Sr**	**Si**	**Ba**
AL-FS	(1)	109	1147	0.14	891	366	6.5	200	17	0.30
	(2)	7.6	0.44	0	0.09	0	0	0	9.0	0
AL-DS	(1)	145	1459	0.97	1434	386	9.4	328	336	0.94
	(2)	90	300	0.54	963	237	6.3	223	45	0.49

Note: (1) Membrane sample after the scaling testing; (2) Membrane sample which was cleaned by DI water flushing after scaling testing.

collected and soaked in the acid immediately after the scaling testing. Ca^{2+} is one of the most abundant elements in the scaling layer. The presence of Mg and Sr was due to the incorporation of these elements into aragonite when the crystal formed and grew (Fernandez-Diaz *et al.* 1996; Brinza *et al.* 2013).

3.3.3.4 Reversibility of scaling and recovery of permeate water flux by hydraulic flushing

After the scaling testing, the membrane was flushed by DI water *in situ* at a crossflow velocity of 16.7 $cm \cdot s^{-1}$ for 20 minutes. The scaling layer formed in AL-FS mode was much easier to remove by hydraulic cleaning. The membrane surface was almost clean and only a few particles were observed on the membrane surface after hydraulic cleaning (Figure 3.3.4), and the recovery of DI water flux was up to 81.8% (Figure 3.3.3a). This result is consistent with previous study of gypsum scaling on the same membrane with AL-FS mode, in which the flux recovery was up to 96% after hydraulic flushing with higher crossflow velocity (21 $cm \cdot s^{-1}$) (Mi & Elimelech, 2010a). However, SEM image revealed that the scaling layer was still present on the membrane surface collected from the AL-DS mode and showed a more distinct needle/acicular crystal form after the hydraulic cleaning (Figure 3.3.4). The recovery of DI water flux was 36% with flushing (Figure 3.3.3b). This scaling layer could not be removed by mechanical brushing using a sterile brush. ICP-MS analysis confirmed that scaling in AL-DS mode is more irreversible than AL-FS mode. Most elements in the scaling layer formed in AL-DS mode were still present after the hydraulic cleaning (Table 3.3.2). This is different from the previous study of gypsum scaling on CTA FO membrane with AL-DS mode. Arkhangelsky *et al.* reported that the diffusion and accumulation of salt in the membrane was the main cause limiting the flux recovery after surface flushing while most scalants were removed from the membrane surface (Arkhangelsky *et al.* 2012).

3.3.3.5 Mechanism of scaling formation in seawater desalination using NH_3/CO_2 FO process

According to membrane surface properties (Figure 3.3.1) and the difference of the reversibility of scaling in AL-DS and AL-FS modes, the membrane surface morphology may play a crucial role in the formation and reversibility of $CaCO_3$ scaling when the HTI CTA FO membrane is used. To confirm this, the AFM adhesion force measurement was performed using a commercial $CaCO_3$ probe (Figure 3.3.5) with SDSW which was a 2% NaCl solution. Figure 3.3.5 shows that the adhesion forces for the AL and SL of membrane were different. When the AL faced the $CaCO_3$ probe (simulating AL-FS mode in FO process), the rupture distance was much shorter than that with SL facing the $CaCO_3$ probe (simulating AL-DS mode in FO process). The larger rupture distance indicated the extensive long range attraction and stronger physical contact between $CaCO_3$ probe and the

SL of membrane in the AL-DS mode (Mi & Elimelech, 2010a, b). Therefore, the scaling on the SL of the FO membrane showed much less reversibility by hydraulic cleaning than that on the AL of the FO membrane. It has been proven in pressure driven membrane processes (e.g., NF, RO) that membrane surface roughness can significantly reduce electrostatic repulsion between foulants and the surface, and the valley regions experienced a greater adhesion force (Elimelech *et al.* 1997; Bowen & Doneva, 2000; Vrijenhoek *et al.* 2001). In addition, the $CaCO_3$ scaling was not observed inside the membrane inner structure even though AL-DS mode was applied. This is attributed to the unique structure of HTI CTA FO membrane. An embedded mesh provides the mechanical support to this membrane while a porous support layer is used in conventional RO and NF membranes. The relatively dense surfaces on both sides of this FO membrane prevent the scalant crystals from entering into the membrane. Similar phenomenon was also observed in the study of gypsum scaling during FO process using the same membrane in AL-DS mode (Arkhangelsky *et al.* 2012).

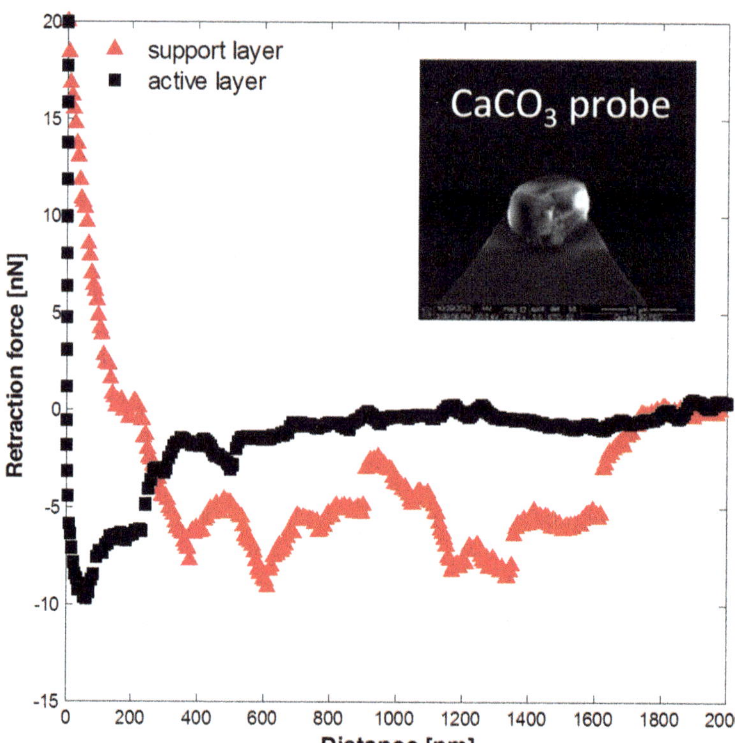

Figure 3.3.5 AFM force measurement using a $CaCO_3$ probe in 2% NaCl solution.

$CaCO_3$ was not present in the pre-filtered seawater because of its low solubility. In addition, the concentration of Ca^{2+} in DSW (50% seawater) was much lower than the calcium saturation concentration which was studied in other FO scaling research (Mi & Elimelech, 2010a; Arkhangelsky *et al.* 2012; Chen *et al.* 2013; Zhang *et al.* 2014). However, there can be a counter direction flow of salts that may interact with the water flux through the membrane during the FO process since FO membrane separates the DS from FS (Phillip *et al.* 2010; She *et al.* 2012; Irvine *et al.* 2013). Solute-solute interactions can have a dramatic impact on the ion selective permeation in the FO process (Irvine *et al.* 2013). The formation of $CaCO_3$ scaling in this study was caused by the interaction between reverse salt flux (i.e., NH_4HCO_3) from the DS and $CaCl_2$ from DSW as the FS to form the insoluble $CaCO_3$ crystal and particle (Hernández-Hernández *et al.* 2008; Chen & Xiang, 2009), followed by the deposition or growth of these crystals and particles on the membrane surface at the feed side. The mechanism of scaling formation in the present study is different from RO and other FO processes reported, in which the scaling is due to the supersaturation of salts present in the feed water. In a previous study, calcium scaling was absent on the membrane surface during a seawater FO desalination when the NaCl solution was used as DS (Li *et al.* 2012). The absence of calcium scaling in the previous research can also be a proof for the scaling mechanism dominated by the formation of insoluble calcium salt via interaction between CO_3^{2-} from the DS and Ca^{2+} from the FS in the present study.

The scaling discovered in the present study can occur over a short time of FO process. It may cover the membrane surface in two different ways. The first, CO_3^{2-} carried by the reverse salt flux interacts with Ca^{2+} to form insoluble $CaCO_3$ crystal in the FS, then deposits onto the membrane surface by the drag force induced by the permeate flux. In this case, the scaling layer can be removed easily. The second, the solute-solute interaction with counter diffusion of CO_3^{2-} and Ca^{2+} occurs on the membrane surface and causes scaling on the membrane, subsequently, newly formed scaling ($CaCO_3$ crystals) with sharp edge penetrates into membrane matrix. Therefore, the scaling under this condition is more difficult to clean hydraulically. The penetration of scaling crystals into membrane matrix can damage the membrane, consequently reducing the membrane performance for the rejection of solutes (Li *et al.* 2014). In order to further demonstrate the mechanism of $CaCO_3$ scaling formation, the particle size was monitored along the FO process in AL-FS and AL-DS mode, respectively. Additional scaling tests (with DSW as the FS) were conducted and terminated when 15 mL permeate was collected, to observe the scaling development on the membrane surface in the initial stage of the FO process. It was found that the membrane surface was clean and no crystals were observed in the AL-FS mode while crystals started to grow on the membrane surface and form clusters in the AL-DS mode (Figure 3.3.6). The development of scaling may result in the different flux patterns between AL-FS and AL-DS

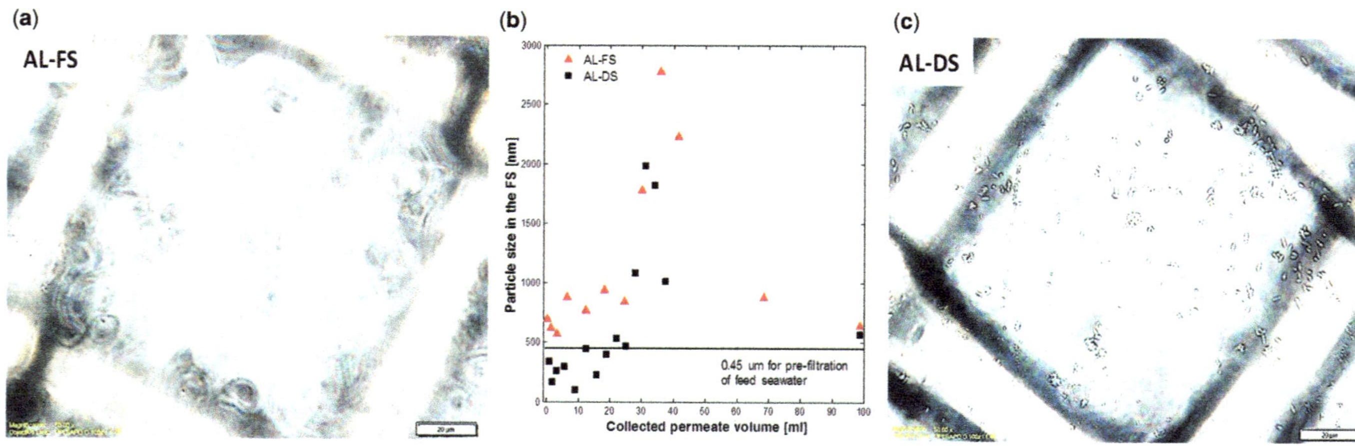

Figure 3.3.6 The particle size in the feed seawater in FO process and the membrane surface in the early stage (15 mL permeate collected) of FO process.

mode when DSW was FS (as shown in Figure 3.3.3). The flux with DSW as the FS approximated to SDSW during initial stage of FO in the AL-FS mode, which can be attributed to the slow formation and deposition of $CaCO_3$ scaling on the membrane active layer (Figure 3.3.6). In addition, the scaling is more reversible in the AL-FS mode (Figure 3.3.4). Conversely, a dramatic decline of flux with DSW as the FS can be observed from the beginning of FO process in the AL-DS mode (Figure 3.3.3). The particle size in the FS under AL-FS mode was larger than AL-DS mode from the beginning of the FO process. This could indicate that the formation of insoluble $CaCO_3$ crystals occurred in the FS under AL-FS mode while the crystals were initially formed on the membrane surface under AL-DS mode. It should be noted that the particle size in the FS was rapidly increased when 30 mL permeate was collected in both AL-FS and AL-DS modes. At this stage, membrane surfaces in both membrane orientations were fully covered by $CaCO_3$ crystals. The formed/deposited $CaCO_3$ crystal could be detached from the top of the scaling layer by the shear stress induced by the crossflow and suspended to the FS. Consequently, the particle size was quickly increased and was not much different afterward in both AL-FS and AL-DS modes.

3.3.4 SUMMARY

Forward Osmosis (FO) is an osmotically driven membrane process, where the membrane separates a draw solution (DS) with high salinity from a feed solution (FS) with low salinity. There can be a counter direction flow of salt (i.e., salt leakage) that may interact with the water flux through the FO membrane. For the first time reported, this study describes a new calcium carbonate scaling phenomenon in the seawater FO desalination process using ammonium bicarbonate as the DS. The scaling on the membrane surface at the feed side is caused by the interaction between an anion reversely diffused from the DS and a cation present in the FS, causing a significant decline of the water flux. The composition of the scaling layer is dominated by the solubility (represented as solubility product constant, K_{sp}) of salt formed by the paired anion and cation. Membrane surface morphology plays a crucial role in the reversibility of the scaling. If the scaling occurs on the active layer of the FO membrane, hydraulic cleaning (increasing crossflow velocity) efficiency to restore the water flux is up to 82%. When scaling occurs on the support layer of the FO membrane, the hydraulic cleaning efficiency is strongly reduced, with only 36% of the water flux recovered. The present study reveals the risk of scaling induced by the interaction of feed solute and draw solute, which is different from the scaling caused by the supersaturation in reverse osmosis and other FO studies reported. The scaling investigated in this study can occur with a very low solute concentration at an early stage of the FO process. This finding provides an important implication for selection of draw solution and development of new membranes in the FO process.

3.3.5 REFERENCES

Al-Amoudi A. S. (2010). Factors affecting natural organic matter (NOM) and scaling fouling in NF membranes: a review. *Desalination*, **259**, 1–10.

Antony A., Low J. H., Gray S., Childress A. E., Le-Clech P. and Leslie G. (2011). Scale formation and control in high pressure membrane water treatment systems: a review. *Journal of Membrane Science*, **383**, 1–16.

Arkhangelsky E., Wicaksana F., Chou S., Al-Rabiah A. A., Al-Zahrani S. M. and Wang R. (2012). Effects of scaling and cleaning on the performance of forward osmosis hollow fiber membranes. *Journal of Membrane Science*, **415–416**, 101–108.

Atkins P. and Jones L. (2010). Chemical Principles. W.H. Freeman and Company, New York, U.S.

Bowen R. W. and Doneva T. A. (2000). Atomic force microscopy studies of membranes: effect of surface roughness on double-layer interactions and particle adhesion. *Journal of Colloid and Interface Science*, **229**, 544–549.

Brinza L., Quinn P. D., Schofield P. F., Mosselmans J. F. W. and Hodson M. E. (2013). Incorporation of strontium in earthworm-secreted calcium carbonate granules produced in strontium-amended and strontium-bearing soil. *Geochimica et Cosmochimica Acta*, **113**, 21–37.

Cath T. Y., Childress A. E. and Elimelech M. (2006). Forward osmosis: principles, applications, and recent developments. *Journal of Membrane Science*, **281**, 70–87.

Chen J. and Xiang L. (2009). Controllable synthesis of calcium carbonate polymorphs at different temperatures. *Powder Technology*, **189**, 64–69.

Chen S. C., Su J., Fu F.-J., Mi B. and Chung T.-S. (2013). Gypsum ($CaSO_4 \cdot 2H_2O$) scaling on polybenzimidazole and cellulose acetate hollow fiber membranes under forward osmosis. *Membranes*, **3**, 354–374.

Chung T.-S., Zhang S., Wang K. Y., Su J. and Ling M. M. (2012). Forward osmosis processes: yesterday, today and tomorrow. *Desalination*, **287**, 78–81.

Coday B. D. and Cath T. Y. (2014). Forward osmosis: novel desalination of produced water and fracturing flowback J. *AWWA*, **106**, 37–38.

Elimelech M., Xiaohua Z., Childress A. E. and Seungkwan H. (1997). Role of membrane surface morphology in colloidal fouling of cellulose acetate and composite aromatic polyamide reverse osmosis membranes. *Journal of Membrane Science*, **127**, 101–109.

Falini G., Gazzano M. and Ripamonti A. (1994). Crystallization of calcium carbonate in presence of magnesium and polyelectrolytes. *Journal of Crystal Growth*, **137**, 577–584.

Fernandez-Diaz L., Putnis A., Prieto M. and Putnis C. V. (1996). The role of magnesium in the crystallization of calcite and aragonite in a porous medium. *Journal of Sedimentary Research*, **66**(3), 482–491.

Hancock N. T. and Cath T. Y. (2009). Solute coupled diffusion in osmotically driven membrane processes. *Environmental Science & Technology*, **43**, 6769–6775.

Hernández-Hernández A., Vidal M. L., Gómez-Morales J., Rodríguez-Navarro A. B., Labas V., Gautron J., Nys Y. and García Ruiz J. M. (2008). Influence of eggshell matrix proteins on the precipitation of calcium carbonate (CaCO3). *Journal of Crystal Growth*, **310**, 1754–1759.

Irvine G. J., Rajesh S., Georgiadis M. and Phillip W. A. (2013). Ion selective permeation through cellulose acetate membranes in forward osmosis. *Environmental Science & Technology*, **47**, 13745–13753.

Jawor A. and Hoek E. M. V. (2009). Effects of feed water temperature on inorganic fouling of brackish water RO membranes. *Desalination*, **235**, 44–57.

Koseoglu H., Kabay N., Yüksel M., Sarp S., Arar Ö. and Kitis M. (2008). Boron removal from seawater using high rejection SWRO membranes – impact of pH, feed concentration, pressure, and cross-flow velocity. *Desalination*, **227**, 253–263.

Lee S. and Lee C. (2005). Scale formation in NF/RO: mechanism and control. *Water Science & Technology*, **51**, 267–275.

Li Z.-Y., Yangali-Quintanilla V., Valladares-Linares R., Li Q., Zhan T. and Amy G. (2012). Flux patterns and membrane fouling propensity during desalination of seawater by forward osmosis. *Water Research*, **46**, 195–204.

Li Z., Valladares R., Linares, Abu-Ghdaib M. and Amy G. (2014). The Process Performance and Fate of Boron in a Multi-Stage Forward Osmosis Process for Integrated Water Management. In: EDS Conference and Exhibition on Desalination for the Environment: Clean Water and Energy, Limassol, Cyprus.

McCutcheon J. R., McGinnis R. L. and Elimelech M. (2005). A novel ammonia – carbon dioxide forward (direct) osmosis desalination process. *Desalination*, **174**, 1–11.

McCutcheon J. R., McGinnis R. L. and Elimelech M. (2006). Desalination by ammonia–carbon dioxide forward osmosis: influence of draw and feed solution concentrations on process performance. *Journal of Membrane Science*, **278**, 114–123.

McGinnis R. L., Hancock N. T., Nowosielski-Slepowron M. S. and McGurgan G. D. (2013). Pilot demonstration of the NH3/CO2 forward osmosis desalination process on high salinity brines. *Desalination*, **312**, 67–74.

Mi B. and Elimelech M. (2010a). Gypsum scaling and cleaning in forward osmosis: measurements and mechanisms. *Environmental Science & Technology*, **44**, 2022–2028.

Mi B. and Elimelech M. (2010b). Organic fouling of forward osmosis membranes: fouling reversibility and cleaning without chemical reagents. *Journal of Membrane Science*, **348**, 337–345.

Mi B. and Elimelech M. (2013). Silica scaling and scaling reversibility in forward osmosis. *Desalination*, **312**, 75–81.

Phillip W. A., Yong J. S. and Elimelech M. (2010). Reverse draw solute permeation in forward osmosis: modeling and experiments. *Environmental Science & Technology*, **44**, 5170–5176.

Rahardianto A., McCool B. C. and Cohen Y. (2008). Reverse osmosis desalting of inland Brackish water of high gypsum scaling propensity: kinetics and mitigation of membrane mineral scaling. *Environmental Science & Technology*, **42**, 4292–4297.

Rice G., Barber A. R., O'Connor A. J., Stevens G. W. and Kentish S. E. (2011). Rejection of dairy salts by a nanofiltration membrane. *Separation and Purification Technology*, **79**, 92–102.

She Q., Jin X., Li Q. and Tang C. Y. (2012). Relating reverse and forward solute diffusion to membrane fouling in osmotically driven membrane processes. *Water Research*, **46**, 2478–2486.

Sheikholeslami R. (2003). Mixed salts – scaling limits and propensity. *Desalination*, **154**, 117–127.

Valladares Linares R., Li Z., Abu-Ghdaib M., Wei C.-H., Amy G. and Vrouwenvelder J. S. (2013). Water harvesting from municipal wastewater via osmotic gradient: an evaluation of process performance. *Journal of Membrane Science*, **447**, 50–56.

Vrijenhoek E. M., Hong S. and Elimelech M. (2001). Influence of membrane surface properties on initial rate of colloidal fouling of reverse osmosis and nanofiltration membranes. *Journal of Membrane Science*, **188**, 115–128.

Wang Y.-N., Wei J., She Q., Pacheco F. and Tang C. Y. (2012). Microscopic characterization of FO/PRO membranes – a comparative study of CLSM, TEM and SEM. *Environmental Science & Technology*, **46**, 9995–10003.

Wei J., Qiu C., Tang C. Y., Wang R. and Fane A. G. (2011). Synthesis and characterization of flat-sheet thin film composite forward osmosis membranes. *Journal of Membrane Science*, **372**, 292–302.

Zhang M., Hou D., She Q. and Tang C. Y. (2014). Gypsum scaling in pressure retarded osmosis: experiments, mechanisms and implications. *Water Research*, **48**, 387–395.

Zhao S. and Zou L. (2011). Relating solution physicochemical properties to internal concentration polarization in forward osmosis. *Journal of Membrane Science*, **379**, 459–467.

Zhao S., Zou L. and Mulcahy D. (2011). Effects of membrane orientation on process performance in forward osmosis applications. *Journal of Membrane Science*, **382**, 308–315.

Chapter 3.4

Impact of spacer thickness on biofouling in forward osmosis*

3.4.1 INTRODUCTION

Osmosis is a natural low energy water transport process not exploited by the drinking water and wastewater industry. The use of forward osmosis (FO) to reclaim water from wastewater is known in applications for oil/gas well drilling wastewater, landfill leachate treatment and water recycling in space missions (Cath *et al.* 2005; Adham *et al.* 2007; Holloway *et al.* 2007; Hickenbottom *et al.* 2013), which are all recent technologies with only a few years of existence. More recently, the use of osmosis to reclaim treated municipal wastewater has been demonstrated in bench-scale experiments (Valladares Linares *et al.* 2013a; Werner *et al.* 2013), where wastewater recovery is integrated with seawater desalination (designated as *indirect* FO desalination). Osmotic membrane bioreactors have been studied as well (Achilli *et al.* 2009). FO is a process that increases water reuse at a lower energy consumption compared to high-energy membrane processes (i.e. reverse osmosis (RO) and nanofiltration (NF)), and therefore, a considerable cost reduction is feasible (Yangali-Quintanilla *et al.* 2011).

When microorganisms are present in a membrane system, biofilm formation is inevitable due to the availability of nutrients in the water flow (Flemming *et al.* 1997). Biofilm formation is the accumulation of microorganisms, including extracellular polymeric substances (EPS) produced by microorganisms, on a surface due to either deposition and/or growth. When biofilm formation causes an unacceptable operational problem such as pressure drop increase, flux reduction and/or salt passage increase, it is called biofouling (Characklis & Marshall, 1990).

---

* This chapter is based, with permission from the copyright holder, on the previously published paper. (Valladares Linares R., Bucs Sz. S., Li Z., Abu-Ghdeeb M., Amy G. and Vrouwenvelder J. S. (2014). Impact of spacer thickness on biofouling in forward osmosis. *Water Research,* **57**, 223–233. doi:10.1016/j.watres.2014.03.046)

All membrane filtration processes suffer from fouling, so the membrane elements have to be cleaned and eventually replaced. Biofouling has been identified as one of the major problems in spiral wound nanofiltration and reverse osmosis membrane operation (Winters & Isquith, 1979; Paul, 1991; Tasaka *et al.* 1994; Flemming *et al.* 1997; Khedr 2000; Saeed *et al.* 2000; Vrouwenvelder *et al.* 2008; van Loosdrecht *et al.* 2012). Feed spacers, separating the membrane sheets in cross-flow systems, are essential for membrane performance and play an important role in biofouling (Vrouwenvelder *et al.* 2009a).

Biofouling assessment in FO membrane studies is very limited in applications involving wastewater effluents. A study by Yoon *et al.* (2013) investigated biofouling in the FO process compared to the RO process. The extent of biofouling, demonstrated by the permeate flux decline, was less severe in the FO process than the RO process. Zou *et al.* (2013) studied the use of feed spacers in forward osmosis systems when using microalgae as a foulant, showing that spacer use not only improved the initial FO flux, but also reduced membrane fouling propensity.

The objective of this study is to determine the influence of feed spacer thickness (28, 31 and 46 mil) on membrane performance and biofouling development in a cross-flow FO system. Experiments consisted of consecutive 5 filtration cycles using similar water flow conditions in parallel systems with different feed spacers. To the authors knowledge this is the first study on the impact of spacer thickness and biofouling in cross-flow FO systems.

3.4.2 MATERIALS AND METHODS

3.4.2.1 Membrane, spacers and cell configuration

The FO membranes used in this study were provided by Hydration Technology Innovations (HTI, Scottsdale, AZ), made of cellulose triacetate embedded in a polyester mesh support. The FO membrane has a thickness of 30 to 50 μm and a structural parameter S of 595 μm, defined as the ratio of the support layer thickness (Ts) and the tortuosity (τ) over the porosity (ε) of the membrane ($S = Ts\ \tau/\varepsilon$). A regular thin film composite (TFC) NF/RO membrane has a higher S value ($S = 2155$ μm), which translates into a lower water flux through the RO membrane compared to the FO membrane using the osmotic power difference as the driving force (Lee *et al.* 2010a; Yip *et al.* 2010).

The membrane cell was a custom-made cross-flow cell made of poly (methyl methacrylate) (PMMA), designed to fit a 20 cm^2 sized membrane sheet. All of the experiments were performed with the active layer (AL) of the membrane facing the FO feed solution (FS), leaving the support side in contact with the draw solution (DS). This configuration has been proven to be the most effective in preventing FO membrane fouling (Cornelissen *et al.* 2008; Mi & Elimelech, 2008; Wang *et al.* 2010). The layout of the FO system is described in previous studies (Li *et al.* 2012).

Three spacers of different thickness were used to grow the biofilm (Figure 3.4.1). Two DOW diamond-shaped spacers of 28 and 31 mil (1 mil = 0.025 mm) thickness (Dow-Filmtec, Midland, MI) used in commercial spiral wound RO modules, and one modified FO spacer produced by HTI with a thickness of 46 mil (HTI, Scottsdale, AZ), were used in commercially available modified FO spiral wound modules. Table 3.4.1 lists details concerning feed flow and characteristics for each spacer, used on both the feed and draw solution side of the membrane.

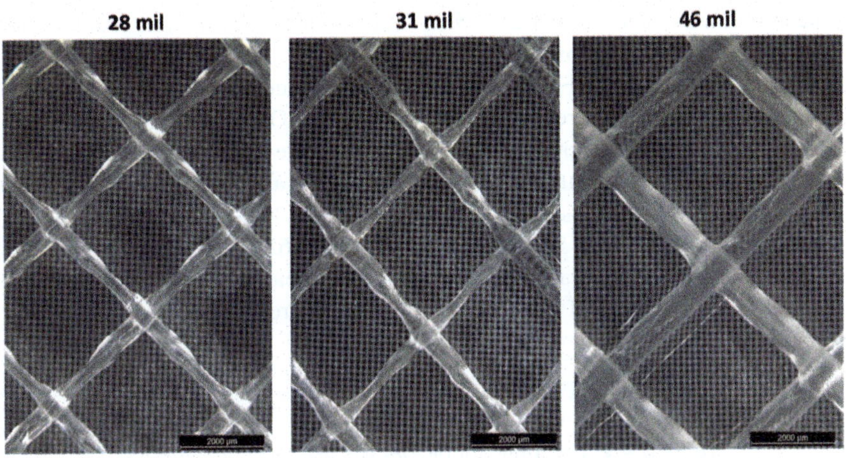

Figure 3.4.1 Different spacer thicknesses (28, 31 and 46 mil; 1 mil = 0.025 mm) on forward osmosis (FO) membrane. The small grid shown inside the FO membrane is the embedded polyester mesh support. All figures have the same scale.

Table 3.4.1 Feed spacers and flow conditions used in the FO cross-flow channel.

Spacer Type (Both Sides#)	Spacer Thickness (mm)	Flow Rate ($L \cdot h^{-1}$)	Linear Flow Velocity ($m \cdot s^{-1}$)
28 mil Dow	0.711	7.2	0.141
31 mil Dow	0.787	7.2	0.120
46 mil HTI	1.168	7.2	0.086

#The same feed spacer was used on both sides of the FO membrane. The study focus was on the fouling at the membrane feed side.

3.4.2.2 Water sources

Two water types were used directly, one after each other. Firstly, a municipal secondary wastewater effluent was used for 20 hours to condition the FO system with bacteria. Subsequently, the system was fed with synthetic municipal water (not containing bacteria) for 5 filtration cycles.

A municipal secondary wastewater effluent was collected from Al-Ruwais wastewater treatment plant in the city of Jeddah, Saudi Arabia. The chemical oxygen demand (COD) was 25 mg · L^{-1}, the total organic carbon (TOC) concentration was 4.84 mg · L^{-1}, the pH was 7.1, the conductivity was 2.33 mS · cm^{-1} and the osmotic pressure was 0.38 bar. The total suspended solids (TSS) and volatile suspended solids (VSS) of the effluent were 2.25 and 2.15 mg · L^{-1}, respectively. The ATP concentration of the effluent was 215 pg · mL^{-1}, which corresponds to a bacterial cell concentration of approximately 10^7 cells · mL^{-1}.

For the subsequent 5 FO filtration cycles, the feed solution was a synthetic municipal wastewater already used in previous FO experiments (Valladares Linares *et al.* 2013a), described in Table 3.4.2. All the chemicals were purchased from Sigma Aldrich (Munich, Germany) and had a purity grade higher than 99.5%. The COD of the synthetic wastewater was 360 mg · L^{-1} before the FO process, with a TOC of 115 mg · L^{-1} and a pH of 5.66. For the draw solution, 4% sodium chloride (NaCl) was used to simulate Red Sea water, with a conductivity of 55 mS · cm^{-1} and pH of 5.74. All solutions were prepared with deionized (DI) ultra-pure water.

Table 3.4.2 Synthetic municipal wastewater constituents.

Compound	Concentration (mg · L^{-1})
Urea	91.74
NH_4Cl	12.75
Sodium acetate	79.37
Peptone	17.41
$MgHPO_4 \cdot 3H_2O$	29.02
KH_2PO_4	23.40
$FeSO_4 \cdot 7H_2O$	5.80
Starch	122.00
Skimmed milk power	116.19
Yeast extract	52.24
$Cr(NO_3)_3 \cdot 9H_2O$	0.77
$CuCl_2 \cdot 2H_2O$	0.54
$MnSO_4 \cdot H_2O$	0.11
$NiSO_4 \cdot 6H_2O$	0.34
$PbCl_2$	0.10
$ZnCl_2$	0.21

3.4.2.3 Biofilm formation

The following procedure was used to grow a biofilm on the membrane feed side surface: (i) inoculation for 20 h with municipal secondary wastewater effluent, and (ii) 5-cycle experiment with recirculation of synthetic wastewater effluent as FS and clean 4% NaCl as DS; each cycle recovered 30% of the initial FS volume (300 mL) from the feed water, after which both FS and DS were replaced with fresh solutions.

After the 5 cycles the FO cell was opened and the membrane along with the stacked spacer of the feed side were conserved for an autopsy. For TOC analysis, DI ultra-pure water was used to dilute the samples. Before using the imaging and elemental characterization techniques, membrane samples were kept hydrated and sealed at 4 °C.

To remove organic matter and disinfect the system, the following cleaning procedure, adapted from Herzberg and Elimelech (2007), was used before starting a new experiment: (i) recirculation of 0.5% sodium hypochlorite for 2 h, (ii) recirculation of 5 mM EDTA at pH 11 for 30 min, (iii) recirculation of 2 mM SDS at pH 11 for 30 min, and (iv) sterilizing the unit by recirculation of 95% ethanol for 10 min. In-between each cleaning agent and at the end, the system was rinsed with DI ultra-pure water for 10 min.

3.4.2.4 Analytical methods

A Shimadzu TOC-V CPH total organic carbon analyzer (Japan) was used to determine the organic carbon present on the membrane and feed spacer surface. The membrane and spacer were taken from the system, placed in a tube containing 20 mL of DI ultra-pure water and subsequently treated in an ultrasonic bath 3 times for 2 minutes. The obtained homogeneous solution was analyzed for TOC.

A Celsis Advance Luminometer (Belgium) was used to measure the amount of Adenine 5′ Triphosphate (ATP) on membrane (and spacer) samples. The tubes with the samples of the fouling material from the membrane sections with a total area of approximately 4 cm^2, diluted in 50 mL sterile water, were placed in an ultrasonic cleaning bath (Bransonic Model 5510; 40 KHz, USA) for 2 minutes, and then mixed on a vortex mixer (Fisher Scientific, 230 V, USA) for 10 seconds at speed 7. This operation was repeated three times. Active biomass was determined in duplicate by measuring the ATP concentration from 50 μL samples, and then related to the area of membrane and spacer.

A stereomicroscope (Leica M205 FA, Leica Microsystems, Wetzlar, Germany) was used to obtain optical images from the FO membrane and spacer before and directly after the fouling experiments.

Confocal Laser Scanning Microscopy (CLSM) (LSM710 upright confocal microscope, Zeiss, Germany) was used to identify and characterize biofouling on the membrane surface after the experiments. The samples were stained with the following dyes based on methods described in the literature (Chen *et al.*

2006; Lee *et al.* 2010b): (i) 4′,6-Diamidino-2-Phenylindole (DAPI) dye (emission wavelength of 358 nm; excitation wavelength of 461 nm) was used to stain DNA to identify total bacterial cells (incubation in dark room for 20 min), (ii) Fluorescein Isothiocyanate (FITC) (emission wavelength of 480 nm; excitation wavelength of 520 nm) was applied to stain the amine-reactive compound-like proteins and amino sugars (incubation time of 60 min); a 0.1 M sodium bicarbonate buffer was used to retain the amine groups so the dying was effective, and (iii) Calcofluor White (emission wavelength of 433 nm; excitation wavelength of 355 nm) was used to stain β-D-glucopyranose polysaccharides (incubation time of 1 min). After each of these three staining stages, the sample was washed twice by phosphate-buffered saline (PBS) pH 7 to remove excess stain and a wipe was used to remove excess dye or buffer remaining.

The fouling on the membrane surface was analyzed by Scanning Electron Microscopy Energy Dispersive X-ray Apparatus (SEM-EDX, Magellan, FEI). The samples were freeze dried for environmental SEM.

An overview of the experimental studies and conditions is shown in Table 3.4.3.

Table 3.4.3 Overview of the forward osmosis studies performed with different spacer thicknesses.

Study	Feed Spacer Thickness (mil)*	Parameter Evaluated/ Equipment Used	Section
Effect of spacer thickness			
on performance	28, 31, 46	flux	3.1
on biomass accumulation	28, 31, 46	ATP and TOC	3.2
on fouling localization	28, 31, 46	stereomicroscope	3.3
on fouling composition	28, 31, 46	CLSM and SEM-EDX	3.4

*1 mil = 0.025 mm.

3.4.3 RESULTS

The development of performance (flux) and fouling amount, spatial localization and composition were investigated in cross flow FO cells with three feed spacers differing in thickness (28, 31 and 46 mil). To relate the flux decline with accumulated fouling material, autopsies were performed on the feed side membrane and spacer sheets.

3.4.3.1 Effect of spacer thickness on performance

The FO flux patterns using the 28, 31 and 46 mil spacers are shown in Figure 3.4.2a, 3.4.2b and 3.4.2c, respectively for 5 filtration cycles. The reduction of water flux through the membrane is due to two processes: (i) loss of osmotic pressure difference resulting from the simultaneous process of concentration of the FS and the dilution of DS, (ii) fouling and biofilm formation on the membrane and spacer surface. However, when a new cycle is started, and the FS and DS are replaced, restoring the osmotic pressure difference, and thus the driving force for the FO process, the effect on the reduction of the flux compared to the previous cycle is only due the effect of the biofilm layer.

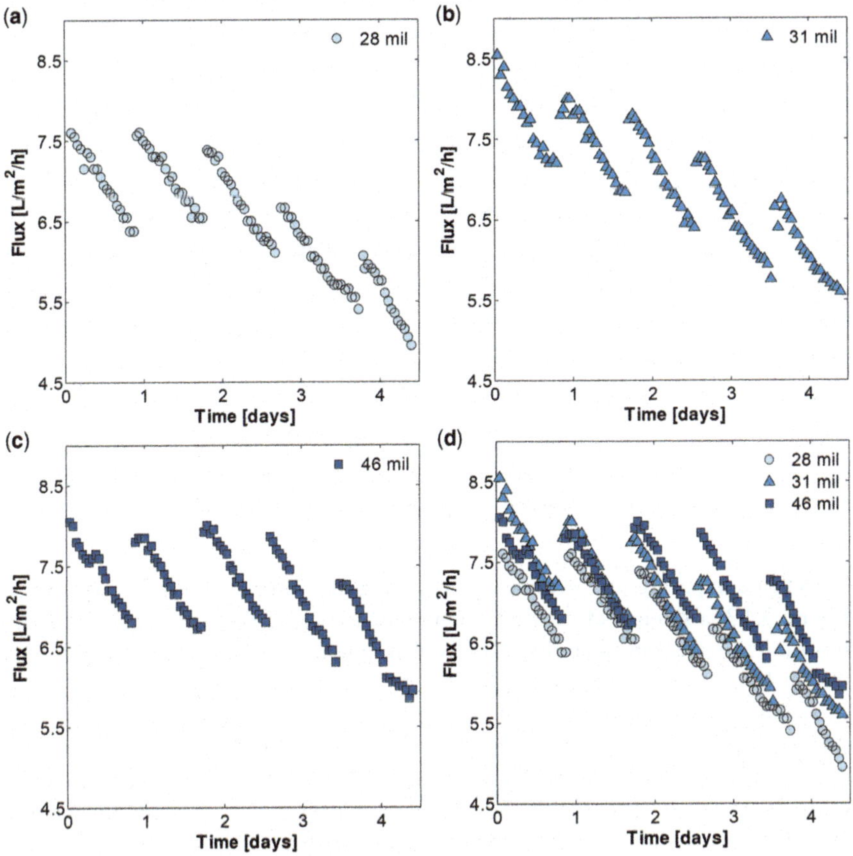

Figure 3.4.2 Flux ($L \cdot m^{-2} \cdot h^{-1}$) patterns over time for FO process with different spacer thicknesses (a) 28 mil, (b) 31 mil, (c) 46 mil and (d) combined graph. Each FO filtration cycle recovers 30% from feed solution to draw solution, after which both feed and draw solutions are replaced.

Results show that the 46 mil feed spacer has the best performance. With the 28 mil spacer, flux is rapidly reduced by the biofilm growth. Figure 3.4.2d compares the fluxes for the three different spacer thicknesses. The thickest spacer (46 mil) clearly outperforms the thinner spacers.

Flux decline calculations show that a lower flux decline can be obtained by the 46 mil thick spacer (Figure 3.4.3). The 28 mil spacer has the lowest performance of all spacer thicknesses throughout the experiment. The exponential curves denote the attachment and growth of a biofilm layer.

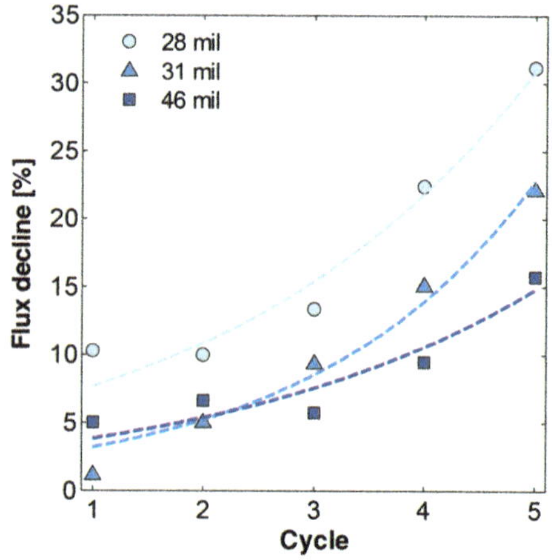

Figure 3.4.3 Flux decline (%) development for each filtration cycle for different spacer thicknesses (28, 31 and 46 mil); dashed lines show the exponential fit for the flux decline due to membrane surface fouling.

When comparing flux decline curves of the 31 mil and 46 mil feed spacers in Figure 3.4.3, a shadow effect (the membrane area pressed under the spacer shape has limited water permeability (Kim & Elimelech, 2012)) can be observed. With the thickest spacer the lowest (≈15%) flux decline was achieved.

Figure 3.4.4 compares the flux decline for the last filtration cycle for the 3 different spacer thicknesses. Running the system with the 28 mil spacer for 5 cycles translates into a flux decline of more than 30%; if the system is run at a lower cross-flow velocity (keeping the same flow rate) using a 46 mil spacer, flux decline is significantly reduced. A thicker spacer performs better for biofouling control.

3.4.3.2 Effect of spacer thickness on biomass accumulation

After running 5 filtration cycles, recovering 30% of the water in each cycle from the feed solution to the draw solution, for each spacer thickness, the biomass concentration on a measured area of the membrane surface (including the spacer section) in contact with the FS (active layer) was calculated.

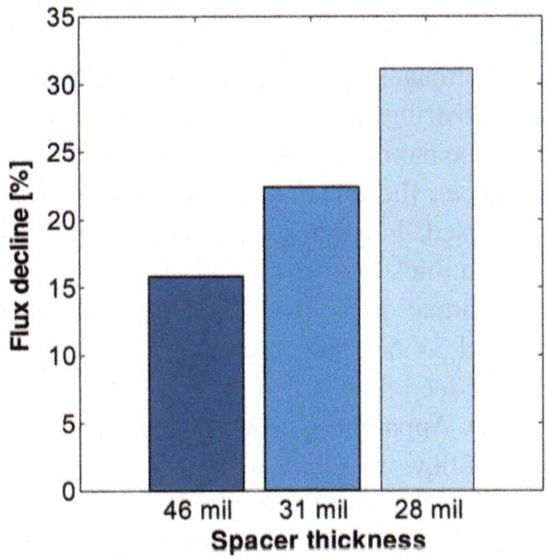

Figure 3.4.4 Flux decline (%) after the fifth filtration cycle for three different spacer thicknesses. The thickest spacer has the lowest flux decline, and thus, gives the best performance.

ATP values (Figure 3.4.5a) were found to be between 10^8 and 10^9 pg · cm^{-2}. Figure 3.4.5b shows the results for TOC concentration, which range from 0.15 to 0.18 mg · cm^{-2}. Spacer thickness had no significant impact on the amount of accumulated bacteria and organic matter including extracellular polymeric substances (EPS) on the membrane surface and the corresponding feed spacer for the same FO operation time.

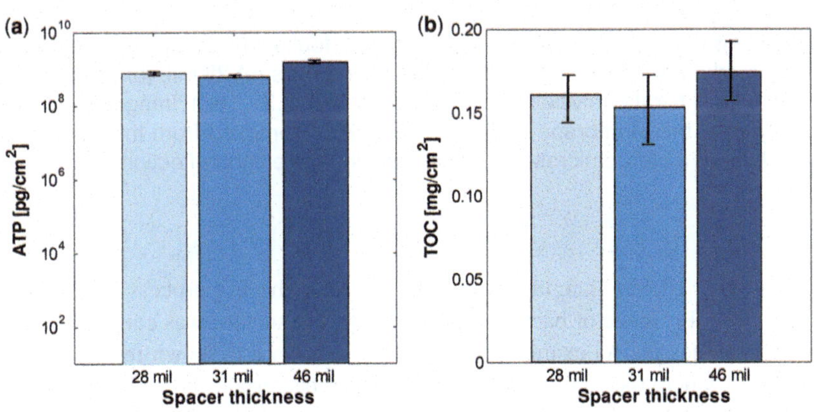

Figure 3.4.5 Biomass concentrations ATP (in pg ATP · cm^{-2}, a) and TOC (in mg · cm^{-2}, b) on the membrane and spacer in the flow cell operated at the same feed flow (7.2 L · h^{-1}) after 5 filtration cycles for different spacer thicknesses.

3.4.3.3 Effect of spacer thickness on fouling localization

Optical stereomicroscopy images of a fouled section of membrane and spacer show the difference in spatial distribution of biofouling for the three spacer thicknesses (Figure 3.4.6). Although the biomass concentration showed no significant difference between the 3 feed spacers, the location of the biofilm changes varied with the thickness of the spacer used. For the 28 mil spacer, both membrane and spacer filaments are covered by a thick biofilm (Figure 3.4.6a). For the 31 mil and 46 mil feed spacers, some membrane areas and spacer filaments are not covered by the biofilm (Figure 3.4.6b and 3.4.6c. The consequence of using different feed spacer thicknesses with the same feed flow is that the cross-flow velocity varies in the FO flow channel (Table 3.4.1). Apparently, growth of biofilm occurs in a less compact structure when a low cross-flow velocity is used (46 mil spacer) compared to a higher flow velocity (28 mil spacer). Spacer thickness and cross-flow velocity have an impact on the localization of the biofilm on the membrane surface and feed spacer filaments.

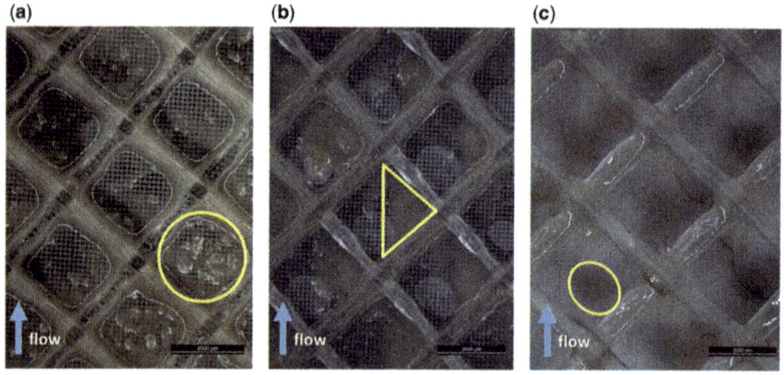

Figure 3.4.6 Varying localization patterns of fouling material (indicated within yellow marked areas) for different spacer thicknesses (a) 28 mil, (b) 31 mil and (c) 46 mil after 5 filtration cycles at constant flow rate (7.2 L · h^{-1}). Images were taken directly after the membrane and the spacer were removed from the FO cell and placed under a stereomicroscope. The white lines are light reflections.

3.4.3.4 Effect of spacer thickness on fouling composition

Results from CLSM imaging of the biofilm after the FO process are shown in Figure 3.4.7. Presence of bacteria on the membrane surface was confirmed for all spacer thicknesses. As example the 28 mil spacer data is shown in Figure 3.4.7a and 3.4.7b, where the blue spots represent stained bacterial cells with a DNA target dye (DAPI). Figure 3.4.7b shows the EPS (matrix of polysaccharides (purple) and proteins (green)), and bacteria (blue) on the fouling layer. Bacterial cells shown in Figure 3.4.7, together with operational performance decline demonstrated in section 3.1, confirm biofouling occurrence.

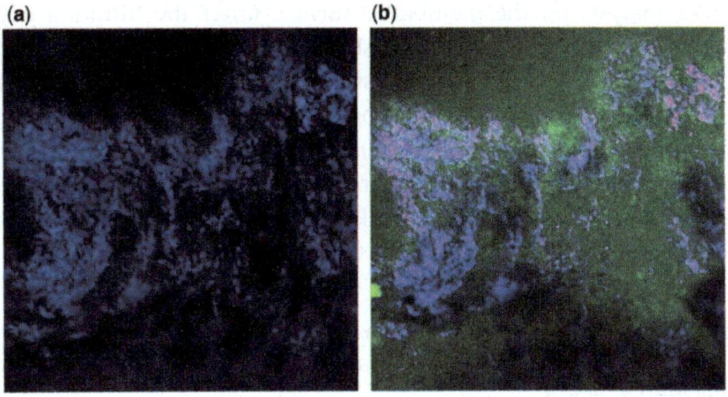

Figure 3.4.7 CLSM images indicating presence of (a) bacterial cells (blue) and (b) accumulation of polysaccharides (purple), proteins (green) and bacterial cells (blue) attached to the membrane surface (feed side) (425 μm^2) after 5 FO filtration cycles. The presented 28 mil spacer data is representative for all spacers thicknesses.

SEM images obtained from the fouled membrane surface for all spacer thicknesses do show similar structures resembling bacteria, EPS and some salt crystals. Figure 3.4.8 illustrates SEM images of a fouled membrane (using a 28 mil feed spacer) at different magnifications after 5 filtration cycles. Bacterial cells can be identified in combination with organic deposits and sodium chloride crystals, most probably coming from the DS side due to reverse draw solute flow.

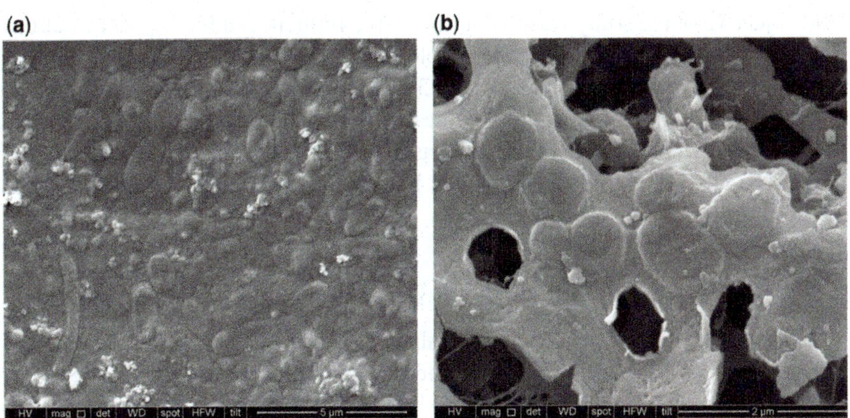

Figure 3.4.8 SEM images of the accumulated material (at magnification 18,000× (a) and 52,000× (b)) on the FO membrane feed side showing structures resembling bacteria, EPS material and salt crystals. The presented 28 mil spacer data is representative for all spacer thicknesses.

An EDX analysis on the membrane surface (after the filtration experiment) for the different spacer thicknesses is shown in Table 3.4.4. The surface of the fouled membrane has a different composition than the virgin membrane due to the presence of EPS and some inorganic deposits, shown as an increase in the amount of sodium and chloride ions (Table 3.4.4). Deposited material composition showed no significant difference among the three spacer thicknesses used. Apparently, based on the few elements detected (Table 3.4.4) only biofouling and limited sodium chloride scaling occurred.

Table 3.4.4 Elemental composition (energy dispersive x-ray analysis) of the fouled surface of an FO membrane (feed side) using the 28, 31 and 46 mil feed spacers.

Element	Virgin Membrane	28 mil	31 mil	46 mil
	[Wt%]			
C	71.31	52.90	44.09	50.41
O	15.50	19.68	16.89	13.84
Na	3.58	3.47	5.63	4.50
Cl	–	2.71	5.83	6.80

3.4.4 DISCUSSION

The objective of the study was to determine the influence of feed spacer thickness on FO performance and biofouling development and their potential for biofouling control. For all spacer thicknesses (28, 31 and 46 mil), operated at the same feed flow and the same run time, the same amount of biomass was found (Figure 3.4.5), while the flux reduction decreased with thicker spacers (Figures 3.4.3 and 3.4.4). The thicker spacer reduced the impact of biofilm on FO membrane performance.

3.4.4.1 Thickest spacer provides the best performance

The use of a thick spacer (46 mil) translates into a better performance of the FO filtration process due to the reduced effect of biofouling. A lower flux decline is achieved when the system is run at a low linear flow velocity (use of a thick spacer). Spacer thickness influences the spatial localization of biofilm in FO. Flux patterns change and due to the less compact biofilm formed, a slower flux decline was obtained when the 46 mil spacer was used. Araújo *et al.* (2012a, 2012b) obtained similar results for RO experiments, reporting that similar amounts of biomass accumulated during the experiment for different spacer thicknesses, but a lower feed channel pressure increase was observed when the thickest spacer was used. Bucs *et al.* (2013) found the same trend (lower impact biomass at thicker spacer) using numerical modeling. Thick spacers were demonstrated to be a feasible control strategy for biofouling.

The shadow effect (limited water permeability under spacer area) for thick spacers shows a smaller influence in FO than in other osmotic membrane processes that require hydraulic pressure (i.e. pressure retarded osmosis) (Kim & Elimelech, 2012). Loose biofouling structures obtained with a thick spacer (46 mil) showed a higher positive impact (improved performance) on reducing flux decline than the negative effect (reduced performance) of the shadow effect. Results show that biofouling in FO is better controlled with thick spacers, limiting the shadow effect due to the lack of hydraulic pressure.

When no spacer is used with different types of fouling in FO membrane systems, an opposite effect for a low cross-flow velocity was reported. Lee *et al.* (2010a) established that when three different initial cross-flow velocities (0.171, 0.256, and 0.342 $m \cdot s^{-1}$) were employed in FO fouling filtration, the flux curve for the lowest cross-flow velocity showed the highest flux decline, whilst at the highest cross-flow velocity, almost no flux decline was observed. FO feed spacers play an important role in biofouling control.

The consequence of using different feed spacer thicknesses in the FO flow channel with the same feed flow is that the cross-flow velocity varies (Table 3.4.1). Spacer thickness and cross-flow velocity both impacted the (i) localization of the biofilm on the membrane surface and feed spacer filaments and the (ii) effect of accumulated biomass on membrane performance. For FO, at low linear velocity the impact of biomass on flux decline is reduced, which is in agreement with results of studies on spiral wound nanofiltration and reverse osmosis systems (Vrouwenvelder *et al.* 2009b; Vrouwenvelder *et al.* 2010; Ying *et al.* 2013).

3.4.4.2 FO and RO show similar biofouling patterns

FO and RO systems differ from each other in several aspects: (i) membrane water flux (up to 10 times higher for RO compared to FO), (ii) driving force (applied hydraulic pressure versus osmotic pressure difference), (iii) biofouling potential in both sides of the membrane for FO systems, and (iv) existence of a reverse draw solute flux in FO processes.

The effect of biofouling on the FO system performance in relation with different feed spacer thicknesses (different cross-flow velocities) is similar to reported data for RO processes. Vrouwenvelder *et al.* (2009b) stated that the impact of biomass on the pressure drop increase will be lower when a low cross-flow velocity is used in the membrane system, inferring that the thickest spacer shows the best performance, and even suggesting modified spacer use as an approach to control biofouling. The 46 mil spacer used in FO filtration (lowest cross-flow velocity) controls biofouling and reduces the impact on flux decline.

This study on FO is in good agreement with results of earlier studies on RO systems concerning the impact of spacer thickness on biomass accumulation and impact of biomass on performance. Previous studies by Vrouwenvelder *et al.* (2010) reported that the same biomass concentrations showed different effects on pressure drop for RO. The same trend is observed in FO filtration, where similar amounts of biomass (ATP and TOC) do not determine performance decline.

As far as the authors' knowledge, similarities on biofouling patterns for FO and RO have not been reported before.

3.4.4.3 Future studies

This study focused on biofouling on the feed side of the FO membrane when a wastewater effluent is used as feed water. Nevertheless, the seawater biofouling potential on the draw solution side may present new insights into FO indirect desalination. Special attention should be given to the transport of small molecular weight carbon and nitrogen compounds from the FS to the DS, which may act as substrate for bacteria to grow and represent a problem for long-term operation of this type of system.

Modeling can be a useful tool to predict and prevent biofouling in FO membranes. Along with this, future studies should focus on the use of modified spacers and novel cleaning strategies, which may be in agreement with cleaning protocols already developed for FO membranes (Valladares Linares *et al.* 2013b), to mitigate biofouling in FO indirect desalination systems.

3.4.5 SUMMARY

Forward osmosis (FO) indirect desalination systems integrate wastewater recovery with seawater desalination. Niche applications for FO systems have been reported recently, due to the demonstrated advantages compared to conventional high-pressure membrane processes such as nanofiltration (NF) and reverse osmosis (RO). Among them, wastewater recovery has been identified to be particularly suitable for practical applications. However, biofouling in FO membranes has rarely been studied in applications involving wastewater effluents. Feed spacers separating the membrane sheets in cross-flow systems play an important role in biofilm formation. The objective of this study was to determine the influence of feed spacer thickness (28, 31 and 46 mil) on biofouling development and membrane performance in a FO system, using identical cross-flow cells in parallel studies. Flux development, biomass accumulation, fouling localization and composition were determined and analyzed. For all spacer thicknesses, operated at the same feed flow and the same run time, the same amount of biomass was found, while the flux reduction decreased with thicker spacers. These observations are in good agreement with biofouling studies for RO systems, considering the key differences between FO and RO. Our findings contradict previous cross-flow studies on particulate/colloidal fouling, where higher cross-flow velocities improved system performance. Thicker spacers reduced the impact of biofouling on FO membrane flux.

3.4.6 REFERENCES

Achilli A., Cath T. Y., Marchand E. A. and Childress A. E. (2009). The forward osmosis membrane bioreactor: a low fouling alternative to MBR processes. *Desalination*, **239**, 10–21.

Adham S., Oppenheimer J., Liu L. and Kumar M. (2007). WateReuse Dewatering Reverse Osmosis Concentrate from Water Reuse Applications using Forward Osmosis. WateReuse Foundation, Alexandria, VA, USA.

Araújo P. A., Kruithof J. C., Van Loosdrecht M. C. M. and Vrouwenvelder J. S. (2012a). The potential of standard and modified feed spacers for biofouling control. *Journal of Membrane Science*, **403–404**, 58–70.

Araújo P. A., Miller D. J., Correia P. B., van Loosdrecht M. C. M., Kruithof J. C., Freeman B. D., Paul D. R. and Vrouwenvelder J. S. (2012b). Impact of feed spacer and membrane modification by hydrophilic, bactericidal and biocidal coating on biofouling control. *Desalination*, **295**, 1–10.

Cath T. Y., Adams D. and Childress A. E. (2005). Membrane contactor processes for wastewater reclamation in space: II. Combined direct osmosis, osmotic distillation, and membrane distillation for treatment of metabolic wastewater. *Journal of Membrane Science*, **257**, 111–119.

Characklis W. G. and Marshall K. C. (1990). Biofilms. John Wiley & Sons , New York City, USA.

Chen M.-Y., Lee D.-J., Yang Z., Peng X. F. and Lai J. Y. (2006). Fluorecent staining for study of extracellular polymeric substances in membrane biofouling layers. *Environmental Science & Technology*, **40**, 6642–6646.

Cornelissen E. R., Harmsen D., de Korte K. F., Ruiken C. J., Qin J.-J., Oo H. and Wessels L. P. (2008). Membrane fouling and process performance of forward osmosis membranes on activated sludge. *Journal of Membrane Science*, **319**, 158–168.

Flemming H. C., Schaule G., Griebe T., Schmitt J. and Tamachkiarowa A. (1997). Biofouling – the Achilles heel of membrane processes. *Desalination*, **113**, 215–225.

Herzberg M. and Elimelech M. (2007). Biofouling of reverse osmosis membranes: role of biofilm-enhanced osmotic pressure. *Journal of Membrane Science*, **295**, 11–20.

Hickenbottom K. L., Hancock N. T., Hutchings N. R., Appleton E. W., Beaudry E. G., Xu P. and Cath T. Y. (2013). Forward osmosis treatment of drilling mud and fracturing wastewater from oil and gas operations. *Desalination*, **312**, 60–66.

Holloway R. W., Childress A. E., Dennett K. E. and Cath T. Y. (2007). Forward osmosis for concentration of anaerobic digester centrate. *Water Research*, **41**, 4005–4014.

Khedr M. G. (2000). Membrane fouling problems in reverse-osmosis desalination applications. *International Desalination and Water-Reuse Quarterly*, **10**, 8–17.

Kim Y. C. and Elimelech M. (2012). Adverse impact of feed channel spacers on the performance of pressure retarded osmosis. *Environmental Science & Technology*, **46**, 4673–4681.

Lee S., Boo C., Elimelech M. and Hong S. (2010a). Comparison of fouling behavior in forward osmosis (FO) and reverse osmosis (RO). *Journal of Membrane Science*, **365**, 34–39.

Lee W., Ahn C. H., Hong S., Kim S., Lee S., Baek Y. and Yoon J. (2010b). Evaluation of surface properties of reverse osmosis membranes on the initial biofouling stages under no filtration condition. *Journal of Membrane Science*, **351**, 112–122.

Li Z.-Y., Yangali-Quintanilla V., Valladares-Linares R., Li Q., Zhan T. and Amy G. (2012). Flux patterns and membrane fouling propensity during desalination of seawater by forward osmosis. *Water Research*, **46**, 195–204.

Mi B. and Elimelech M. (2008). Chemical and physical aspects of organic fouling of forward osmosis membranes. *Journal of Membrane Science*, **320**, 292–302.

Paul D. (1991). Reverse osmosis: scaling, fouling and chemical attack. *Desalination Water Reuse*, **1**, 8–11.

Saeed M. O., Jamaluddin A. T., Tisan I. A., Lawrence D. A., Al-Amri M. M. and Chida K. (2000). Biofouling in a seawater reverse osmosis plant on the Red Sea coast, Saudi Arabia. *Desalination*, **128**, 177–190.

Tasaka K., Katsura T., Iwahori H. and Kamiyama Y. (1994). Analysis of RO elements operated at more than 80 plants in Japan. *Desalination*, **96**, 259–272.

Valladares Linares R., Li Z., Abu-Ghdaib M., Wei C.-H., Amy G. and Vrouwenvelder J. S. (2013a). Water harvesting from municipal wastewater via osmotic gradient: an evaluation of process performance. *Journal of Membrane Science*, **447**, 50–56.

Valladares Linares R., Li Z., Yangali-Quintanilla V., Li Q. and Amy G. (2013b). Cleaning protocol for a FO membrane fouled in wastewater reuse. *Desalination and Water Treatment*, **51**, 4821–4824.

van Loosdrecht M. C., Bereschenko L., Radu A., Kruithof J. C., Picioreanu C., Johns M. L. and Vrouwenvelder H. S. (2012). New approaches to characterizing and understanding biofouling of spiral wound membrane systems. *Water Science and Technology*, **66**, 88.

Vrouwenvelder J. S., Manolarakis S. A., van der Hoek J. P., van Paassen J. A. M., van der Meer W. G. J., van Agtmaal J. M. C., Prummel H. D. M., Kruithof J. C. and van Loosdrecht M. C. M. (2008). Quantitative biofouling diagnosis in full scale nanofiltration and reverse osmosis installations. *Water Research*, **42**, 4856–4868.

Vrouwenvelder J. S., Graf von der Schulenburg D. A., Kruithof J. C., Johns M. L. and van Loosdrecht M. C. M. (2009a). Biofouling of spiral-wound nanofiltration and reverse osmosis membranes: a feed spacer problem. *Water Research*, **43**, 583–594.

Vrouwenvelder J. S., Hinrichs C., Van der Meer W. G. J., Van Loosdrecht M. C. M. and Kruithof J. C. (2009b). Pressure drop increase by biofilm accumulation in spiral wound RO and NF membrane systems: role of substrate concentration, flow velocity, substrate load and flow direction. *Biofouling*, **25**, 543–555.

Vrouwenvelder J. S., Buiter J., Riviere M., van der Meer W. G. J., van Loosdrecht M. C. M. and Kruithof J. C. (2010). Impact of flow regime on pressure drop increase and biomass accumulation and morphology in membrane systems. *Water Research*, **44**, 689–702.

Wang Y., Wicaksana F., Tang C. Y. and Fane A. G. (2010). Direct microscopic observation of forward osmosis membrane fouling. *Environmental Science & Technology*, **44**, 7102–7109.

Werner C. M., Logan B. E., Saikaly P. E. and Amy G. L. (2013). Wastewater treatment, energy recovery and desalination using a forward osmosis membrane in an air-cathode microbial osmotic fuel cell. *Journal of Membrane Science*, **428**, 116–122.

Winters H. and Isquith I. R. (1979). In-plant microfouling in desalination. *Desalination*, **30**, 387–399.

Yangali-Quintanilla V., Li Z., Valladares R., Li Q. and Amy G. (2011). Indirect desalination of Red Sea water with forward osmosis and low pressure reverse osmosis for water reuse. *Desalination*, **280**, 160–166.

Ying W., Gitis V., Lee J. and Herzberg M. (2013). Effects of shear rate on biofouling of reverse osmosis membrane during tertiary wastewater desalination. *Journal of Membrane Science*, **427**, 390–398.

Yip N. Y., Tiraferri A., Phillip W. A., Schiffman J. D. and Elimelech M. (2010). High performance thin-film composite forward osmosis membrane. *Environmental Science & Technology*, **44**, 3812–3818.

Yoon H., Baek Y., Yu J. and Yoon J. (2013). Biofouling occurrence process and its control in the forward osmosis. *Desalination*, **325**, 30–36.

Zou S., Wang Y.-N., Wicaksana F., Aung T., Wong P. C. Y., Fane A. G. and Tang C. Y. (2013). Direct microscopic observation of forward osmosis membrane fouling by microalgae: critical flux and the role of operational conditions. *Journal of Membrane Science*, **436**, 174–185.

Chapter 3.5

Effect of cleaning methods to remove organic fouling*

3.5.1 INTRODUCTION

Membrane processes are now commonly used in water reuse and seawater desalination. Forward osmosis (FO) represents a new opportunity to solve the global water crisis because of its low-energy requirement compared to high-energy processes, such as reverse osmosis (RO) or nanofiltration (NF). Previous studies investigated fouling in FO membranes (Shibutani *et al.* 2011; Yangali-Quintanilla *et al.* 2011), showing that for low pressure osmotic processes, fouling is highly reversible. Previous work by Zhang *et al.* (2006) has determined that extracellular substances (EPS) are the first step of fouling, comprising polysaccharides and proteins mainly, and are considered to be precursors of transparent exopolymer particles (TEP). TEP fouling cannot be controlled by feed water disinfection or microfiltration (Bar-Zeev *et al.* 2009; Villacorte *et al.* 2009; Berman *et al.* 2011), thus the need for an effective cleaning method.

This research focuses on the cleaning efficiency of several methods to remove NOM-fouling on the active layer (AL) of a FO membrane immersed in a municipal secondary wastewater effluent (SWWE). Seawater was used as draw solution (DS). In addition, identification and cleaning of TEP on both the active and the support layer (SL) of the membrane were performed.

3.5.2 MATERIALS AND METHODS

The FO membranes used for this study are made of cellulose triacetate (CTA) (HTI, Scottsdale, AZ, USA). The AL of the membrane was facing the FO feed, leaving the SL in contact with the DS. This configuration (FO mode) has been proven to be the

*This chapter is based, with permission from the copyright holder, on the previously published paper. (Valladares Linares R., Li Z., Yangali-Quintanilla V., Li Q. and Amy G. (2013). Cleaning protocol for a FO membrane fouled in wastewater reuse. *Desalination and Water Treatment*, **51**, 4821–4824. doi: 10.1080/19443994.2013.795345) most effective in preventing fouling of the FO membrane (Cornelissen *et al.* 2008; Mi & Elimelech, 2008; Wang *et al.* 2010; Zou *et al.* 2011). The feed for the FO process was SWWE from the city of Jeddah. For the DS, a batch of Red Sea water was used.

Figure 3.5.1 shows a detailed scheme of the setup to induce fouling in the membrane. This configuration has been used in previous studies with relevant results (Valladares Linares *et al.* 2012). The FO dilution process takes 24 hours recirculating the DS inside the cell, increasing the level in the DS tank, while the SWWE concentrates (volume decreases). The cycle was repeated by replacing the diluted draw solution (DDS) with DS every day, to ensure that a fouling layer was formed on the surface of the membrane.

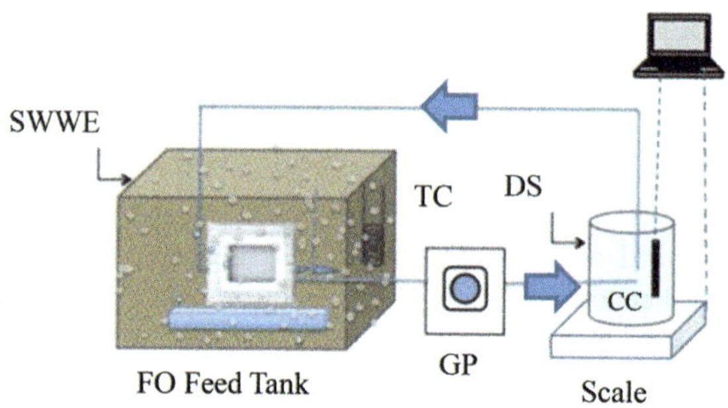

Figure 3.5.1 FO system layout. TC – temperature controller, CC – conductivity controller, GP – gear pump.

An evaporator was used to concentrate the SWWE without involving a membrane process, in order to identify the NOM fraction causing fouling. This sample was referred to as evaporated wastewater effluent (EWWE), and compared to the concentrated feed water after the FO process (CFW). A liquid chromatograph coupled with organic carbon detection analyser (LC-OCD) was used to analyse the organic carbon content of the water samples, following the procedure established by Huber *et al.* (2011).

Two hydrodynamic cleaning procedures were used to remove fouling from the surface of the FO membrane. First, osmotic backwash was performed replacing the feed with 4% NaCl solution and recirculating DI water inside the cell. Afterwards, cleaning with air scouring was performed for 15 minutes with air-stone diffusers at a rate of 12.5 $L \cdot h^{-1}$ of air per liter of SWWE. This last procedure has been proved to be very effective to remove foulants in an FO membrane.

Two chemical cleaning solutions were used to clean fouling on the FO membrane after the use of hydrodynamic procedures. The first washing solution was a 1% aqueous solution of NaOCl. The second one contained 0.8% sodium EDTA and 1% Alconox (commercial detergent containing phosphates). Both solutions were tested on both the active and the support layers to remove NOM and TEP fouling.

The Villacorte *et al.* (2009) protocol was used to prepare the cationic dye Alcian Blue, with which the fouled FO membrane was stained in order to assess TEP accumulation (Villacorte *et al.* 2009). Transmitted and reflected light microscopy was used with different objectives to determine the best option for TEP removal.

3.5.3 RESULTS

LC-OCD chromatograms of the CFW, EWWE and DDS suggest that biopolymers (protein-like substances and polysaccharides), present in the SWWE and in the EWWE, but not in the CFW or the DDS (no partitioning through the membrane is observed) were the main constituents of the fouling layer in the active layer of the FO membrane (Figure 3.5.2).

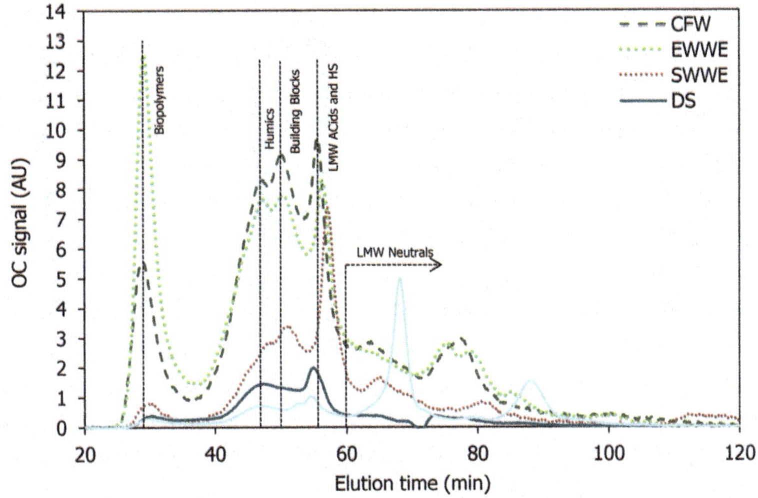

Figure 3.5.2 LC-OCD chromatograms for the CFW, EWWE, SWWE, DS and DDS.

After achieving a flux decline of 25% cleaning was performed. Osmotic backwash did not help to recover any flux. The interaction of the salt in the cleaning solution with the foulants might have enhanced the pore blockage and thus the water flux kept decreasing. A second hypothesis is the reduction of the osmotic pressure difference due to salt accumulation in the fouling layer, which eventually reduce the flux through the membrane. Nevertheless, NOM-fouling showed high hydraulic reversibility (89.5% flux recovery) when air scouring for 15 minutes within the concentrated wastewater effluent was used as a cleaning technique.

TEP forms clusters clearly identifiable with an optical microscope (Figure 3.5.3a and 3.5.3c). After a 10-minute cleaning of the FO membrane with the Alconox + Sodium EDTA solution, TEP was significantly removed (Figure 3.5.3b and 3.5.3d); nevertheless, traces of these particles can still be observed on both layers of the membrane after the cleaning.

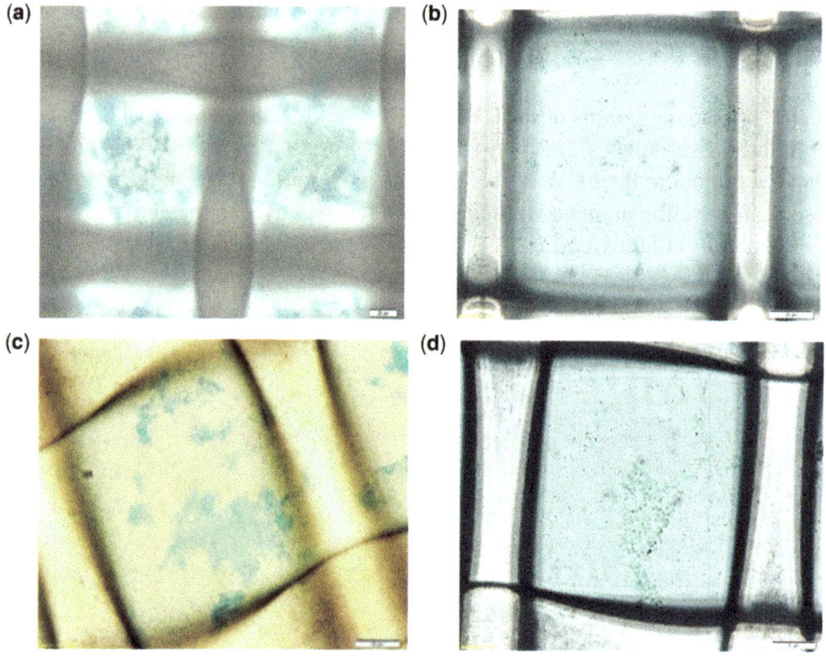

Figure 3.5.3 a) TEP fouling in the AL of an FO membrane (60×); b) AL of a chemically cleaned FO membrane (100×); c) TEP fouling in the SL of an FO membrane (100×); d) SL of a chemically cleaned FO membrane (100×).

Chemical cleaning of the AL with Alconox + Sodium EDTA proved to slightly increase the reversibility (93.6%). Chemical cleaning of the SL increased the chemically reversible fouling up to 94.5%. The irreversible fouling in this set of experiments was 5.5%, which may be attributed to biopolymers that cannot be removed from the membrane surface, along with the trace TEP left on the membrane surface.

Cleaning with a 1% aqueous solution of NaOCl compromised the integrity of the membrane as proven by conductivity and flux measurements, even though the pH of the solution was kept bellow 8, as recommended by the manufacturer.

3.5.4 SUMMARY

Forward osmosis (FO) is an emerging technology which can be applied in water reuse applications. Osmosis is a natural process that involves less energy consumption than reverse osmosis (RO), and therefore can be applied as a dilution process before low pressure RO; it is expected to compete favorably against current advanced water reuse technologies that use microfiltration/ultrafiltration and RO. The focus of this research was to assess the efficiency of different cleaning procedures to remove fouling from the surface of a FO membrane during the operation of a submerged system working in FO-mode (active layer (AL) facing feed solution) intended for secondary wastewater effluent recovery, using seawater as draw solution, which will be diluted and can further be fed to a low pressure RO unit to produce fresh water. Natural organic matter (NOM) fouling was expected to affect the AL, while, for the support layer (SL), transparent exopolymer particles (TEP) were used as indicators of fouling due to their stickiness and propensity to enhance the attachment of other foulants in seawater on the membrane surface. The composition of the NOM fouling layer was determined after proper characterization with a liquid chromatograph coupled with organic carbon detection (LC-OCD), showing biopolymers and protein-like substances as the main constituents. NOM fouling showed high hydraulic reversibility after a 25% flux decline was observed, up to 89.5% when *in-situ* air scouring for 15 minutes was used as a cleaning technique. Chemical cleaning with a mixture of Alconox, an industrial detergent containing phosphates, and sodium EDTA showed to increase the reversibility (93.6%). Osmotic backwash using a 4% NaCl solution and DI water proved to be ineffective to recover flux due to the salt diffusion phenomena occurring at the AL. Part of the flux that could not be recovered is attributable to TEP fouling on the SL, which forms clusters clearly identifiable with an optical microscope and TEP-specific dyeing/staining of the membrane. Two solutions were tested to clean in-place the SL: a 1% NaOCl solution and the same detergent solution used to clean the AL. The former agent compromised the integrity of the FO membrane, as proved with flux and conductivity measurements; 94.5% of flux was recovered with the latter agent, showing that the chemically irreversible fouling for the FO membrane is on the order of 5.5%, which might be associated with the adsorption of biopolymers on the AL and some TEP residuals on the SL. Physical cleaning (air scouring) of the AL proved to be the most effective way to control fouling.

3.5.5 REFERENCES

Bar-Zeev E., Berman-Frank I., Liberman B., Rahav E., Passow U. and Berman T. (2009). Transparent exopolymer particles: potential agents for organic fouling and biofilm formation in desalination and water treatment plants. *Desalination and Water Treatment*, **3**, 7.

Berman T., Mizrahi R. and Dosoretz C. G. (2011). Transparent exopolymer particles (TEP): a critical factor in aquatic biofilm initiation and fouling on filtration membranes. *Desalination*, **276**, 184–190.

Cornelissen E. R., Harmsen D., de Korte K. F., Ruiken C. J., Qin J.-J., Oo H. and Wessels L. P. (2008). Membrane fouling and process performance of forward osmosis membranes on activated sludge. *Journal of Membrane Science*, **319**, 158–168.

Huber S. A., Balz A., Abert M. and Pronk W. (2011). Characterisation of aquatic humic and non-humic matter with size-exclusion chromatography – organic carbon detection – organic nitrogen detection (LC-OCD-OND). *Water Research*, **45**, 879–885.

Mi B. and Elimelech M. (2008). Chemical and physical aspects of organic fouling of forward osmosis membranes. *Journal of Membrane Science*, **320**, 292–302.

Shibutani T., Kitaura T., Ohmukai Y., Maruyama T., Nakatsuka S., Watabe T. and Matsuyama H. (2011). Membrane fouling properties of hollow fiber membranes prepared from cellulose acetate derivatives. *Journal of Membrane Science*, **376**, 102–109.

Valladares Linares R., Yangali-Quintanilla V., Li Z. and Amy G. (2012). NOM and TEP fouling of a forward osmosis (FO) membrane: foulant identification and cleaning. *Journal of Membrane Science*, **421–422**, 217–224.

Villacorte L. O., Kennedy M. D., Amy G. L. and Schippers J. C. (2009). The fate of Transparent Exopolymer Particles (TEP) in integrated membrane systems: removal through pre-treatment processes and deposition on reverse osmosis membranes. *Water Research*, **43**, 5039–5052.

Wang Y., Wicaksana F., Tang C. Y. and Fane A. G. (2010). Direct Microscopic Observation of Forward Osmosis Membrane Fouling. *Environmental Science & Technology*, **44**, 7102–7109.

Yangali-Quintanilla V., Li Z., Valladares R., Li Q. and Amy G. (2011). Indirect desalination of Red Sea water with forward osmosis and low pressure reverse osmosis for water reuse. *Desalination*, **280**, 160–166.

Zhang H.-M., Xiao J.-N., Cheng Y.-J., Liu L.-F., Zhang X.-W. and Yang F.-L. (2006). Comparison between a sequencing batch membrane bioreactor and a conventional membrane bioreactor. *Process Biochemistry*, **41**, 87–95.

Zou S., Gu Y., Xiao D. and Tang C. Y. (2011). The role of physical and chemical parameters on forward osmosis membrane fouling during algae separation. *Journal of Membrane Science*, **366**, 356–362.

Part 4
Rejection of Pollutants

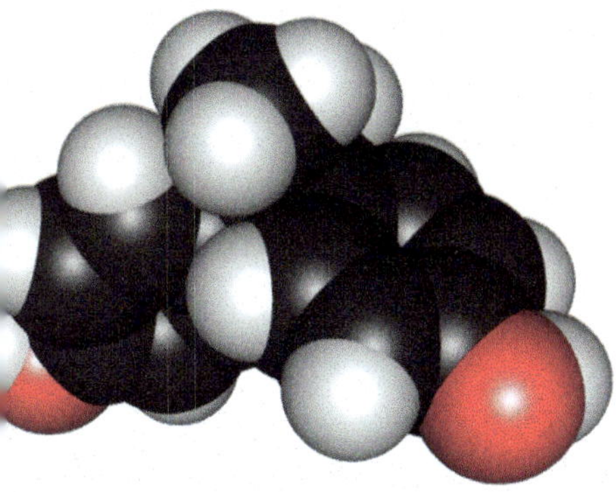

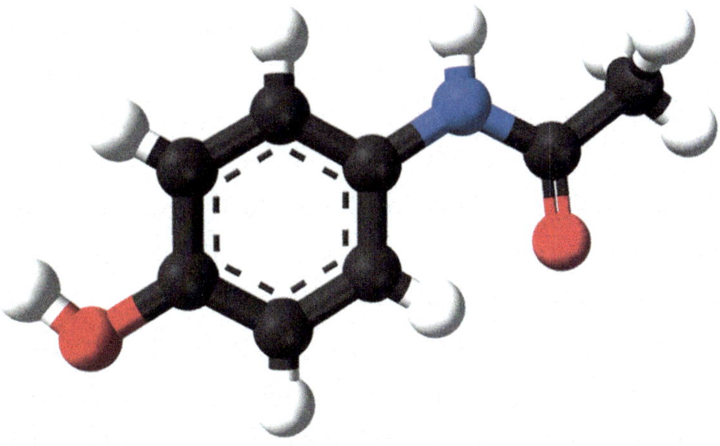

Chapter 4.1

Rejection of micropollutants by FO membranes*

4.1.1 INTRODUCTION

Organic micropollutants, or emerging organic contaminants, are substances that include, but are not limited to, pharmaceutically active compounds (PhACs), endocrine disrupting compounds (EDCs), disinfection by-products and pesticides, that are present in the environment and, due to the increasing concentration detected in recent studies, are arising concern among researchers and regulatory agencies, because most of them are not yet regulated and their impacts on human life are not quite known (Snyder *et al.* 2003).

There is an urging need to understand their partitioning, accumulation and removal from water, soil, air and biota; unfortunately, there is not a clear idea of the exact risks of chronic exposure to a mixture of organic micropollutants, including quantities ingested through drinking water, leaving a big question yet to be answered (Yangali-Quintanilla *et al.* 2010b). There are current technologies that can remove these organic micropollutants from wastewaters, but generally they are energy intensive processes, such as nanofiltration (NF) and reverse osmosis (RO) (Kimura *et al.* 2003; Bellona *et al.* 2004). Conventional wastewater treatment plants will not remove completely these chemicals, which will partition mainly into the sludge produced and the water effluent, generating a threat for the discharge site and the downstream areas (Kümmerer *et al.* 2000; Snyder *et al.* 2003; Ternes *et al.* 2004).

Findings indicate that fouling significantly impacts on membrane performance and varies the removal pattern of neutral trace organic contaminants (Agenson & Urase, 2007). Xu *et al.* found that membrane fouling facilitated the transport

---

* This chapter is based, with permission from the copyright holder, on the previously published paper. (Valladares Linares R., Yangali-Quintanilla V., Li Z. and Amy G. (2011). Rejection of micropollutants by clean and fouled forward osmosis membrane. *Water Research*, **45**, 6737–6744. doi:10.1016/j.watres.2011.10.037)

of hydrophobic and hydrophilic organic contaminants through a cellulose triacetate (CTA) RO membrane (Xu *et al.* 2006). For trace organic contaminants (hormones), colloidal fouling in an RO membrane (CTA) caused a significant increase in the permeate concentration of these substances (Ng & Elimelech, 2004). Three mechanisms have been identified to govern the influence of fouling in the rejection capacity of a NF membrane: modification of the membrane charge surface, pore restriction, and cake enhanced concentration polarization (Nghiem & Hawkes, 2007). Fouling can change the surface characteristics of the membrane, either to improve or degrade the rejection capability and flux of it. There are studies that can prove both situations for NF and RO membranes (Bellona *et al.* 2004; Xu *et al.* 2006; Nghiem & Coleman, 2008), but there is no literature on the precise effects of organic contaminants and their rejection by FO membranes.

Forward osmosis (FO) coupled with RO membranes have been proposed in the last years to be an efficient double barrier system to reject most of emerging contaminants (Cornelissen *et al.* 2008; Cath *et al.* 2011). Cath *et al.* studied the removal of micropollutants by a FO-RO system using a membrane biological reactor (MBR) and the effluent from an activated sludge tank as feed waters for the spiral wound FO membrane, and synthetic seawater as draw solution (DS). They measured the concentration of the micropollutants present in the MBR and the effluent (no micropollutants were spiked in the water), varying from 2 to 400 $ng \cdot L^{-1}$ (in some cases the compounds were below the limit of quantification). The rejections were mostly high with the FO membrane, except for Bisphenol A. When the hybrid process was considered, very high percentages of rejection (>99%) were achieved (Cath *et al.* 2011).

This study focuses on the study of the difference in the rejection of a cocktail of 13 micropollutants that are known to occur in aquatic environments (Heberer, 2002) (5 hydrophilic nonionic, 3 hydrophobic nonionic and 4 hydrophilic ionic micropollutants) by a clean FO membrane (inorganic solution with ionic-strength used as feed) and a fouled FO membrane (secondary wastewater effluent (SWWE) used as feed). Significantly higher-than-normal concentrations of the micropollutants are used in the experiment to mainly account for membrane adsorption after steady-state saturation (Yangali-Quintanilla *et al.* 2009).

4.1.2 MATERIALS AND METHODS

4.1.2.1 FO membrane

Hydration Technology Innovations (HTI, Scottsdale, AZ) FO membranes were used for this study. The original coupons, measuring 4" × 6", were cut into flat-sheet membranes to fit the custom-made plate and frame cell with an area of 202 cm^2 for each side made out of poly (methyl methacrylate), forming a channel where the DS recirculates.

4.1.2.2 RO membrane

The RO membrane used is an aromatic polyamide membrane produced for desalination of brackish water by Dow-Filmtec (Midland, MI, USA) under the name BW30. The RO system consists of a positive displacement pump (Hydra-Cell, MN, USA), a crossflow filtration cell accommodating a 139 cm^2 membrane (SEPA CF II, Sterlitech, Kent, WA, USA), needle valves, pressure gauges, a proportional pressure relief valve and stainless steel tubing (Swagelok BV, Netherlands).

4.1.2.3 Source waters

For the feed water of the FO process, a SWWE was used; the effluent was collected from the Al-Ruwais wastewater treatment plant in Jeddah, Saudi Arabia, after sedimentation of the effluent of the activated sludge aeration tanks (secondary treatment). The SWWE has a pH of 7.2 and a conductivity of 3300 $\mu S \cdot cm^{-1}$. The biological oxygen demand (BOD_5) of the wastewater effluent was 20 $mg \cdot L^{-1}$, and the DOC was 5 $mg \cdot L^{-1}$. The effluent received no pretreatment, except that it was kept at 4 °C to prevent changes in the composition.

Seawater from the Red Sea was taken from the feeding line of the desalination plant at KAUST, Thuwal, Saudi Arabia, and was used as DS for the FO process. The water has a pH of 7.8, the dissolved organic carbon (DOC) was determined to be 1 $mg \cdot L^{-1}$, the total dissolved solids (TDS) are 40,500 $mg \cdot L^{-1}$, and the conductivity was 57,500 $\mu S \cdot cm^{-1}$. The raw seawater was pretreated with a 0.45 μm filter.

An inorganic synthetic solution (SS) was used to simulate the ionic-strength of a wastewater effluent, avoiding the presence of suspended solids and fouling agents. It was calculated based on the results of the analysis done to the wastewater effluent from Jeddah following the Standard Methods (Eaton *et al.* 2005) for common inorganic elements such as ammonium, nitrate, nitrite, total nitrogen and phosphate.

Both feed waters and the DS were kept at 20 ± 0.5 °C during the FO experiments with a chilling/heating device.

For the RO experiment to be compared to FO-LPRO, the RO feed water was prepared diluting prefiltered seawater with DI water, reaching the conductivity of the diluted draw solution (DDS) after the FO process. For the FO-LPRO, the same DDS coming from the FO process with SWWE as feed water is used in the LPRO system.

4.1.2.4 Experimental setup and procedure

The FO membrane cell was connected, via a gear pump (Coleparmer), to the DS container. A balance (TE6101, Sartorius AG, Germany) was used as flow recorder when connected to a computer; based on the difference in weight and the effective membrane area, the flux can be calculated. The feed water tank contained 18 L of SWWE and the DS tank held 0.5 L of pretreated seawater. The recirculation flow was set at 50 mL · min^{-1}. The recording of the weight data was activated immediately after the membrane was submerged in the feed tank; at the same time the osmotic dilution process started, because the tubing system and the cell were prefilled with DS before being in contact with the feed water. A stirring device was used to create turbulence in the feed water and achieve proper mixing. A schematic is presented in Figure 4.1.1.

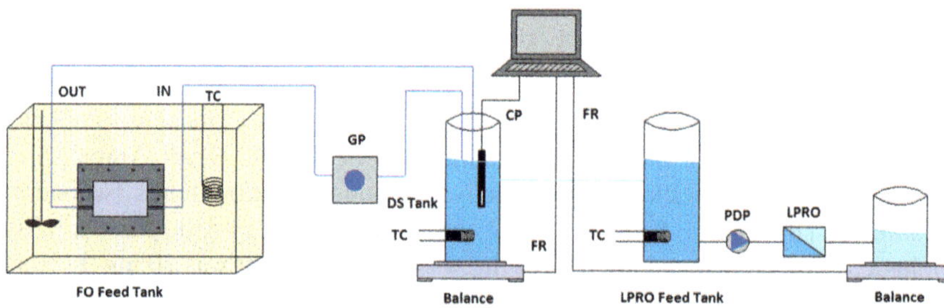

Figure 4.1.1 FO-RO system layout. TC = temperature controller, GP = gear pump, CP = conductivity probe, FR = flow recorder, PDP = positive displacement pump.

The dilution experiment was performed for 24 hours; the DS increased its volume due to continuous osmosis between the feed water and the DS recirculating in the cells. The dilution cycle was repeated for 5 days by replacing the fresh DS at the beginning of each new day.

The same procedure was followed with a new membrane and the SS as feed water to test the capability of the clean membrane to reject the micropollutants (no fouling agents present in the feed water).

The operating parameters for the RO system were set as follows: feed pressure of 15 bar, permeate flow of 1.6 mL · min^{-1}, concentrate flow of 80 mL · min^{-1} and a RO flux of 7 L · m^{2} · h^{-1}. The recovery of the process was stabilized to 2%.

4.1.2.5 Micropollutants stock preparation and analyses

All the micropollutants were purchased from Sigma Aldrich (Munich, Germany). The list of micropollutants used to prepare the stock solutions is presented in Table 4.1.1. Compounds were classified according to their speciation in water as hydrophobic when the logarithm of the distribution coefficient (log D), which refers to the ratio of the sum of the concentrations of all forms of the compound in each of the two immiscible phases forming a mixture, generally water and octanol, is higher than 2.6, and hydrophilic when log D $\leq$ 2.6. Ionic compounds shown in the table are negatively charged at pH 7, which was determined by using ADME/Tox Web Software. Physicochemical properties were calculated using Molecular Modeling Pro.

Table 4.1.1 Properties of the micropollutants spiked to the feed water.

Name	ID	MW ($g \cdot mol^{-1}$)	log D[a] (pH 7)	Mol. Length (nm)[b]	Width (nm)[b,c]	Depth (nm)[b]	Equiv. Width (nm)[b]
1,4-dioxane	DIX	88	−0.17	0.71	0.66	0.52	0.59
Acetaminophen	ACT	151	0.23	1.14	0.68	0.41	0.53
Metronidazole	MTR	171	−0.27	0.93	0.90	0.48	0.66
Phenazone	PHZ	188	0.54	1.17	0.78	0.56	0.66
Caffeine	CFN	194	−0.45	0.98	0.87	0.56	0.70
Bisphenol A	BPA	228	3.86	1.25	0.83	0.75	0.79
Carbamazepine	CBM	236	2.58	1.20	0.92	0.58	0.73
17α-ethynilestradiol	EE2	296	3.98	1.48	0.87	0.84	0.85
Ibuprofen	IBF	206	3.97	1.39	0.73	0.55	0.64
Naproxen	NPX	230	0.34	1.37	0.78	0.75	0.76
Fenoprofen	FNP	242	0.38	1.16	0.93	0.74	0.83
Gemfibrozil	GFB	250	2.30	1.58	0.94	0.65	0.78
Ketoprofen	KTP	254	−0.13	1.16	0.92	0.74	0.83

[a]ADME/Tox Web Software.
[b]Molecular Modeling Pro.
[c]Equivalent width = $(\text{width} \times \text{depth})^{0.5}$.

All of the compounds, except for the 17α-ethynilestradiol, were prepared in a stock solution of $10\ mg \cdot L^{-1}$ with Milli-Q water. This organic micropollutant was diluted in 1% wt. ethanol solution, and then diluted into a concentration of $10\ mg \cdot L^{-1}$ with Milli-Q water.

The organic micropollutants were spiked into the SS, SWWE and RO feed from the stock solution prepared with a concentration of $10\ mg \cdot L^{-1}$. The target concentration in the FO and RO feed of each micropollutant independently was $10\ \mu g \cdot L^{-1}$. Samples of the FO and RO feed water were taken before the experiment started, and control samples were taken before the addition of the micropollutants.

For the comparison of rejections between the SWWE and the SS, the 2nd and 3rd day samples of DDS were mixed to form a composite. Afterwards, rejection over time for the SWWE was tested between 2nd and 3rd day and 4th and 5th day composites. A sample of blank DI water was taken as control.

The micropollutants in water samples were analysed by Technologiezentrum Wasser, (TZW, Karlsruhe, Germany). The information about the procedures for analyses of micropollutants in water samples were referenced in a previous study (Yangali-Quintanilla *et al.* 2010a). The uncertainty of measurement was ±20%, not determined for each compound individually sampled, but determined during method validation, for all of the compounds listed in Table 4.1.1.

4.1.2.6 FO membrane characterization

An Anton Paar Zeta Potential Analyser (Austria) was used to determine the zeta potential (ZP) of the FO membrane. It uses a clamping cell where two pieces of membrane are used to create a channel of 25 mm of length and 5 mm width, with the active layers facing each other, and then the charge of the membrane in mV is measured when an electrolyte flows through it. For this analysis, two electrolytes were used, 10 mM KCl and the SWWE previously used as feed water. The ZP is measured in the pH range in which the membrane can operate (4 to 8), so the proper injection of acid (0.1 M HCl) or base (0.1 M NaOH) is added in the titration process. The results presented in this study were calculated with the Helmholtz-Smoluchowski equation.

The contact angles of clean and fouled FO membranes were measured with a goniometer CAM200 (KSV, Finland) by using the sessile drop method. The fouled membrane samples were dried for 24 hours at room temperature (20 °C) before measurements.

Confocal laser scanning microscopy (CLSM) (LSM710 uplight confocal microscope, Zeiss, Germany) was used to compare the surface of the FO membrane when different feed waters were used in the FO process.

4.1.3 RESULTS AND DISCUSSION

4.1.3.1 Zeta potential and contact angle

The ZP of the new membrane (virgin membrane) and the fouled membrane (after being used in the 5-day experiment with SWWE as feed water) was determined to account for the membrane surface charge. Figure 4.1.2 shows the graph for the ZP in mV for each value of pH, using different electrolytes. The new membrane has a negative charge for the experiments carried out in this study, because the pH of the SWWE is 7.6, compared to the isoelectric point of the new membrane, which is below pH 3 (using the SWWE as electrolyte).

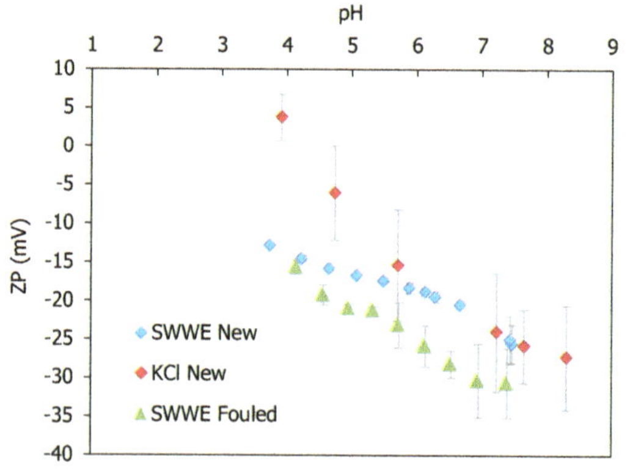

Figure 4.1.2 Zeta Potential of the FO membrane for varying pH values and electrolytes.

The fouling layer after the FO process increased the negative charge of the membrane. When the membrane was in contact with the wastewater effluent, the ZP was negative, which means that the ionic micropollutants were rejected due to electrostatic repulsion. This rejection was higher when the membrane was fouled; the charge became more negative due to the presence of negatively charged radicals in the foulants, mainly coming from the NOM acids (carboxylic radicals) and polysaccharides, as described in the literature (Cho *et al.* 1998; Fan *et al.* 2001; Shim *et al.* 2002).

There was also an increase in the hydrophilicity of the fouled membrane, based on the measurements of the contact angle of the FO membrane in both conditions, being 58.8° ± 0.3 and 49° ± 3 for the clean and fouled membrane, respectively. The foulants increased the adsorption capacity of the membrane for hydrophilic compounds, and thus, the rejection of hydrophobic neutral micropollutants was higher when the membrane was fouled.

4.1.3.2 Rejection of micropollutants by FO

Rejections of the FO and RO membranes were calculated with Equation 4.1.1.

$$\text{Rejection} = \left(1 - \left(\frac{C}{C_0}\right)\right) \times 100 \qquad (4.1.1)$$

For FO rejection when the feed water (SWWE or SS) was spiked with the micropollutants, C_0 is the concentration of the feed water, and C is the concentration of the DDS. For the RO rejection, the C_0 is the concentration of the RO feed water, and the C is the concentration of the RO permeate. For the FO-LPRO process, the initial FO feed water concentration is C_0 and C is the concentration of the RO permeate.

Figure 4.1.3 shows the comparison for the rejection of the 13 micropollutants by a clean FO membrane and a fouled FO membrane for the composite samples of the 2nd and 3rd day of experiments. Figure 4.1.4 illustrates and summarizes the rejection mechanisms of the FO membrane. It can be noticed that for the hydrophilic neutral compounds the rejection with the SS as feed water (clean membrane) is higher than the rejection when SWWE is used as feed (fouled membrane), ranging from 2% difference for Dioxane to 6% difference for Phenazone, except for Caffeine. Two hypothesis might explain these results: (i) the increase in surface charge might potentially result in a higher molecular weight cut-off (MWCO) due to membrane swelling, as described by Xu *et al.* for NF and RO membranes, reducing the rejection of the hydrophilic neutral compounds, and (ii) higher hydrophilicity generated by the fouling layer on the membrane may allow a higher amount of micropollutants to partition through the membrane, and eventually, decrease the rejection (shown in Figure 4.1.4a, b) (Xu *et al.* 2006). This hypothesis is more significant than the first one, because it is based independently from the type of membrane used and focuses on the fouling layer instead, which is certainly occurring. In the case of Caffeine, the rejection for the fouled membrane increases 15% compared to the clean membrane; this can be explained by the high hydrophilicity of the compound (log D – 0.45), being absorbed by the cake layer formed in the membrane (mainly hydrophilic) and thus, preventing the compound from partitioning through the membrane into the DS, forming a double-barrier-like system. Along with this hypothesis, the MWCO of the membrane (≈200 Da) is really close to the MW of this compound, and the foulants can be blocking the pores of the membrane, increasing the rejection of Caffeine with the fouled membrane. For the hydrophobic neutrals, the increase in rejections goes from 8.7% for Bisphenol A, to a 9.8% for Carbamazepine. In the case of the ionic contaminants, the increase goes from 0.4% for Naproxen, to 6% for Ibuprofen when the membrane is fouled.

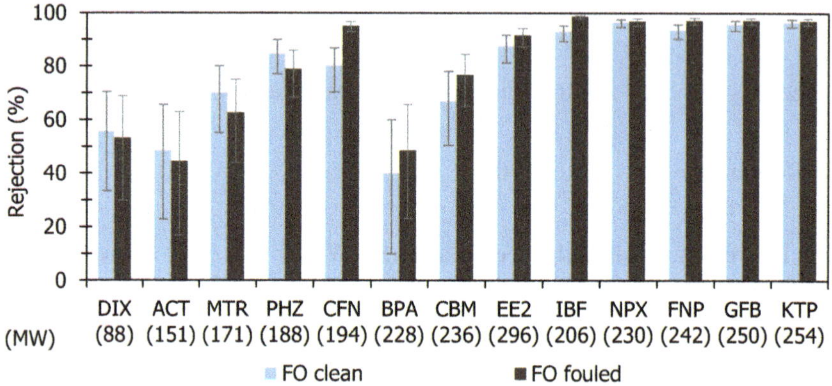

Figure 4.1.3 Rejection of micropollutants by a clean and fouled FO membrane.

The increment in the rejection of hydrophobic neutral compounds is due to the higher hydrophilicity of the membrane when the fouling cake layer is present; this phenomenon is also associated with an increased adsorption capacity (for hydrophilic compounds) and reduced mass transport capacity (diffusion and partitioning) of hydrophobic compounds through the fouled membrane (Figure 4.1.4c, d). For the ionic compounds, the negative charge of the membrane is greater when fouled (described in Section 4.1.3.1), increasing eventually the electrostatic repulsion (Donnan exclusion) between the negative charge of the membrane surface (Figure 4.1.4e, f) as mentioned by Verliefde *et al.* (2008) and Xu *et al.* (2006) and the negative charge of the compound at pH 7.3. Nevertheless, the rejection is still low for hydrophobic neutral compounds, being Bisphenol A again the lowest with 48.7% rejection with the fouling layer.

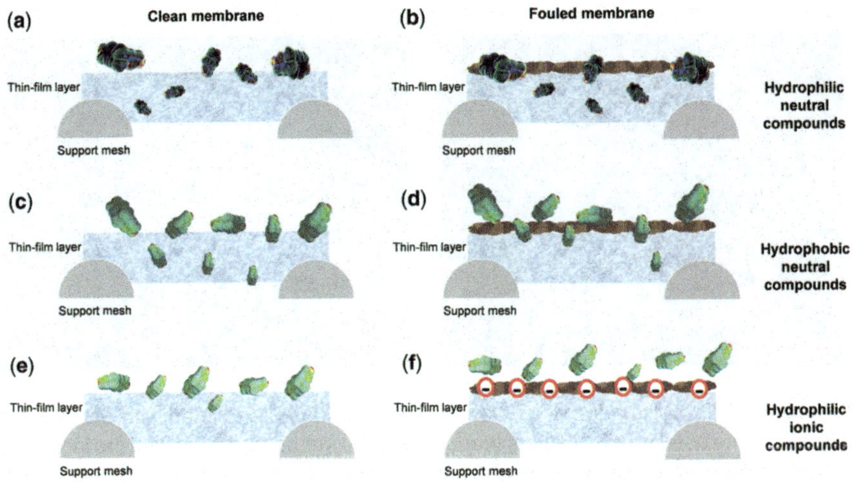

Figure 4.1.4 Rejection mechanisms of different types of micropollutants by a clean and fouled FO membrane. (a and b) For the hydrophilic neutral substances, the fouled membrane is able to reject less efficiently the contaminants due to membrane swelling. (c and d) For the hydrophobic compounds, rejection by the fouled membrane is increased due to reduced mass transport capacity of hydrophobic compounds. (e and f) For the hydrophilic ionic compounds, the fouling layer increases the negative charge of the membrane, increasing the rejection of these compounds.

Figure 4.1.5 shows images of the active layer of a FO membrane in different feed waters. Figure 4.1.5a corresponds to a CLSM image of the FO membrane used with SS as feed water, where the support mesh can be clearly distinguished and no fouling is present on the surface of the membrane. Nevertheless, when SWWE is used as feed for the FO process, the foulants start to accumulate on the surface of the membrane, forming a cake layer clearly identifiable in the CLSM image on Figure 4.1.5b, covering almost completely the support mesh. Consequently, rejection hypothesis based on the fouling layer formation are encouraged in this study.

Figure 4.1.6 compares the rejection percentages for the RO process and the coupled FO-RO process. For the RO process alone, the rejection of most compounds is higher than 99%, except for the two smallest micropollutants, Dioxane (67.9%) and Acetaminophen (89.4%), which have a MW lower or equal than the MWCO of the RO membrane, which can be estimated as 150 Da. When the RO processes is coupled to the FO, the results for rejection can go as high as 89.1% for Dioxane, 96.3% for Acetaminophen, 98.9% for Metronidazole and for the rest of the compounds the rejection was higher than 99%.

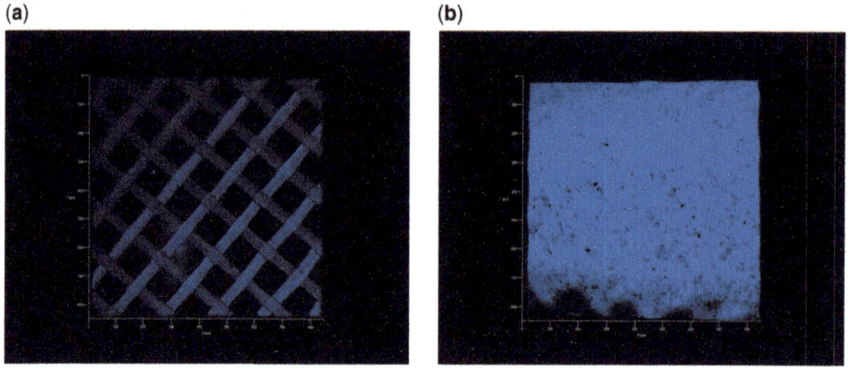

Figure 4.1.5 CLSM images of a FO membrane. (a) Membrane used with SS as feed; (b) membrane used with SWWE as feed. For the FO membrane used with the inorganic solution (SS), no fouling can be seen on the surface of the membrane; however, when the membrane is used with SWWE as feed, foulants accumulate on the surface.

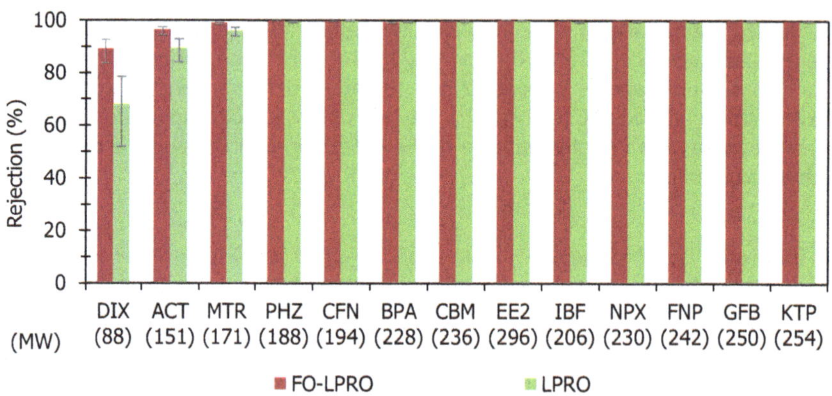

Figure 4.1.6 Rejection of micropollutants by an RO membrane and a coupled FO-LPRO system.

These results can be compared with the results presented by Cath *et al.* (2011) described in Section 4.1.3.1. In the case of the experiments mentioned in the present study, the concentrations of the spiked MPs were in the range of 1 to 10 $\mu g \cdot L^{-1}$, at least 2 orders of magnitude higher than those used by Cath *et al.* and there was no crossflow velocity on the feed side, which means a higher concentration polarization effect, leading to the accumulation of micropollutants in the membrane surface; thereby, the rejections achieved from the single FO process were lower. Nevertheless, when the coupled process was considered, the results of this work and that of Cath *et al.* were similar, achieving very high percentages of rejection (>99%).

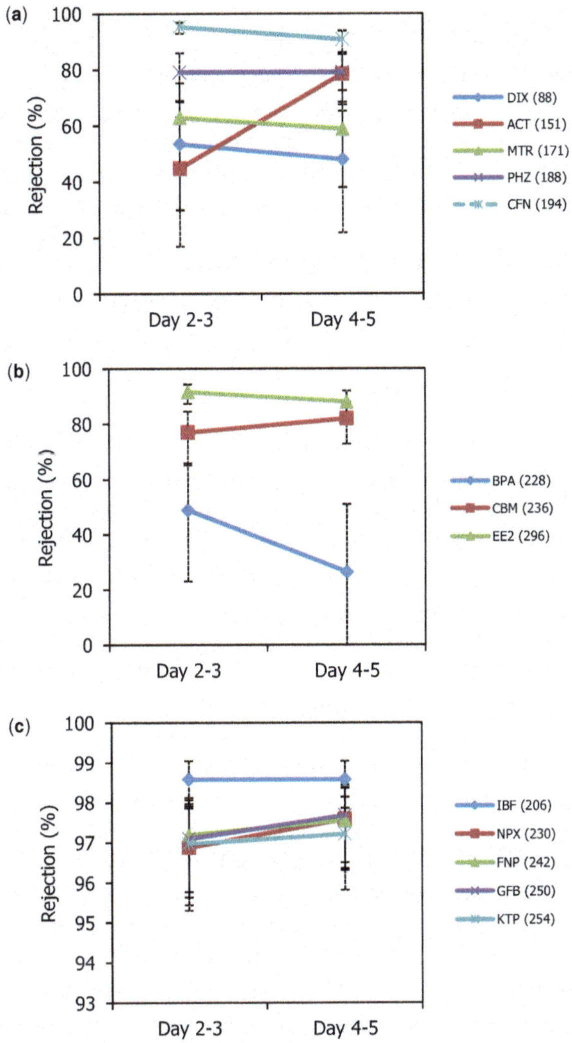

Figure 4.1.7 Rejection of micropollutants on the 2nd and 3rd day and rejection on the 4th and 5th day of the FO process. (a) Hydrophilic neutrals, (b) hydrophobic neutrals, and (c) hydrophilic ionic.

Figure 4.1.7 shows the rejection of micropollutants for a composite sample of the 2nd and 3rd day DDS of the FO process with SWWE as feed, and the rejection percentage for a composite sample of the 4th and 5th day DDS with the same feed water. Figure 4.1.7a shows the hydrophilic neutral pollutants; note that for all of them, excluding Acetaminophen, the rejection decreases over time due to the formation of the fouling layer, which increases the hydrophilicity of the surface and promotes partitioning of these contaminants into the DS side. For Figure 4.1.7b, rejection of hydrophobic neutrals decreases over time, even for Carbamazepine, which can have the same decreasing tendency once the experimental error is considered. The cake layer formed in the surface of the membrane increases the effect of external concentration polarization (ECP), and even though the rejection of the 4th and 5th day composite is still higher than the rejection of the clean membrane (SS as feed water), ECP plays an important role in the reduction of rejection over time for the hydrophobic neutral compounds. Figure 4.1.7c shows an increase in the rejection for hydrophilic ionic micropollutants over time, which was expected due to the formation of the fouling layer in the membrane and the change in the surface charge over time.

4.1.4 SUMMARY

As forward osmosis (FO) gains attention as an efficient technology to improve wastewater reclamation processes, it is fundamental to determine the influence of fouling in the rejection of emerging contaminants (micropollutants). This study focuses on the rejection of 13 selected micropollutants, spiked in a secondary wastewater effluent, by a FO membrane, using Red Sea water as draw solution (DS), differentiating the effects on the rejection caused by a clean and fouled membrane. The resulting effluent was then desalinated at low pressure with a reverse osmosis (RO) membrane, to produce a high quality permeate and determine the rejection with a coupled forward osmosis – low pressure reverse osmosis (FO-LPRO) system. When considering only FO with a clean membrane, the rejection of the hydrophilic neutral compounds was between 48.6% and 84.7%, for the hydrophobic neutrals the rejection ranged from 40.0% to 87.5%, and for the ionic compounds the rejections were between 92.9% and 96.5%. With a fouled membrane, the rejections were between 44.6% to 95.2%, 48.7% to 91.5% and 96.9% to 98.6%, respectively. These results suggest that, except for the hydrophilic neutral compounds, the rejection of the micropollutants is increased by the presence of a fouling layer, possibly due to the higher hydrophilicity of the FO fouled membrane compared to the clean one, the increased adsorption capacity of hydrophilic compounds and reduced mass transport capacity, membrane swelling, and the higher negative charge of the membrane surface, related to the foulants composition, mainly NOM acids (carboxylic radicals) and polysaccharides or polysaccharide-like substances. However, when coupled with RO, the rejections in both cases increased above 96%. The coupled FO-LPRO system was an effective double barrier against the selected micropollutants.

4.1.5 REFERENCES

Agenson K. O. and Urase T. (2007). Change in membrane performance due to organic fouling in nanofiltration (NF)/reverse osmosis (RO) applications. *Separation and Purification Technology*, **55**, 147–156.

Bellona C., Drewes J. E., Xu P. and Amy G. (2004). Factors affecting the rejection of organic solutes during NF/RO treatment – a literature review. *Water Research*, **38**, 2795–2809.

Cath T., Hancock N., Xu P. and Heil D. (2011). A Comprehensive Study of Micropollutants Rejection by Forward Osmosis and Hybrid FO-RO. In: 2012 Membrane Technology Conference American Water Works Association, Glendale, Arizona.

Cho J., Amy G., Pellegrino J. and Yoon Y. (1998). Characterization of clean and natural organic matter (NOM) fouled NF and UF membranes, and foulants characterization. *Desalination*, **118**, 101–108.

Cornelissen E. R., Harmsen D., de Korte K. F., Ruiken C. J., Qin J.-J., Oo H. and Wessels L. P. (2008). Membrane fouling and process performance of forward osmosis membranes on activated sludge. *Journal of Membrane Science*, **319**, 158–168.

Eaton A. D., Clesceri L. S., Rice E. W. and Greenberg A. E. (2005). Standard Methods for the Examination of Water and Wastewater, 21st edn. Copublished by American Public Health Association, Water Environment Federation, and American Water Works Association (AWWA), New York.

Fan L., Harris J. L., Roddick F. A. and Booker N. A. (2001). Influence of the characteristics of natural organic matter on the fouling of microfiltration membranes. *Water Research*, **35**, 4455–4463.

Heberer T. (2002). Occurrence, fate, and removal of pharmaceutical residues in the aquatic environment: a review of recent research data. *Toxicology Letters*, **131**, 5–17.

Kimura K., Amy G., Drewes J. E., Heberer T., Kim T.-U. and Watanabe Y. (2003). Rejection of organic micropollutants (disinfection by-products, endocrine disrupting compounds, and pharmaceutically active compounds) by NF/RO membranes. *Journal of Membrane Science*, **227**, 113–121.

Kümmerer K., Al-Ahmad A. and Mersch-Sundermann V. (2000). Biodegradability of some antibiotics, elimination of the genotoxicity and affection of wastewater bacteria in a simple test. *Chemosphere*, **40**, 701–710.

Ng H. Y. and Elimelech M. (2004). Influence of colloidal fouling on rejection of trace organic contaminants by reverse osmosis. *Journal of Membrane Science*, **244**, 215–226.

Nghiem L. D. and Coleman P. J. (2008). NF/RO filtration of the hydrophobic ionogenic compound triclosan: transport mechanisms and the influence of membrane fouling. *Separation and Purification Technology*, **62**, 709–716.

Nghiem L. D. and Hawkes S. (2007). Effects of membrane fouling on the nanofiltration of pharmaceutically active compounds (PhACs): mechanisms and role of membrane pore size. *Separation and Purification Technology*, **57**, 176–184.

Shim Y., Lee H.-J., Lee S., Moon S.-H. and Cho J. (2002). Effects of natural organic matter and ionic species on membrane surface charge. *Environmental Science & Technology*, **36**, 3864–3871.

Snyder S. A., Westerhoff P., Yoon Y. and Sedlak D. L. (2003). Pharmaceuticals, personal care products, and endocrine disruptors in water: implications for the water industry. *Environmental Engineering Science*, **20**, 21.

Ternes T. A., Joss A. and Siegrist H. (2004). Scrutinizing pharmaceuticals and personal care products in wastewater treatment. *Environmental Science & Technology*, **38**, 392a–399a.

Verliefde A. R. D., Cornelissen E. R., Heijman S. G. J., Verberk J. Q. J. C., Amy G. L., Van der Bruggen B. and van Dijk J. C. (2008). The role of electrostatic interactions on the rejection of organic solutes in aqueous solutions with nanofiltration. *Journal of Membrane Science*, **322**, 52–66.

Xu P., Drewes J. E., Kim T.-U., Bellona C. and Amy G. (2006). Effect of membrane fouling on transport of organic contaminants in NF/RO membrane applications. *Journal of Membrane Science*, **279**, 165–175.

Yangali-Quintanilla V., Sadmani A., McConville M., Kennedy M. and Amy G. (2009). Rejection of pharmaceutically active compounds and endocrine disrupting compounds by clean and fouled nanofiltration membranes. *Water Research*, **43**, 2349–2362.

Yangali-Quintanilla V., Maeng S. K., Fujioka T., Kennedy M. and Amy G. (2010a). Proposing nanofiltration as acceptable barrier for organic contaminants in water reuse. *Journal of Membrane Science*, **362**, 334–345.

Yangali-Quintanilla V., Sadmani A., McConville M., Kennedy M. and Amy G. (2010b). A QSAR model for predicting rejection of emerging contaminants (pharmaceuticals, endocrine disruptors) by nanofiltration membranes. *Water Research*, **44**, 373–384.

Chapter 4.2

Rejection of boron*

4.2.1 INTRODUCTION

One of the biggest challenges in reverse osmosis (RO) desalination is boron removal, which is usually obtained with special high-boron-rejection RO membranes (with the drawback of lower water permeability) or a pretreatment step to increase pH; both involve additional cost to achieve drinking water quality standards (Dominguez-Tagle *et al.* 2011). Another solution for boron removal is using a double-pass RO system, which is e.g. more susceptible to severe scaling problems (Hasson *et al.* 2011). The World Health Organization (WHO) lowered its boron guideline value for drinking water down to 2.4 ppm (World Health Organization, 2011); however, most governments and environmental agencies have not applied this new value to regulations.

Boron exists as non-ionic boric acid form at the natural pH level of seawater which is about 7–8.5, due to its high pKa (9.2). Rejection mechanisms are mostly based on size exclusion at these pH values. At a pH higher than 9.2, concentrations of borate and other ionic forms of boron become dominant compared to non-ionic forms, which are easier to remove by RO membranes due to the electrostatic repulsion mechanism. The reason of the low rejection of boric acid (non-ionic) in membrane systems is because of its ability to diffuse through the membranes in a non-ionic form in the way similar to carboxylic acids or water (Nadav, 1999; Koseoglu *et al.* 2008).

Forward osmosis (FO) has the potential to reduce the energy consumption of producing fresh water from seawater when compared to conventional high-pressure membrane processes such as RO. FO can find a niche application as a lower energy process to achieve high quality water (Shaffer *et al.* 2012; Valladares Linares *et al.* 2013), particularly if it used in a hybrid *direct* desalination process. A FO *direct*

---

*This chapter is based, with permission from the copyright holder, on the previously published paper. (Valladares Linares R., Li Z. Y., Sarp S., Park Y. G., Amy G. and Vrouwenvelder J. S. (2015). Higher boron rejection with a new TFC forward osmosis membrane. *Desalination and Water Treatment*, doi:10.1080/19443994.2014.940220) desalination process uses saline water as the feed solution (FS) and an osmotic reagent as the draw solution (DS). The DS gets diluted with the water permeating through the membrane. This diluted solution can be converted in high quality water by an additional treatment step involving low pressure RO, enabling a robust lower cost desalination system compared to traditional high pressure seawater RO systems.

The rejection of boric acid by cellulose triacetate (CTA) FO membranes has been studied in the past with debatable results. (Hancock & Cath, 2009) demonstrated that the rejection of boric acid was between 12% to 33%, depending on the flux conditions. Other studies with the same membrane (Jin *et al.* 2011; Jin *et al.* 2012; Kim *et al.* 2012) showed that boron rejection can be as high as 70%, depending on the draw solution used in the process, the pH of the feed water and the flux of water through the membrane. This highlights the importance of reverse salt diffusion in FO as one of the key mechanisms governing the boron flux. Independently, the use of non-charged draw solutes has been proposed to reduce the reverse salt diffusion through the membrane (Yong *et al.* 2012), but has not yet been shown to reduce the boron solute flux in a FO *direct* desalination process.

In this study, a new generation of thin-film composite (TFC) FO membranes was tested and compared to a commercially available cellulose triacetate (CTA) FO membrane. The study focused on determining the boron rejection of the membranes, where a deionized (DI) water and 5 mM H_3BO_3 solution was used as feed solution (FS). Two solutions (0.32 M NaCl and 1 M glucose) with the same approximate osmotic pressure were used as draw solution (DS) independently to assess the interaction of charged solutes with the membrane in terms of reverse draw solute flux, and their role in the rejection of boron.

4.2.2 MATERIALS AND METHODS

4.2.2.1 FO membranes

Two FO membranes were used for this study: (i) a cellulose triacetate (CTA) membrane with an embedded polyester mesh and (ii) a thin-film composite (TFC) membrane with an identical embedded polyester mesh on the support layer. Both membranes were purchased from Hydration Technologies Innovation (HTI, Scottsdale, AZ, USA).

Tests were made with both membrane orientations: active layer (AL) facing the feed solution (AL-FS) and active layer facing the draw solution (AL-DS). Tests on pure water flux and draw solution rejection were completed in the AL-FS orientation, which has been proven to be the most effective in preventing fouling of the FO membrane (Cornelissen *et al.* 2008; Mi & Elimelech, 2008; Wang *et al.* 2010; Zou *et al.* 2011).

The zeta potential (ZP) of both FO membranes was calculated with an Anton Paar Zeta Potential Analyser (Austria). For this analysis a solution of 10 mM KCl was used as electrolyte. The ZP is measured in the pH range in which the membrane can be used safely (4 to 8 for the CTA membrane and 2 to 10 for the TFC), so the proper injection of acid (0.1 M HCl) or base (0.1 M KOH) is added in the titration process to assess the zeta potential.

4.2.2.2 Experimental setup and procedure

The system consisted of a cross-flow FO cell accommodating a 2 cm × 10 cm membrane coupon. Experiments were performed at a flow rate of 200 mL · min^{-1}, equivalent to a cross-flow velocity of 0.084 m · s^{-1}. 1 L of feed and draw solution were used for each condition. This configuration has been used in previous FO studies (Li *et al.* 2012).

Two feed solutions were used: deionized (DI) water and DI water spiked with 0.002 M H_3BO_3. For the draw solution, two solutes were used at a concentration that could generate the same osmotic power, determined with osmolality measurements: a 0.32 M NaCl solution (corresponding to 1.8% NaCl solution), and a 1 M glucose solution. Each experiment recovered 100 mL from the FS to the DS.

Water temperature was kept constant with a heating/chilling device at 20 ± 1 °C. The pH of the feed water was maintained at 7.2. No dissociation of B is expected due to the stability of boric acid at this pH (Jin *et al.* 2012).

Three fluxes were measured/calculated and reported in the results and discussion section: (i) water flux through the FO membrane (from FS to DS), (ii) boron flux through the FO membrane (from FS to DS), and (iii) reverse solute flux (from DS to FS) which refers to the draw solution molecules being transported through the membrane in opposite direction of the water flux.

An additional experiment was run with a pre-fouled membrane to study the effect of fouling caused by natural organic matter (NOM). A SEPA CF II filtration cell (Sterlitech, Kent, WA, USA) accommodating a 139 cm² membrane sheet was used in dead-end mode for 24 h at a pressure of 100 psi to foul the membrane with a solution containing 20 mg · L^{-1} of Xanthan gum to simulate a fouling layer of biopolymer.

4.2.3 RESULTS AND DISCUSSION

An overview of the studies is given in Table 4.2.1.

Table 4.2.1 Scheme of experimental studies on CTA and TFC forward osmosis membranes.

Study	Figure	Table	Results Section
membrane characterization			
– SEM observation	1		3.1
– zeta potential measurement	2		3.1
membrane performance			
– water flux	3		3.2.1
– draw solute rejection		2	3.2.2
boron flux			
– impact membrane orientation	4, 5		3.3
– impact of draw solution	4, 5		3.3
– impact fouled membrane	4, 5		3.3

4.2.3.1 Membrane characterization

Figure 4.2.1 shows scanning electron microscopy (SEM) images of both CTA and TFC membranes. The CTA membrane has a smooth uniform layer on both sides, and no pores can be identified on its surface. On the other hand, the support layer of the TFC membrane (Figure 4.2.1f) shows a highly porous structure, one of the reasons why the membrane presents a higher water flux due to reduced concentration polarization effect (as specified by the membrane supplier, and found as well in this study). For the TFC membrane, the polyamide thin-film layer (Figure 4.2.1e) serves as the separation layer (active layer). The polyester mesh filaments can be seen in the cross section for both CTA and TFC membranes (Figure 4.2.1a and 4.2.1d).

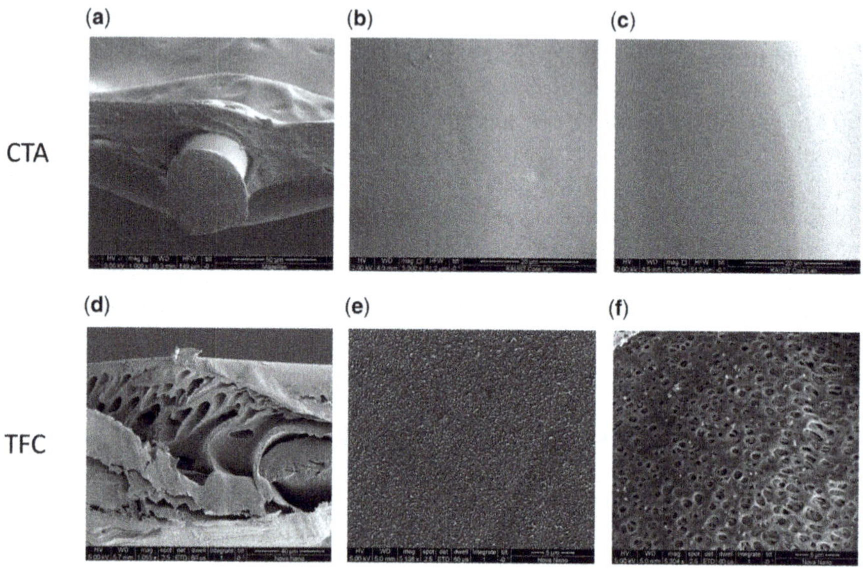

Figure 4.2.1 SEM images showing the different structure of a cellulose triacetate (CTA) and a thin-film composite (TFC) forward osmosis membrane: (a and d) cross section (magnification 1000×) showing the polyester embedded mesh (fibers), (b and e) active layer (magnification 5000×), and (c and f) support layer (magnification, 5000×).

The results for zeta potential presented in Figure 4.2.2 were calculated using the Helmholtz-Smoluchowski equation (Valladares Linares *et al.* 2011). The CTA membrane has a slight negative charge after pH 4.5, for both AL and SL. The TFC membrane has a negative charge for the AL, and a significantly stronger negative charge for the SL.

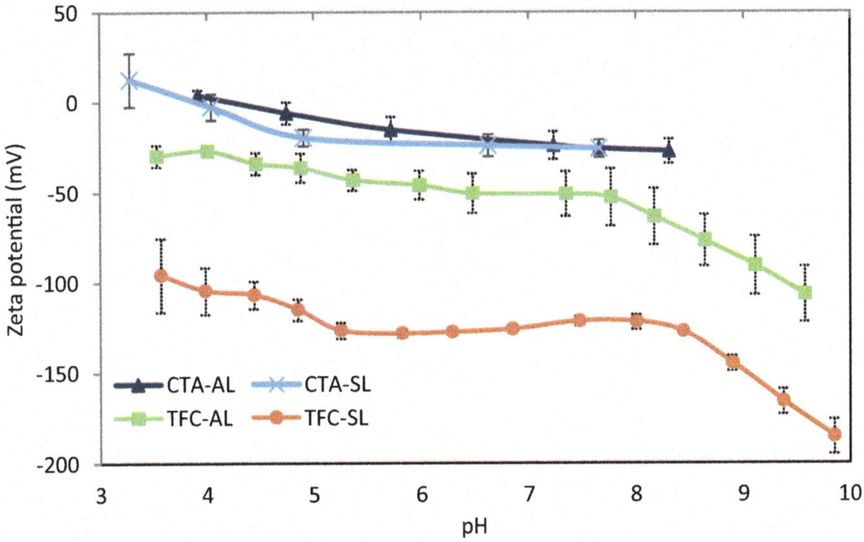

Figure 4.2.2 Zeta potential measured with a 10 mM KCl electrolyte for the CTA (same composition for active layer (AL) and support layer (SL)) and TFC (polyamide AL and polysulfone SL) forward osmosis membranes. The TFC FO membrane has a more negatively charged surface on both sides, compared to the CTA membrane.

In summary, the new TFC membrane has higher water permeability than the CTA membrane. The stronger negative charge of the TFC membrane on both SL and AL is expected to have an effect on the reverse draw solute flux compared to the less negatively charged CTA membrane. Such effect is discussed further in Section 4.2.3.3 in relation with boron flux results.

4.2.3.2 Membrane performance

4.2.3.2.1 Water flux

Figure 4.2.3 shows the water flux (permeate flux) patterns for the two membranes after the extraction of a fixed water volume (100 mL) through the FO membrane. Figure 4.2.3a represents the water flux obtained with DI water as feed solution, whereas Figure 4.2.3b shows the results obtained with DI water spiked with 0.002 M H_3BO_3 as feed solution. There is no significant difference in the water flux when boron is added to the feed solution. Nevertheless, there is a clear difference between using NaCl and glucose as DS, considering that both solutions have similar osmotic pressure. The lower water flux obtained with glucose may be explained by a higher viscosity of the DS, which is a phenomenon known to occur with concentrated solutions of saccharides such as glucose or sucrose (Garcia-Castello *et al.* 2009). When comparing the performance of the two membranes, the new TFC membrane has about 20% higher water flux than the CTA membrane, regardless of the draw solution used under the experimental conditions in this study.

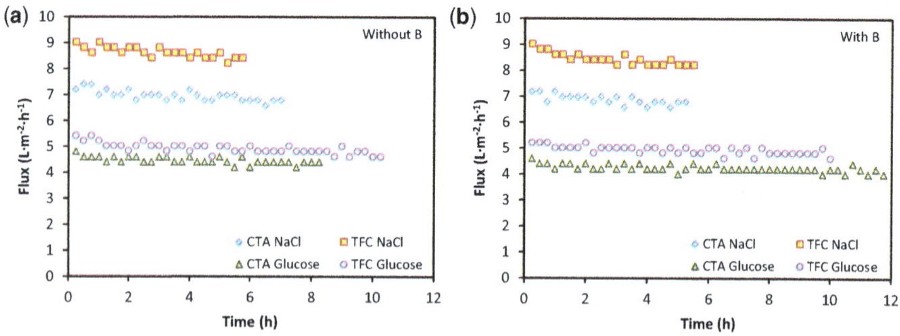

Figure 4.2.3 Water flux patterns for different FO membranes. (a) Flux patterns for deionized water as feed solution for both CTA and TFC membrane using different draw solutions (NaCl and glucose); (b) flux patterns for DI water spiked with boron as feed solution for both CTA and TFC membranes using different draw solutions. All experiments were performed in AL-FS orientation.

4.2.3.2.2 Draw solute rejection

The rejection percentage of draw solution is summarized in Table 4.2.1. The rejection for both, NaCl and glucose molecules, is equally high for both CTA and TFC membrane types. When results in Figure 4.2.3 are compared to rejection results (Table 4.2.2), the TFC forward osmosis membrane shows to have a better performance (higher water flux at the same driving force) without compromising the low solute flux characteristic of membranes designed for osmotic processes.

Table 4.2.2 Total rejection of solutes from draw solution to the feed by FO membranes.

	Rejection* (%)			
Membrane Type	**NaCl**	**Glucose**	**NaCl + B****	**Glucose + B****
CTA	99.2	99.9	99.2	99.9
TFC	99.1	99.8	99.1	99.9

*Experiments conducted in AL-FS orientation.
**Boron was added as 5 mM boric acid (H_3BO_3) to the feed solution.

Based on the results for water flux and draw solute rejection, both TFC and CTA FO membranes are suitable for desalination in terms of solute rejection for charged (NaCl) and neutral (glucose) osmotic agents, even when boron is present in the feed water, which had no impact on the water flux or reverse draw solute flux.

In conclusion, the new TFC FO membrane shows a better performance than the first generation CTA FO membrane for water permeability.

4.2.3.3 Boron flux

The boron flux in mg · m^{-2} · h^{-1} for the different membranes and draw solutions used can be seen in Figure 4.2.4a and 4.2.4c. The interaction with a charged and non-charged molecule (NaCl and glucose, respectively) as draw solution does not directly impact the boron flux. This might be related to the neutral charge of boric acid at pH 7. On the other hand, the membrane orientation does play a role in boron flux, which increases when the active layer faces the draw solution.

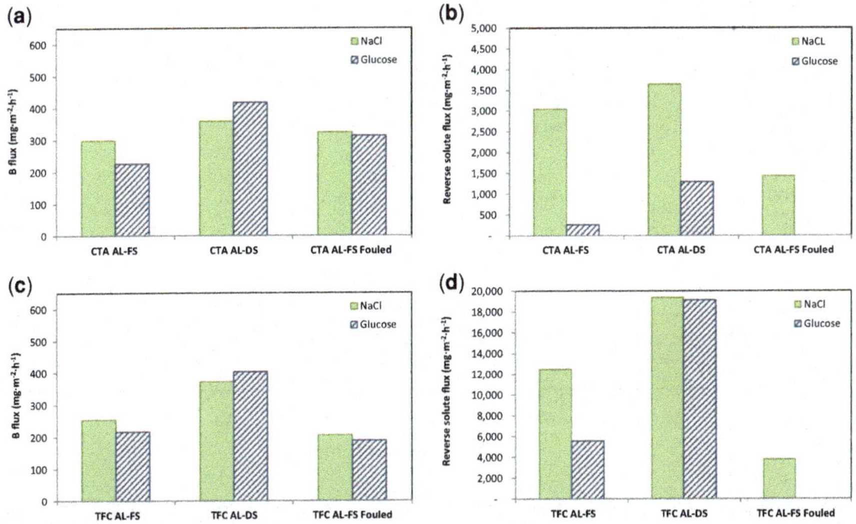

Figure 4.2.4 Boron flux (a and c) and reverse solute flux (b and d) in mg · m^{-2} · h^{-1} for CTA and TFC forward osmosis membranes using NaCl and glucose as draw solution for three conditions: active layer facing feed solution (AL-FS), active layer facing draw solution (AL-DS) and a fouled membrane in AL-FS orientation.

The effect of NOM membrane fouling on boron removal was studied with pre-fouled CTA and TFC membranes. The NOM fouling layer, composed of a biopolymer, does not seem to play an important role in promoting or decreasing boron transport through the membrane, as suggested by Jin *et al.* in previous studies (Jin *et al.* 2012).

Figure 4.2.4b and 4.2.4d show the reverse solute flux for both membranes under different membrane orientations. When glucose is used as draw solution, the reverse draw solute flux is significantly lower for both membranes in both orientations due to the larger molecule size compared to NaCl, except for the TFC membrane operated in AL-DS mode, which might be explained by the structure of the active layer with its lower negative charge compared to the support layer, and the highly porous support which allows a higher number of solute molecules inside its structure.

A comparison between the membrane types, permeate water flux and boron flux is shown in Figure 4.2.5. A linear positive correlation can be observed between the boron flux and the water flux (both boron and water fluxes have the same direction: from FS to DS). As the water flux increases, the boron flux increases, for both TFC and CTA membranes.

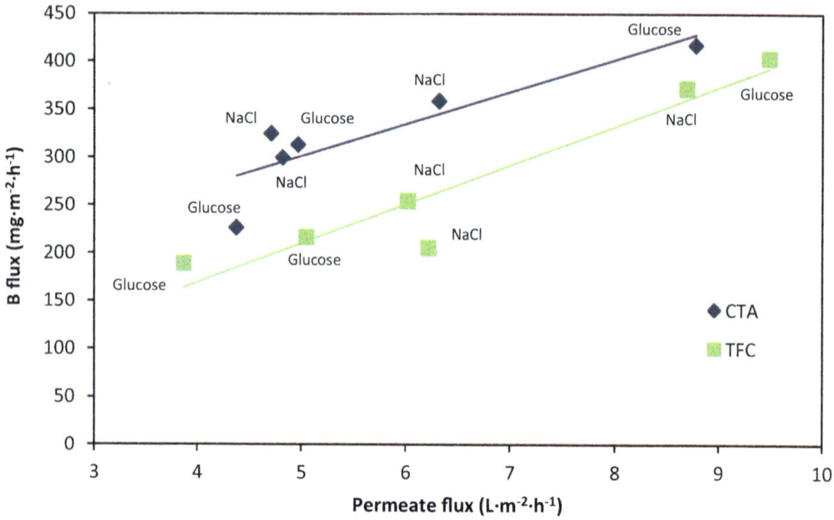

Figure 4.2.5 Linear relation between boron flux and water permeate flux for CTA and TFC forward osmosis membrane. Independent from the draw solute used (NaCl or glucose), boron flux is directly proportional to the permeate flux. The TFC forward osmosis membrane enabled a lower boron flux than the CTA forward osmosis membrane.

In essence, this study shows that the new generation TFC forward osmosis membrane exhibits compared to the first generation CTA membrane: (i) a 40% lower boron flux, and (ii) 20% higher water flux for both types of draw solutes tested (NaCl and glucose). This demonstrates the potential for boron removal for new generation TFC FO membranes, even at high water fluxes compared to CTA FO membranes.

4.2.4 SUMMARY

Due to the stringent limits for boron in drinking and irrigation water, water treatment facilities have to incur additional treatment to remove boron down to a safe concentration. Forward osmosis (FO) is a membrane technology that may reduce the energy required to remove boron present in seawater. In *direct* FO desalination hybrid systems, fresh water is recovered from seawater using a recoverable draw solution, FO membranes are expected to show high boron rejection. This study focuses on determining the boron rejection capabilities of a new generation thin-film composite (TFC) FO membrane compared to a first generation cellulose triacetate (CTA) FO membrane. The effects of water permeate flux, membrane structure, draw solute charge and reverse solute flux on boron rejection were determined. For TFC and CTA FO membranes, experiments showed that when similar operating conditions are applied (e.g. membrane type, draw solute type) boron rejection decreases with increasing permeate flux. Reverse draw solute flux and membrane fouling have no significant impact on boron rejection. Compared to the first generation CTA FO membrane operated at the same conditions, the TFC FO membrane showed a 40% higher boron rejection capability and a 20% higher water flux. This demonstrates the potential for boron removal for new generation TFC FO membranes.

4.2.5 REFERENCES

Cornelissen E. R., Harmsen D., de Korte K. F., Ruiken C. J., Qin J.-J., Oo H. and Wessels L. P. (2008). Membrane fouling and process performance of forward osmosis membranes on activated sludge. *Journal of Membrane Science,* **319**, 158–168.

Dominguez-Tagle C., Romero-Ternero V. J. and Delgado-Torres A. M. (2011). Boron removal efficiency in small seawater Reverse Osmosis systems. *Desalination,* **265**, 43–48.

Garcia-Castello E. M., McCutcheon J. R. and Elimelech M. (2009). Performance evaluation of sucrose concentration using forward osmosis. *Journal of Membrane Science,* **338**, 61–66.

Hancock N. T. and Cath T. Y. (2009). Solute coupled diffusion in osmotically driven membrane processes. *Environmental Science & Technology,* **43**, 6769–6775.

Hasson D., Shemer H., Brook I., Zaslavschi I., Semiat R., Bartels C. and Wilf M. (2011). Scaling propensity of seawater in RO boron removal processes. *Journal of Membrane Science,* **384**, 198–204.

Jin X., Tang C. Y., Gu Y., She Q. and Qi S. (2011). Boric acid permeation in forward osmosis membrane processes: modeling, experiments, and implications. *Environmental Science & Technology,* **45**, 2323–2330.

Jin X., She Q., Ang X. and Tang C. Y. (2012). Removal of boron and arsenic by forward osmosis membrane: influence of membrane orientation and organic fouling. *Journal of Membrane Science,* **389**, 182–187.

Kim C., Lee S., Shon H. K., Elimelech M. and Hong S. (2012). Boron transport in forward osmosis: measurements, mechanisms, and comparison with reverse osmosis. *Journal of Membrane Science,* **419–420**, 42–48.

Koseoglu H., Kabay N., Sarp M., Yüksel S., Arar Ö. and Kitis M. (2008). Boron removal from seawater using high rejection SWRO membranes – impact of pH, feed concentration, pressure, and cross-flow velocity. *Desalination,* **227**, 253–263.

Li Z.-Y., Yangali-Quintanilla V., Valladares-Linares R., Li Q., Zhan T. and Amy G. (2012). Flux patterns and membrane fouling propensity during desalination of seawater by forward osmosis. *Water Research,* **46**, 195–204.

Mi B. and Elimelech M. (2008). Chemical and physical aspects of organic fouling of forward osmosis membranes. *Journal of Membrane Science,* **320**, 292–302.

Nadav N. (1999). Boron removal from seawater reverse osmosis permeate utilizing selective ion exchange resin. *Desalination,* **124**, 131–135.

Shaffer D. L., Yip N. Y., Gilron J. and Elimelech M. (2012). Seawater desalination for agriculture by integrated forward and reverse osmosis: improved product water quality for potentially less energy. *Journal of Membrane Science,* **415–416**, 1–8.

Valladares Linares R., Yangali-Quintanilla V., Li Z. and Amy G. (2011). Rejection of micropollutants by clean and fouled forward osmosis membrane. *Water Research,* **45**, 6737–6744.

Valladares Linares R., Li Z., Abu-Ghdaib M., Wei C.-H., Amy G. and Vrouwenvelder J. S. (2013). Water harvesting from municipal wastewater via osmotic gradient: an evaluation of process performance. *Journal of Membrane Science,* **447**, 50–56.

Wang Y., Wicaksana F., Tang C. Y. and Fane A. G. (2010). Direct microscopic observation of forward osmosis membrane fouling. *Environmental Science & Technology,* **44**, 7102–7109.

World Health Organization (2011). Guidelines for Drinking-Water Quality, 4th edn, WHO Press, Geneva, Switzerland.

Yong J. S., Phillip W. A. and Elimelech M. (2012). Coupled reverse draw solute permeation and water flux in forward osmosis with neutral draw solutes. *Journal of Membrane Science,* **392–393**, 9–17.

Zou S., Gu Y., Xiao D. and Tang C. Y. (2011). The role of physical and chemical parameters on forward osmosis membrane fouling during algae separation. *Journal of Membrane Science,* **366**, 356–362.

Part 5
Draw Solution and Membranes

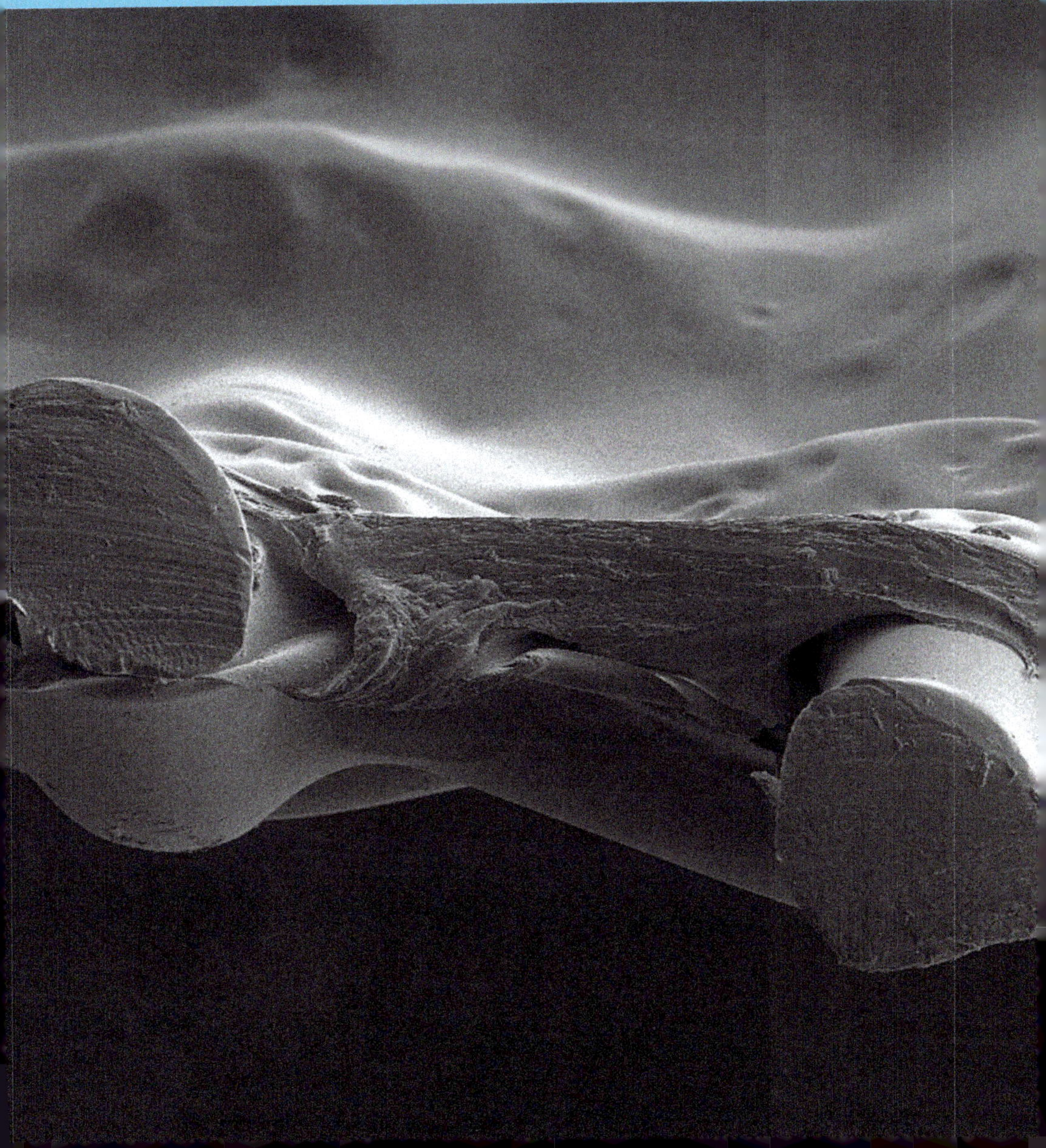

Chapter 5.1

Draw solution*

5.1.1 INTRODUCTION

The rapid growth of population, economics, and industrialization has led to serious concerns on global sustainability in terms of water, energy, environment, and climate (Shao *et al.* 2009; Escobar, 2010; Elimelech & Phillip, 2011; Lee *et al.* 2011; Chung *et al.* 2012a, 2012b; Zhao *et al.* 2012b). Clean water scarcity, energy crisis, environmental pollution, and climate change have plagued many communities. Adequate water supply and affordable energy sources become essential in today's world to sustain public health and national prosperity. Presently, over one-third of the world's population face clean water shortage and 2.6 billion people lack adequate sanitation (Montgomery & Elimelech, 2007; Shannon *et al.* 2008). As estimated, the world population will increase to 9 billion by 2050 from current 7 billion (Elimelech & Phillip, 2011). The problem of clean water shortage will be aggravated with this exponential population growth. Therefore, one of the most crucial challenges of the 21st century is to meet the increasing demand for clean water.

However, water production and energy development are co-dependent. The production of drinkable water is an energy-intensive process, while power generation often requires a large amount of water (King & Webber, 2008; McGinnis & Elimelech, 2008; Sivertsen *et al.* 2012). Since current technologies to produce clean water are energy-intensive processes, any fluctuation in oil price and energy cost will directly affect water production. Concerns on water security and the search for low-energy separation technologies to produce clean water have gained much attention in recent years (Greenlee *et al.* 2009; Choi *et al.* 2010; Elimelech &

*This chapter is based, with permission from the copyright holder, on the previously published paper. (Ge Q., Ling M. and Chung T. (2013). Draw solutions for forward osmosis processes: Developments, challenges, and prospects for the future. *Journal of Membrane Science*, **442**, 225–237. doi:10.1016/j.memsci.2013.03.046)

Phillip, 2011; Lee *et al.* 2011; Subramani *et al.* 2011; Chon *et al.* 2012; Chung *et al.* 2012a, 2012b; Zhao *et al.* 2012b). Methods to alleviate the water shortage stresses have been developed rapidly, in which membrane process is considered as a very promising technology. Among a variety of membrane processes, reverse osmosis (RO) is currently the most widely used desalination technology for clean water production (Greenlee *et al.* 2009; Lee *et al.* 2011). However, RO is an energy intensive process because a high hydraulic pressure is required to overcome the osmotic pressure generated by the source water. Other problems in RO, such as significant concentration polarization and membrane fouling, lower the feed water recovery rate. Moreover, the discharge of concentrated brine after RO is a serious environmental issue. In summary, the highly fluctuated oil price, uncertainty of energy cost, issues of water security and brine disposal have led every thirsty country to explore alternative technologies for clean water production.

Forward osmosis (FO) has been considered as an emerging membrane technology for water reuse and desalination (Cath *et al.* 2006; Chung *et al.* 2012a; Chung et al. 2012b; Zhao *et al.* 2012b). Unlike pressure-driven membrane processes, FO is a naturally occurring osmosis-driven process involving a semipermeable membrane and a draw solution. Ideally, the semipermeable membrane performs as a barrier that rejects salts or unwanted elements but allows water to pass through, while the draw solution has a higher osmotic pressure that extracts water from the feed across the membrane. The prominent advantages of FO over others include high water recovery rate, brine discharge minimization, low fouling propensity, and low energy requirements (Cath *et al.* 2006; Mi & Elimelech, 2010; Zhang *et al.* 2010; Cai *et al.* 2011; Zhang *et al.* 2011; Chung *et al.* 2012a; Liu & Mi, 2012; Zhao *et al.* 2012b). These advantages have resulted in rapidly growing interests in FO from various disciplines. Not only can it potentially alleviate the stresses of water supply (Cath *et al.* 2006; Shannon *et al.* 2008; Chung *et al.* 2012a; Crow, 2012; Su *et al.* 2012b; Zhao *et al.* 2012b) and promote power generation (Lee *et al.* 1981; Loeb, 2001; Gerstandt *et al.* 2008; McGinnis & Elimelech, 2008; Skilhagen *et al.* 2008; Achilli *et al.* 2009a; Thorsen & Holt, 2009; Zhang *et al.* 2013), but also facilitate the areas of life science, such as protein or pharmaceutical enrichment (Yang *et al.* 2009b; Ling & Chung, 2011b; Wang *et al.* 2011), controlled drug release and food processing (Santus & Baker, 1995; Thombre *et al.* 1999; Petrotos & Lazarides, 2001; Wang *et al.* 2005; Garcia-Castello *et al.* 2009).

Since the FO process requires the both presence of a semipermeable membrane and an osmotic pressure difference across the membrane, remarkable progresses have been made in not only FO membranes but also draw solutions. An ideal FO membrane is capable of providing high water permeability, high rejection of solutes, substantially reducing internal concentration polarization (ICP), and has high chemical stability and mechanical strength. Likewise, a good draw solution must have the following characteristics: high water flux, minimal reverse draw solute flux, no toxicity, reasonably low cost, and easy recovery. Meanwhile, it is required to be compatible with the FO membrane. A variety of draw solutions have been developed in the past few decades (McCutcheon *et al.* 2005; Cath *et al.* 2006; Chung *et al.* 2012a, 2012b; Zhao *et al.* 2012b); however, reviews on draw solution are quite rare to date (Cath *et al.* 2006; Chekli *et al.* 2012). With a number of novel draw solutes proposed in recent years (Ge *et al.* 2010; Ling *et al.* 2010; Yen *et al.* 2010; Li *et al.* 2011a; Ling *et al.* 2011; Phuntsho *et al.* 2011; Carmignani *et al.* 2012; Ge *et al.* 2012a; Iyer 2012; Noh *et al.* 2012; Stone *et al.* 2013), providing an updated and comprehensive review on draw solution is highly warranted. This work will not only conduct a comprehensive and critical review on the existing draw solutions and their applications in FO integrated processes, but also address the strategies and prospects to design better draw solutes.

5.1.2 FUNDAMENTALS OF FO PROCESSES

According to the 2nd law of thermodynamics, chemical potential tends to equilibrate in an isolated system. FO processes follow the 2nd law of thermodynamics where solvent molecules move naturally from a low concentration solution across a semipermeable membrane to a high concentration solution in order to equilibrate their overall chemical potential. The separation of feed solvent and solutes takes place in the low concentration solution due to the difference in osmotic pressure ($\Delta\pi$) across the membrane and the unique characteristics of the semipermeable membrane which only allows solvent molecules to pass though. Therefore, $\Delta\pi$ is the driving force in the transport of solvent molecules across the membrane, while the semipermeable characteristics of FO membrane control salt rejection or water/solute selectivity. In accordance with the literature (McCutcheon *et al.* 2005; Cath *et al.* 2006; Chung *et al.* 2012a, 2012b; Crow, 2012; Zhao *et al.* 2012b), the terms "draw solution" and "feed solution" represent the high concentration solution and low concentration solution, respectively. When an asymmetric membrane is used, there are usually two modes of FO processes according to the membrane orientation; namely, pressure retarded osmosis (PRO) mode and forward osmosis (FO) mode. In the PRO mode, the draw solution flows against the selective layer while the feed solution flows against the porous support layer. The reverse holds true for the FO mode. Unless otherwise stated, the general term of FO processes in this work includes both PRO and FO modes.

The water flux, J_w ($L\cdot m^{-2}\cdot h^{-1}$ or LMH), in an FO process can be described by the equation below (Lee *et al.* 1981; Loeb, 2001; Cath *et al.* 2006; Zhao *et al.* 2012b):

$$J_w = A\sigma\Delta\pi \tag{5.1.1}$$

where A is the water permeability coefficient, σ is the reflection coefficient. Generally, the effective $\Delta\pi$ across the selective layer of the membrane is lower than the bulk $\Delta\pi$, thus the real water flux across the membrane is lower than the theoretical value (Gray *et al.* 2006; McCutcheon & Elimelech, 2006; Phillip *et al.* 2010; Su *et al.* 2010a; Su & Chung, 2011; Su *et al.* 2012b; Zhao *et al.* 2012b).

As illustrated in Figure 5.1.1 (Phillip *et al.* 2010; Su & Chung, 2011), the decline in water flux is associated with complicated phenomena during water transport across the membrane such as concentration polarization (Cath *et al.* 2006; Gray *et al.* 2006; McCutcheon & Elimelech, 2006; Su *et al.* 2010a). Concentration polarization is caused by the dilutive effect of the draw solution and the concentrative effect of the feed solution. As a result, the draw solute concentration at the membrane surface is lower than its bulk concentration, whereas the feed concentration at the membrane surface is higher than its bulk concentration with water permeating to the draw solution from the feed side.

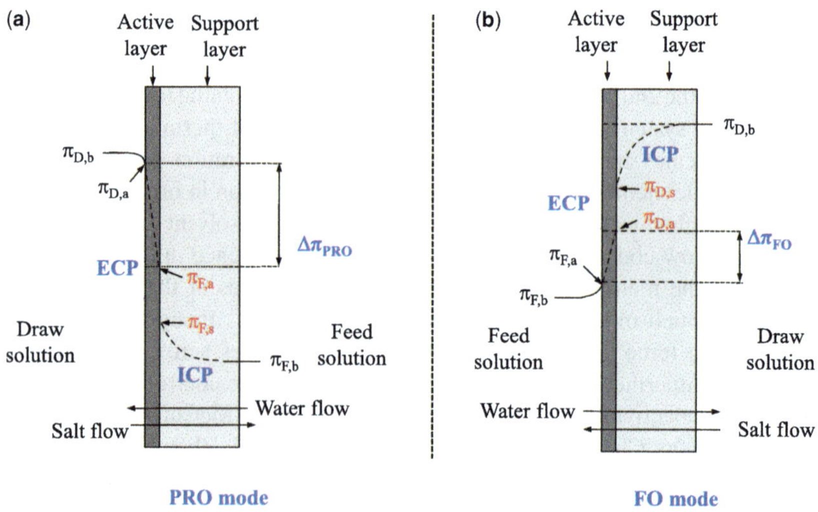

Figure 5.1.1 Illustration of driving force and concentration polarization for (a) the PRO mode and (b) the FO mode in FO processes. Figure adapted from Phillip *et al.* (2010), Su and Chung (2011).

Because of concentration polarization, the effective osmotic pressure difference across the membrane declines, as shown below,

$$\pi_{D,a} - \pi_{F,a} < \pi_{D,b} - \pi_{F,b} \tag{5.1.2}$$

where $\pi_{D,a}$, $\pi_{F,a}$, $\pi_{D,b}$ and $\pi_{F,b}$ represent the osmotic pressures of the draw and feed solutions at the membrane active layer surfaces and bulky solutions, respectively, as indicated in Figure 5.1.1. As a result, Equation 5.1.1 in real cases must be modified as:

$$J_w = A\sigma\Delta\pi = A\sigma(\pi_{D,a} - \pi_{F,a}) \tag{5.1.3}$$

According to the location where concentration polarization takes place, there are two categories of concentration polarization: external concentration polarization (ECP) and ICP (Figure 5.1.1). ECP occurs at the outer surface of the membrane selective layer. Both concentrative ECP and dilutive ECP may exist in an FO process depending on the membrane orientation. Concentrative ECP occurs in the FO mode, while dilutive ECP occurs in the PRO mode (Gray *et al.* 2006; McCutcheon & Elimelech, 2006; Phillip *et al.* 2010; Su *et al.* 2010a; Su & Chung, 2011). It is notable that the effect of ECP on permeate flux can be alleviated by increasing flow turbulence (Cath *et al.* 2006). Unlike ECP, ICP takes place within the porous support layer. It is the most troublesome phenomenon in FO processes because it cannot be easily eliminated. Therefore, the decline in water flux in FO is primarily caused by ICP (Mehta & Loeb, 1979; Gray *et al.* 2006; McCutcheon & Elimelech, 2006; Phillip *et al.* 2010; Su *et al.* 2010a; Su & Chung, 2011; Fang *et al.* 2012) and the percentage of flux reduction can be as high as 80% (Mehta & Loeb, 1979; Su *et al.* 2012b).

Similar to ECP, ICP also has two types: dilutive ICP and concentrative ICP. Dilutive ICP occurs under the FO mode due to the dilution of draw solutes within the membrane support, while concentrative ICP happens under the PRO mode owing to the accumulation of feed solutes inside the porous support. In a given membrane orientation, both ICP and ECP occur simultaneously. However, for most FO processes, only dilutive ICP needs to be considered if the operation is in the FO mode. In this case, the influence of concentrative ECP may be negligible ($\pi_{F,a} \approx \pi_{F,b}$) especially when pure DI water is used as the feed, feed flow rate is high, and reverse salt flux is insignificant. Similarly, only dilutive ECP should be considered in the PRO mode and the influence of concentrative ICP may be negligible ($\pi_{F,a} \approx \pi_{F,b}$) if DI water is used as the feed and reverse salt flux is insignificant (Su *et al.* 2010a).

Many efforts have been made to model the effects of ECP and ICP on water flux in the FO process (Loeb *et al.* 1997; McCutcheon & Elimelech, 2006; McCutcheon & Elimelech, 2007; Tan & Ng, 2008; Su & Chung, 2011). McCutcheon and Elimelech have developed models considering the effects of both ICP and ECP on water flux. The water flux in the FO mode can be described (McCutcheon & Elimelech, 2006; McCutcheon & Elimelech, 2007) as:

$$J_w = A\left[\pi_{D,a}\exp(-J_wK) - \pi_{F,a}\exp\left(\frac{J_w}{k}\right)\right] \tag{5.1.4}$$

where $\pi_{D,m}$ and $\pi_{F,m}$ are the osmotic pressures of the draw and feed solutions, respectively, k is the mass transfer coefficient, and K is called the solute resistivity. The water flux in the PRO mode can be described as Equation 5 by considering the effects of both concentrative ICP and dilutive ECP on water flux (McCutcheon & Elimelech, 2006; McCutcheon & Elimelech, 2007):

$$J_w = A\left[\pi_{D,a}\exp\left(-\frac{J_w}{k}\right) - \pi_{F,a}\exp(J_wK)\right] \tag{5.1.5}$$

The solute resistivity within the porous support layer (K) is defined by

$$K = \frac{t\tau}{\varepsilon D} = \frac{S}{D} \tag{5.1.6}$$

where t, τ, and ε are the thickness, tortuosity, and porosity of the support layer, respectively, and D is the diffusion coefficient of a draw solution. The term $t\tau/\varepsilon$ is often referred to as the membrane structural parameter (S) (Zhang *et al.* 2013). As aforementioned, ICP is the primary factor in reducing water flux. Investigations have shown that ICP is influenced by the structure (thickness, tortuosity, and porosity) of the membrane support layer, and the diffusion coefficient of the draw solution. An FO membrane with a high S value and/or a draw solution with a low diffusion coefficient may cause significant ICP in the FO process.

5.1.3 IDEAL DRAW SOLUTION FOR THE FO PROCESS

The availability of a suitable draw solution is crucial for advancing FO technology. An appropriate draw solution not only promotes the efficiency of the FO process, but also saves costs of the subsequent steps in recovering and replenishing the draw solute. In addition to have minimal toxicity and low cost, an ideal draw solution needs to fulfill the following requirements. Firstly, it is able to generate a high osmotic pressure. As the osmotic pressure difference between the draw solution and feed solution across the membrane is the driving force for the FO process, the osmotic pressure of a draw solution must be higher than that of the feed solution to ensure a positive permeate flux. According to the Morse equation derived from the Van't Hoff equation by considering dilute ionic solutions (Wikipedia), the osmotic pressure of a solution, π, can be expressed as follows:

$$\pi = iMRT = i\left(\frac{n}{V}\right)RT \tag{5.1.7}$$

where i is the Van't Hoff factor, M is the molarity of the solute which is equal to the ratio of the number of solute moles (n) to the volume of the solution (V), R is the gas constant of $8.3145\ \mathrm{J \cdot K^{-1} \cdot mol^{-1}}$, T is the absolute temperature. Hence, to achieve a high osmotic pressure, a good solubility of the draw solute in water is required to get a high n or M value. In addition, an ionic compound which is able to fully dissociate to produce more ionic species is preferred because it may result in a high i value. This indicates that multivalent ionic solutes are the most favorable. Therefore, compounds with a high water solubility and a high degree of dissociation are potential candidates as draw solutes.

Secondly, the reverse flux of the draw solute must be minimal. As most FO membranes are not ideally semipermeable, the draw solute is also likely to diffuse from the draw solution into the feed solution (Cath *et al.* 2006; Gray *et al.* 2006; McCutcheon & Elimelech, 2006; Hancock & Cath, 2009; Phillip *et al.* 2010;

Su *et al.* 2010a, 2010b; Su & Chung, 2011; Chung *et al.* 2012a, 2012b; Zhao *et al.* 2012b). This reverse flux occurs due to the large concentration difference between the draw solution and the feed solution. The reverse flux of a draw solute, J_s (g m^{-2} h^{-1} or gMH), can be determined from the increase of the draw solute in the feed concentration:

$$J_s = \frac{(C_t V_t) - (C_0 V_0)}{\Delta t} \frac{1}{A_m} \tag{5.1.8}$$

where C_0 ($mol \cdot L^{-1}$) and V_0 (L) are the initial salt concentration and feed volume, respectively, while C_t ($mol \cdot L^{-1}$) and V_t (L) are the salt concentration and feed volume over a predetermined time Δt (h), respectively, during the test. A_m (m^2) is the effective membrane area. For draw solutes exhibiting electrical conductivities in their aqueous solutions, their concentrations in the feed can be calculated by the conversion of conductivity when pure DI water as feed. Otherwise, different methods, such as Total Organic Carbon (TOC), are used to determine their concentrations.

The occurrence of reverse draw solute flux adversely affects the FO and subsequent processes. It not only reduces the driving force and contaminates the feed solution, but also increases the replenishment cost of the draw solute. For example, if FO is applied to enrich proteins or pharmaceuticals, the reverse flux may degrade these products and even denature them because they may be sensitive to salt (Yang *et al.* 2009b; Ling & Chung, 2011b; Morão *et al.* 2011; Wang *et al.* 2011). For the production of potable water, it is important that the trace amount of draw solutes in the final product water must be below the maximum allowable level (USEPA, 2009). In an osmotic membrane bioreactor (OMBR) system, the reverse flux makes the draw solute accumulate in the bioreactor which may have toxic effects on the microbial community (Achilli *et al.* 2009b; Alturki *et al.* 2012; Yap *et al.* 2012). Therefore, a minimum reverse flux of the draw solute is vital for an FO process.

Thirdly, easy regeneration of the diluted draw solution after an FO process is required. Usually, FO is coupled with another process to produce clean water. For example, FO based seawater desalination involves at least two steps: (1) extraction of water from the seawater by a draw solution, and (2) separation of the product water from the diluted draw solution by means of RO (Cornelissen *et al.* 2008; Bamaga *et al.* 2009; Achilli *et al.* 2010; Yangali-Quintanilla *et al.* 2011; Yong *et al.* 2012), nanofiltration (NF) (Tan & Ng, 2010; Zhao & Zou, 2011a; Zhao *et al.* 2012a), ultrafiltration (UF) (Ling & Chung, 2011a; Ge *et al.* 2012a), membrane distillation (MD) or other thermal methods (McCutcheon *et al.* 2005; McGinnis *et al.* 2007; Li *et al.* 2011b; Wang *et al.* 2011; Ge *et al.* 2012b). These processes consume energy as either hydraulic pressure or heating is required to reconcentrate the diluted draw solution. Hence, easy regeneration of draw solutions is highly desirable to lower energy consumption and overall operation costs.

Finally, the draw solute is preferred to have a small molecular weight and low viscosity in its aqueous solution. As aforementioned, concentration polarization is inevitable in the FO process. Concentration polarization, especially ICP, is partially caused by the low diffusion coefficient of the draw solute. As the diffusion coefficient of a draw solute is inversely proportional to its molecular weight and solution viscosity, a draw solute with a large molecular weight and high viscosity would have a low diffusion coefficient; hence, significant ICP. Therefore, a draw solute with a small molecular weight and low viscosity is favorable. Other requirements include solid state at ambient temperature for easy handling and compatibility with FO membranes.

5.1.4 LITERATURE REVIEW ABOUT DRAW SOLUTIONS

A great deal of effort has been made on exploring suitable draw solutions for FO processes over the past few decades. Many compounds have been proposed in the literature. An overview of various solutes, their recovery methods and possible drawbacks is summarized in Table 5.1.1. The physicochemical properties and water fluxes of some typical draw solutes used in FO processes are included in Table 5.1.2. Generally, the draw solutes shown in these two tables can be divided into two categories, i.e. commercially available compounds and synthetic compounds. The evolution of these two categories of compounds is given below.

5.1.4.1 Commercially available compounds as draw solutes

Using commercially available compounds as draw solutes prevailed before 2007 (Cath *et al.* 2006; Lee *et al.* 2011). These compounds can be classified into four types according to their physicochemical properties: volatile compounds, nutrient compounds, inorganic salts, and organic salts and polymers. This classification is not strict, they may overlap each other.

5.1.4.1.1 Volatile compounds

As early as in 1964, Neff first proposed the mixture of ammonia and carbon dioxide gases as a draw solute for seawater desalination (Neff, 1964). In his patent, he elaborated the details of system design, the conditions to prepare ammonium bicarbonate (NH_4HCO_3) solution from ammonia and carbon dioxide, the separation of water and the regeneration of NH_4HCO_3 as well as potential applications such as irrigation, sugar and drug concentration. Although it is feasible to use NH_4HCO_3 as draw solute, it produces a limited osmotic pressure due to its comparatively low solubility in water. To overcome this problem, McCutcheon *et al.* (2006) and McGinnis *et al.* (2007) investigated ammonium salts by mixing ammonium hydroxide (NH_4OH) and NH_4HCO_3 at proper proportions or varying the ratio of NH_3 to CO_2 to improve the solubility. Higher osmotic pressures and better performance

Table 5.1.1 Overview of the development and recovery approaches of draw solutes used in FO technology.

Year	Researcher(s)	Draw Solute(s)	Method of Recovery	Drawbacks	Reference
1964	Neff	ammonia and carbon dioxide	Heating	Energy intensive	(Neff, 1964)
1965	Batchelder	Volatile solutes (e.g. SO_2)	Heating or air stripping	Energy intensive, toxic	(Batchelder, 1965)
1965	Glew	Mixture of H_2O and another gas (SO_2) or liquid (aliphatic alcohols)	Distillation	Energy intensive	(Glew, 1965)
1970	Hough	Organic acid and inorganic salts	Temperature variation or chemical reaction	Complicated procedures, many corrosive chemical involvement	(Hough, 1970)
1972	Frank	Al_2SO_4	Precipitation by doping $Ca(OH)_2$	Toxic by-products	(Frank, 1972)
1975	Kravath & Davis	Glucose	None	Not pure water	(Kravath & Davis, 1975)
1976	Kessler & Moody	Glucose–Fructose	None	Not pure water	(Kessler & Moody, 1976)
1989	Stache	Fructose	None	Not pure water	(Stache, 1989)
1992	Yaeli	Glucose	Low pressure RO	Energy intensive	(Yaeli, 1992)
1997	Loeb *et al.*	$MgCl_2$	None	Not pure water	(Loeb *et al.* 1997)
2002	McGinnis	KNO_3 & SO_2	SO_2 is recycled through standard means	Energy intensive, toxic	(McGinnis, 2002)

(*Continued*)

Table 5.1.1 Overview of the development and recovery approaches of draw solutes used in FO technology (*Continued*).

Year	Researcher(s)	Draw Solute(s)	Method of Recovery	Drawbacks	Reference
2005 – 2007	McCutcheon *et al.*	NH_3 & CO_2 (NH_4HCO_3) or NH_4OH-NH_4HCO_3	Moderate heating (~60 °C)	High reverse draw solute flux Insufficient removal of ammonia	(McCutcheon *et al.* 2005, 2006; McGinnis *et al.* 2007)
2007	Adham *et al.*	Magnetic nanoparticles	Captured by a canister separator	Poor performance, agglomeration	(Adham *et al.* 2007)
2007	Adham *et al.*	Dendrimers	Adjusting pH or UF	Not feasible	(Adham *et al.* 2007)
2007	Adham *et al.*	Albumin	Denatured and solidified by heating	Not feasible	(Adham *et al.* 2007)
2008	McCormick *et al.*	Salt, ethanol	pervaporation-based separations	High reverse draw solute flux and low water flux	(McCormick *et al.* 2008)
2010	Yen *et al.*	2-Methylimidazole based solutes	Membrane Distillation (MD)	Materials costly	(Yen *et al.* 2010)
2010 – 2011	Ling *et al.* Ge *et al.*	Magnetic nanoparticles	Recycled by external magnetic field	Agglomeration	(Ge *et al.* 2010; Ling *et al.* 2010)
2011	Li *et al.*	Stimuli-responsive polymer hydrogels	Deswelling of the polymer hydrogels	Energy intensive, Poor water flux	(Li *et al.* 2011a; Li *et al.* 2011b)
2011	Ling & Chung	Hydrophilic nanoparticles	UF	Poor water flux	(Ling & Chung, 2011a)

2011	Phuntsho *et al.*	Fertilizers	None	Only applicable in agriculture	(Phuntsho *et al.* 2011)
2011	Iyer	fatty acid-polyethylene glycol	Thermal method	Poor water flux	(Iyer, 2012)
2012	Su *et al.*	Sucrose	NF	Relatively low water flux	(Su *et al.* 2012a)
2012	Ge *et al.*	Polyelectrolytes	UF	Relatively high viscosity	(Ge *et al.* 2012a)
2012	Noh *et al.*	Thermo-sensitive solute (Derivatives of Acyl-TAEA)	Not studied	Poor water flux	(Noh *et al.* 2012)
2012	Yong *et al.*	urea, ethylene glycol, and glucose	Not studied	Low water flux and high draw solute flux	(Yong *et al.* 2012)
2012	Bowden *et al.*	Organic salts	RO	Low water flux, energy intensive	(Bowden *et al.* 2012)
2012	Carmignani *et al.*	polyglycol copolymers	NF	High viscosity, severe ICP	(Carmignani *et al.* 2012)
2012	Stone *et al.*	hexavalent phosphazene salts	Not studied	Not economical and practical	(Stone *et al.* 2013)

Table 5.1.2 Overview of the physicochemical properties and FO water flux of draw solutes used in FO processes.

Draw Solute(s)	Concentration	Osmotic Pressure (atm)	Molecular Weight	Feed Solution	Water Flux (LMH)	Remark	Reference
NaCl	0.60 M	28	58.5 $g \cdot mol^{-1}$	DI water	9.6	CTA flat sheet membrane, FO mode	(Achilli *et al.* 2010)
$MgCl_2$	0.36 M	28	95 $g \cdot mol^{-1}$	DI water	8.4	CTA flat sheet membrane, FO mode	(Achilli *et al.* 2010)
KCl	2 M	89.3	74.6 $g \cdot mol^{-1}$	DI water	22.6	CTA flat sheet membrane, FO mode	(Phuntsho *et al.* 2011)
NH_4HCO_3	0.67 M	28	79 $g \cdot mol^{-1}$	DI water	7.3	CTA flat sheet membrane, FO mode	(Achilli *et al.* 2010)
Sucrose	1 M	26.7	342.3 $g \cdot mol^{-1}$	DI water	12.9	CA hollow fiber, FO mode	(Su *et al.* 2012a)

PAA-Na 1200	0.72 g/mL	44	1200 Da	DI water	22	CA hollow fiber, PRO mode	(Ge *et al.* 2012a)
PEG-$(COOH)_2$-MNPs 250	0.065 M	73	None	DI water	13	CTA flat sheet membrane, PRO mode	(Ge *et al.* 2010)
1,2,3-Trimethylimidazolium iodide	1M	50	238 $g \cdot mol^{-1}$	DI water	13	CTA flat sheet membrane, PRO mode	(Yen *et al.* 2010)
Sodium formate	0.68 M	28	68 $g \cdot mol^{-1}$	DI water	9.4	CTA flat sheet membrane, FO mode	(Bowden *et al.* 2012)
Polyglycol copolymer	30~70%	40~95	> 500 Da	3.5%NaCl	≥ 4	CTA flat sheet membrane, FO mode	(Carmignani *et al.* 2012)
Sodium hexa-carboxylatophenoxy phosphazene	0.067 M	None	1089 $g \cdot mol^{-1}$	DI water	6	CTA flat sheet membrane, FO mode	(Stone *et al.* 2013)

were achieved when the resultant ammonium mixtures were used as draw solutes. In 1965, sulfur dioxide (SO_2) was used as a draw solute by Batchelder (1965). In his work, SO_2 was added to seawater or freshwater and the resultant solution was used as a draw solution in an FO process to extract water from seawater. The separation of SO_2 from water was carried out by heating or air stripping. Although a positive water flux was observed in the experiments, water flux and salt rejection were not quantified in his patent. The concept of using volatile compounds as draw solutes was further developed by Glew in the same year (Glew, 1965). In his study, a mixture of water and SO_2 or liquid (e.g. aliphatic alcohols) was employed as a draw solution in the FO process. The purpose of adding SO_2 or alcohol is to lower water activity and increase the osmotic pressure of the resultant aqueous solution. The production of water and the recovery of SO_2 were proposed by means of distillation.

Later in 2002, a combination of potassium nitrate (KNO_3) and SO_2 was used as draw solutions by McGinnis in a system consisting of two FO units for seawater desalination (McGinnis, 2002). Hot seawater and saturated KNO_3 solution were used as the respective feed and draw solutions in the first FO unit. With water permeating to the draw solution from the feed seawater, the KNO_3 solution was diluted. The diluted KNO_3 solution was reconcentrated by a saturated SO_2 solution in the second FO unit. The SO_2 solute was then removed by standard means, such as heating. In this way, both KNO_3 and SO_2 were recycled. Although SO_2 was used as a draw solute in several works, processes involving SO_2 need to be very cautious in view of its special physicochemical properties. SO_2 is volatile and corrosive, and its solution is acidic and unstable. Meanwhile, SO_2 is a redox agent. It is susceptible to react with both oxidant and reducing agents. Hence, careful operations are needed for the SO_2 involved FO process and the subsequent posttreatment.

In summary, volatile draw solutes can be separated from product water by means of heating or distillation, and the regeneration of draw solutions can be achieved by dissolving the volatile gases back into water. However, most early patents on volatile compounds as draw solutes had insufficient experimental data to demonstrate their superiority or advantages.

5.1.4.1.2 Nutrient compounds

With the development of FO technology, its application in seawater desalination was extended to other needs. The second type of draw solutes, nutrient compounds, was proposed originally for an emergency water supply in lifeboats, but later extended for food and wine concentrations (Herron *et al.* 1994) and wastewater recycle (Su *et al.* 2012a). Glucose was first explored as a draw solute in the FO process for seawater desalination by Kravath and Davis (1975). Glucose present in the final product is edible and intended for drinking purpose in emergency lifeboats. Therefore, it excluded the need for draw solute recovery. Later, Kessler and Moody (1976), Stache (1989), and Yaeli (1992) further developed the concept of using nutrient solutes in the FO process. Kessler and Moody investigated the possibility of using the combination of glucose-fructose as draw solutes. It was claimed that better FO performance was achieved when compared with a pure glucose solution. Stache refined the idea by using fructose alone as the draw solute because fructose has the merits of a higher efficiency in osmosis. In addition, there is no thirst induced after drinking, making fructose more attractive as a draw solute. Yaeli (1992) extended Kravath and Davis (1975) idea by including glucose regeneration through low pressure RO processes. He was the pioneer introducing the concept of a "hybrid" system consisting of an FO and a low pressure RO. The studies on using nutrient compounds as draw solutes have led to various commercial products today (Herron *et al.* 1994; Jiao *et al.* 2004; Moßhammer *et al.* 2006).

The application of nutrient compounds in FO processes is still attracting much attention recently (Su *et al.* 2012a; Yong *et al.* 2012). Sucrose was used as a draw solute by Su *et al.* in wastewater treatment using double-skinned hollow fiber membranes (Su *et al.* 2012a). It was found that the draw solution of 1 M sucrose was able to produce a comparable water flux with 1 M $MgCl_2$ solution under the PRO mode. This interesting outcome was achieved due to the insignificant reduction in osmotic gradient across the semipermeable membrane since sucrose has a larger molecule size than $MgCl_2$ which results in a negligible reverse sucrose flux.

5.1.4.1.3 Inorganic salts

From the late 1990s, attention on draw solution has shifted to commercially available inorganic salts. To date, various inorganic salts have been used in the FO process. Draw solutions from inorganic salts give reasonably high fluxes and can be readily recovered by RO (Cornelissen *et al.* 2008; Bamaga *et al.* 2009; Achilli *et al.* 2010; Yangali-Quintanilla *et al.* 2011; Yong *et al.* 2012). As early as in 1970, Hough (1970) proposed a number of acids such as valeric acid, manganic acid, and glucuronic acid, and a variety of water soluble inorganic salts such as calcium, sodium, potassium, cesium and barium salts, as draw solutes in FO seawater desalination. By utilizing their solubility change when varying temperature or their reactions with sulfuric acid (H_2SO_4) or carbon dioxide (CO_2), water can be produced by precipitating these draw solutes. A precipitable salt, aluminum sulfate ($Al_2(SO_4)_3$), was also explored as a draw solute for FO by Frank (1972). This salt could then be treated with calcium hydroxide ($Ca(OH)_2$). The resultant precipitates of calcium sulfate ($CaSO_4$) and aluminium hydroxide ($Al(OH)_3$) were removed from the product water. Subsequently, sulfuric acid (H_2SO_4) or carbon dioxide (CO_2) was added to neutralize the excess $Ca(OH)_2$ left in the water. The drawbacks of aforementioned two approaches are obvious in view of the complicated posttreatment procedures and the involvement of many chemicals including both base ($Ca(OH)_2$ and $Al(OH)_3$) and acid (H_2SO_4). In the late 1990s, $MgCl_2$ was employed as a draw solute by Loeb *et al.* (1997) for a fundamental FO study. However, the recovery of $MgCl_2$ from its diluted solution was not discussed.

Later, with studies on inorganic draw solutes extension, more inorganic salts were employed in FO processes. These inorganic salts can be classified into three types as follows according to their recycling methods.

The first type is the thermal recoverable inorganic draw solutes, for example, ammonium bicarbonate (NH_4HCO_3). McCutcheon, Elimelech, McGinnis and their co-workers have extensively studied NH_4HCO_3 for FO seawater desalination since early 2000s (McCutcheon *et al.* 2005; Gray *et al.* 2006; McCutcheon & Elimelech, 2006; McCutcheon *et al.* 2006). A high water flux and a high water recovery were achieved because of the high osmotic pressure generated by the NH_4HCO_3 solution. A thermal method was applied to recover NH_4HCO_3 draw solute. When heated to around 60 °C, NH_4HCO_3 salt decomposed into ammonia and carbon dioxide gases, which were then separated from the product water. The NH_4HCO_3 draw solution could be regenerated by dissolving the gases back into water. In view of the low decomposition temperature of NH_4HCO_3, the energy required in its recovery can be tapped from low-grade waste heat in industries. Therefore, the NH_4HCO_3-promoted FO seawater desalination process was claimed to save up to 85% energy when compared with other desalination technologies (McGinnis & Elimelech, 2007). However, the high reverse flux of NH_4HCO_3 is a serious problem (Achilli *et al.* 2010). The draw solute leakage not only reduces the driving force across the membrane, but also contaminates the feed product and increases the replenishment cost. Meanwhile, insufficient removal of ammonia in product water remains a big obstacle towards its complete acceptance.

The second type of inorganic salts is those compounds which can be functioned as fertilizers. Nine fertilizers were investigated as draw solutes by Shon and his-coworkers (Phuntsho *et al.* 2011). Figure 5.1.2 illustrates a brief diagram of his approach. The FO process in their work is similar to those conventional FO desalination processes (Wang *et al.* 2009; Yang *et al.* 2009a; Su *et al.* 2010a; Wang *et al.* 2010a; Su *et al.* 2012a). As the diluted draw solution after FO is for irrigation, the separation of permeate water from draw solution is unnecessary. Therefore, the only required energy in this approach is to run the two pumps which do not exert much hydraulic pressure. As the draw solute recovery process after FO is regarded as the most energy-intensive step for seawater desalination (Cath *et al.* 2006; McGinnis & Elimelech, 2007; Zhao *et al.* 2012b), the absence of draw solute recovery in their work will ensure energy-savings. However, limitations still exist with this concept. First of all, most of chemical fertilizers exhibit acidic properties in their aqueous solutions. This makes them not appropriate for some membranes which are unable to operate under acidic conditions, for example, cellulose acetate/triacetate (CA/CTA) membranes. Such membranes may alter their structures since they contain ester groups which will undergo hydrolysis when exposed to acidic environments, possibly resulting in inconsistent FO performance (Ge *et al.* 2012a). Secondly, some fertilizers selected in this work only partially dissolve in water. Others are able to totally dissolve in water but partially dissociate in their aqueous solutions. As a result, some draw solutions may not have enough osmotic pressures to extract water from feed seawater through an FO process. High concentrations of fertilizers can fulfill this requirement, but they cause significant concentration polarization which leads to a decline in both water flux and water recovery. Blended fertilizers were also tried as draw solutes, but the bulk osmotic pressure and water flux may be lower than the sum of the osmotic pressures and water fluxes of the two individual draw solutes (Phuntsho *et al.* 2012).

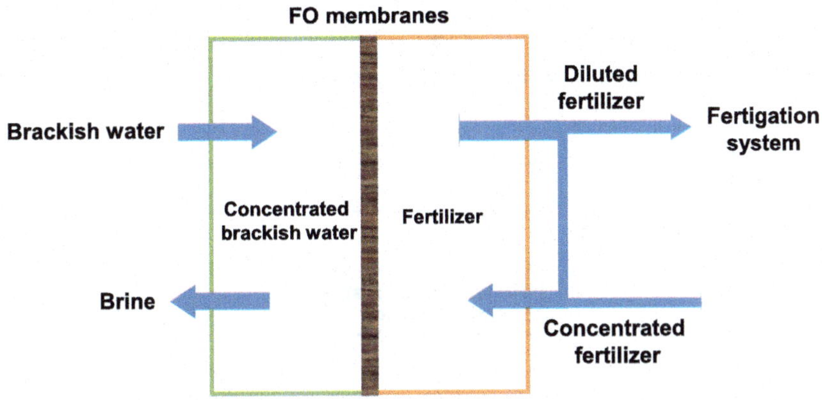

Figure 5.1.2 Design concept of the fertilizer drawn FO desalination process for irrigation.

Except for the aforementioned compounds, the rest of inorganic salts are assigned to the third type of draw solutes due to their similar recycle approach by using pressure-driven processes (McCormick *et al.* 2008; Achilli *et al.* 2010; Tan & Ng, 2010). These inorganic salts have been used in many FO applications. Some protocols were developed to select optimal draw solutes for different applications (Achilli *et al.* 2010; Kim *et al.* 2012). The flow diagram of a protocol built by Achilli *et al.* (2010) for water reuse and desalination is taken as an example and illustrated in Figure 5.1.3 (Achilli *et al.* 2010).

More than 500 inorganic compounds were initially considered and 14 inorganic salts remained at the end of the screened process. Draw solutions from the selected salts were then evaluated by determining their water fluxes and reverse salt fluxes through the FO process, and analyzing the reconcentration of these draw solutions using a RO process design software. Due to ICP influence, different draw solutions produced different water fluxes when they were tested at a concentration generating the same bulk osmotic pressure. As HTI (Hydration Technologies Inc) FO membranes were used in their study which had a negatively charged surface, the reverse salt fluxes were determined by the sizes of hydrated anions. Draw solutions containing larger hydrated anions such as $MgSO_4$, $KHCO_3$, $NaHCO_3$, Na_2SO_4, $(NH_4)_2SO_4$, and K_2SO_4showed lower reverse salt fluxes. However, based on the overall evaluation of water flux, reverse salt flux and RO permeate concentration, none of these draw solutions ranks the highest. Therefore, exploration of novel draw solutes is vital for future FO applications.

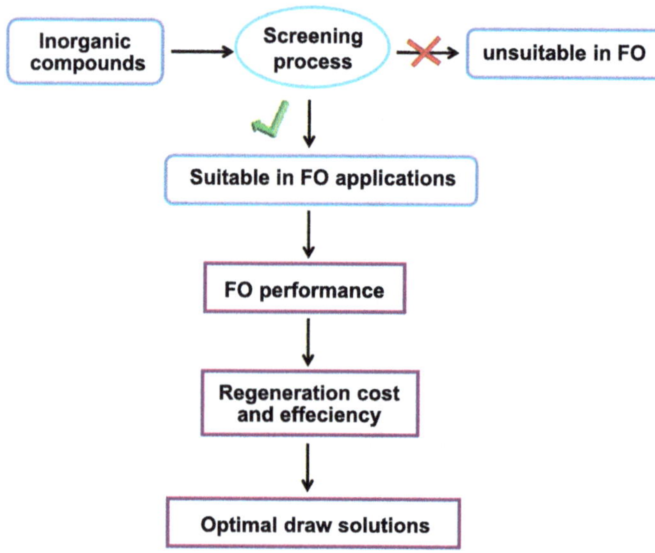

Figure 5.1.3 Illustration of flow diagram for draw solution selection. Figure adapted from Achilli *et al.* (2010).

5.1.4.1.4 Organic salts

Recently, Bowden *et al.* proposed a series of draw solutions from organic salts and used them in an OMBR system for wastewater treatment (Bowden *et al.* 2012). The advantage of using biodegradable organic salts is to mitigate the accumulation of draw solutes in OMBR. As an organic salt is defined as the combination of an organic anion with an organic or inorganic cation, for easier comparison with inorganic salts of NaCl and $MgCl_2$, the organic salts proposed in their work consisted of an organic anion and an inorganic cation of either Na^+ or Mg^{2+}. Ten organic salts derived from the combination of organic anions of formate, acetate, propionate and fumarate/maleate with Na^+ or Mg^{2+} were initially considered and evaluated according to the modified screen procedure developed by Achilli *et al.* (2010). Four organic salts; namely, sodium formate, sodium acetate, sodium propionate and magnesium acetate were selected for further assessments in OMBR, RO reconcentration, and biodegradation. Generally, an organic salt with a shorter carbon chain outperforms that with a longer carbon chain when they have the same cation species at the same osmotic pressure. This is because the former has less ICP than the latter due to a higher diffusion coefficient. Organic salts prove their advantages over their inorganic counterparts in terms of higher salt rejection in the RO reconcentration process and biodegradation potential required in an OMBR system. However, the FO water fluxes of the organic salts are much lower than those of their inorganic analogues generated under the same conditions. For example, the organic sodium salts produce water fluxes of 8.7~9.4 LMH at the concentration generating an osmotic pressure of 28 atm, as compared to ~14 LMH generated by NaCl with the same conditions (Ling *et al.* 2010). This may be because that the organic sodium salts with larger sizes have lower diffusion coefficients and are more affected by ICP than NaCl. Furthermore, using RO processes to recover the draw solutes indicates its drawback in view of the intensive energy consumption required in RO.

5.1.4.2 Synthetic materials as draw solutes

To advance FO technology, many researchers have been exploring the viability of using synthetic materials as draw solutes, and a variety of synthetic compounds have been proposed and applied to FO processes (Adham *et al.* 2007; Ge *et al.* 2010; Ling *et al.* 2010; Yen *et al.* 2010; Bai *et al.* 2011; Li *et al.* 2011a; Ling *et al.* 2011; Phuntsho *et al.* 2011; Ge *et al.* 2012a; Noh *et al.* 2012). Some of them exhibit great potentials as draw solutes.

Hydrophilic magnetic nanoparticles (MNPs) are considered as one of promising draw solutes and have generated a fair amount of interest recently (Adham *et al.* 2007; Warne *et al.* 2008; Ge *et al.* 2010; Ling *et al.* 2010; Bai *et al.* 2011; Ling & Chung, 2011a, 2011b; Ling *et al.* 2011). The concept of using MNPs as draw solute was originally disclosed by Warne *et al.* (2008) where superparamagnetic nanoparticles were synthesized via multi-step reactions. An external magnetic field was proposed to recycle draw solutes in their work. Although the concept is attractive, very little science and experimental data were disclosed to demonstrate the feasibility of MNPs as draw solutes. MNPs were first successfully explored as draw solutes for FO processes by Chung's group in the same year (Ge *et al.* 2010; Ling *et al.* 2010; Ling & Chung, 2011a, 2011b; Ling *et al.* 2011). In their studies, a series of hydrophilic MNPs have been synthesized by one-pot reactions and used as draw solutes in FO since 2010, for example, MNPs capped by polyacrylic acid or poly (ethylene glycol) diacid (PAA-MNPs and PEG-$(COOH)_2$-MNPs, respectively). Moderate performance with water fluxes of ~18 LMH was achieved when they were used as draw solutes (Ge *et al.* 2010; Ling *et al.* 2010). Their recyclability was carried out by an external magnetic field. The effects of surface ligands on particle hydrophilicity and osmotic pressure were studied to reveal the relationship between osmotic pressure generation and the surface chemistry of MNPs. It was also found that the osmotic pressure of MNPs can be enhanced by improving their surface hydrophilicity and/or reducing their particle size. Therefore, a draw solution of nanoparticles is able to produce water fluxes in FO by engineering the particle size and/or surface hydrophilicity. It is worth noting that the reverse fluxes of the MNPs were not detected due to their large particle sizes compared to inorganic species such as NaCl and $MgCl_2$. With this regard, a draw solution made from nanoparticles is superior in some special FO applications where no reverse flux of draw solutes is allowed for the final products, for example, protein or pharmaceutical enrichment where trace impurity may denature the target products (Ling & Chung, 2011b).

The recovery of nanoparticles from their diluted solution can be carried out by various methods including magnetic field, UF, and electric field, etc., depending on the particle size and properties. To recycle MNPs draw solutes, an external magnetic field has been employed to capture MNPs from their diluted draw solutions (Ge *et al.* 2010; Ling *et al.* 2010; Bai *et al.* 2011). Nanoparticle agglomeration was observed under a high-strength magnetic field and therefore FO performance of recycled MNPs dropped below their original MNPs draw solutes. To overcome this problem, ultrasonication was applied to redisperse those agglomerated MNPs (Ling & Chung, 2011a). Unfortunately, ultrasonication was not able to effectively restore MNPs draw solutes without changing their magnetic properties. The characterization of MNPs after ultrasonication showed that the magnetic cores were gradually oxidized and lost response to magnetic force although their particle sizes were reduced. To advance the application of nanoparticle draw solutes, UF was employed to recycle the diluted nanoparticle draw solutions. It was observed that UF could regenerate nanoparticle draw solutions effectively without causing obvious nanoparticle agglomeration, although nanoparticles smaller than the pore size of the UF membrane might leak through the membrane (Zhao & Zou, 2011a). In addition, thermo-responsive MNPs were synthesized as draw solutes where temperature stimulus was applicable to recycle them (Ling *et al.* 2011). When heated above the critical temperature of the thermo-responsive polymer coated on the nanoparticle surface, the nanoparticles would agglomerate spontaneously so that they can be caught and regenerated easily by either a UF or a low-strength magnetic field separation. In reality, thermo-responsive MNPs can be preheated by solar energy and then regenerated repeatedly by an integrated system consisting of a UF membrane and/or a low-strength magnetic field without causing nanoparticle agglomeration and performance deterioration (Ling *et al.* 2011; Crow, 2012).

In view of the low energy consumption, FO processes can be also used to regenerate nanoparticle draw solutes when RO retentate or other highly concentrate solution is employed as the draw solution. This concept has been demonstrated by Ling *et al.* to enrich proteins via a dual-FO process (Ling & Chung, 2011b). Most recently, surface-ionized nanoparticles have been explored as draw solutes in FO (Ling & Chung, 2012). Such nanoparticles exhibited improved FO performance after the surface ionization. After an FO process, the ionic nanoparticles could be regenerated via an integrated process comprising an electric field and a NF system.

Although nanoparticles are a promising class of draw solutes in FO applications; however, disadvantages also exist in such draw solutes. Their FO performance is not so good as inorganic salts due to their large size and suspension state in water, both of which result in a lower osmotic pressure and more significant concentration polarization. To overcome these disadvantages, synthetic materials which are able to dissolve well in water and form homogenous solutions have been explored by Ge *et al.* (2012a) and Yen *et al.* (2010). 2-methylimidazole-based organic compounds are one of them (Yen *et al.* 2010). Due to the good solubility, studies on both neutral and charged 2-methylimidazole-based compounds were carried out in the FO process. Consistent with aforementioned discussions, compounds with charges exhibit superiorities to neutral compounds in terms of higher water fluxes and lower reverse solute fluxes. The tailor-designed 2-methylimidazole-based organic compounds are able to draw water from seawater and can be recycled by FO-MD integrated processes. However, similar to draw solutes from inorganic salts, the reverse flux of draw solutes is an issue. In addition, the high cost of synthesis is another drawback of this kind of draw solutes. Nevertheless, this work brings more insights of using organic compounds as draw solutes in FO for future studies.

To lower the material cost and reduce the reverse flux, polyelectrolytes of polyacrylic acid sodium (PAA-Na) salts were also investigated as draw solutes by Ge *et al.* (2012a), Ge *et al.* (2012b). The characteristics of high water solubility and flexible configuration ensure the suitability of PAA-Na as draw solutes and their relative easy recycle through a pressure-driven UF process. Experimental results confirmed a high water flux with an insignificant reverse salt flux when using PAA-Na salts as the draw solutes. In addition, a high salt rejection was achieved when recycling PAA-Na by UF processes. Clearly, the multi-ionic PAA-Na outperforms conventional ionic salts, such as NaCl, when comparing their FO performance via the same membranes (Gray *et al.* 2006). The recycled PAA-Na exhibits repeatable performance and indicates no aggregation problems which exist in MNPs draw solutions (Ge *et al.* 2010; Ling *et al.* 2010). However, it was observed that an increase in PAA-Na solution concentration resulted in a viscosity enhancement. The high viscosity prevents polyelectrolytes from being practically used as draw solutes at ambient conditions. Fortunately, a temperature increase leads to a significant decline in viscosity. Therefore, a hybrid FO-MD system may be useful to increase the performance of draw solutions from polyelectrolytes since MD works at a relatively high temperature (Ge *et al.* 2012b). In view of these characteristics, a PAA-Na salt was employed as the draw solute in an FO-MD hybrid process to recycle water from the model wastewater containing an acid dye (Ge *et al.* 2012b). The hybrid system worked very well and the details will be discussed in the following section.

Other synthetic materials besides those discussed above were also investigated as draw solutes. Some stimuli-responsive polymer hydrogels have been developed as draw solutes by Wang's group for FO desalination (Li *et al.* 2011a). These polymer hydrogels are able to extract and release water when stimulus by either temperature or pressure, or by light with the incorporation of light-absorbing carbon particles (Li *et al.* 2011b). They can extract water from a feed saline solution in an FO desalination process and then undergo a reversible volume change when exposed to environmental stimulus. However, the reported performance of all hydrogels as draw solutes is poor at room temperature. Even though better performance is achieved at 50 °C, it requires a high hydraulic pressure of 30 bars during the dewatering process. This greatly increases energy consumption and cost, hence their practicability is in doubt. Recently, thermosensitive draw solutes based on N,N′,N″-triacylated tris (2-aminoethyl) amine (acyl-TAEA) derivatives were designed by Lee and his coworkers (Noh *et al.* 2012). Their solutions can draw water from saline water when temperatures are lower than the phase separation temperature, while the recovery of draw solutes takes place at higher temperatures. Similar to the hydrogel system aforementioned, the FO performance of acyl-TAEA draw solutes is still poor.

5.1.5 APPLICATIONS OF TYPICAL DRAW SOLUTIONS IN INTEGRATED FO PROCESSES

Although FO is an important membrane process, however, it is usually coupled with another technique, like NF (Tan & Ng, 2010; Zhao & Zou, 2011a; Zhao *et al.* 2012a), UF (Ling & Chung, 2011a; Ge *et al.* 2012a), MD (Curcio & Drioli, 2005; McCutcheon *et al.* 2005; Wang *et al.* 2011; Ge *et al.* 2012b), RO (Cornelissen *et al.* 2008; Bamaga *et al.* 2009; Achilli *et al.* 2010; Yangali-Quintanilla *et al.* 2011; Yong *et al.* 2012), etc. to obtain the final product and regenerate the draw solutes. Therefore, an integrated hybrid system is essential to optimize FO performance. Promising hybrid systems have been demonstrated in many fields including power generation (Pattle, 1954; Loeb *et al.* 1976; Lee *et al.* 1981; Loeb, 1998; Seppälä & Lampinen, 1999; Curcio & Drioli, 2005; McGinnis *et al.* 2007; Post *et al.* 2007; Yip *et al.* 2011), seawater/brackish water desalination (Cath *et al.* 2006; Choi *et al.* 2009; Martinetti *et al.* 2009; Bamaga *et al.* 2011; Zhao *et al.* 2012b), wastewater treatment (Batchelder, 1965; Cath *et al.* 2005b; Holloway *et al.* 2007; Cornelissen *et al.* 2011; Ge *et al.* 2012b), pharmaceutical drug delivery (Herbig *et al.* 1995; Santus & Baker, 1995; Thombre *et al.* 1999; Lin & Ho, 2003) and protein/pharmaceutical enrichment (Yang *et al.* 2009b; Nayak & Rastogi, 2010; Ling & Chung, 2011b; Wang *et al.* 2011). Some of them are briefly exemplified as follows.

5.1.5.1 Seawater desalination

NH_4HCO_3 has been intensively investigated by Elimelech and his coworkers for FO desalination (McCutcheon *et al.* 2005; Cath *et al.* 2006; Gray *et al.* 2006; McCutcheon & Elimelech, 2006; McCutcheon *et al.* 2006; Phillip *et al.* 2010). Figure 5.1.4 illustrates a representative hybrid process consisting of FO desalination and NH_4HCO_3 recovery (Cath *et al.* 2006). The NH_4HCO_3 solution extracts water from the feed seawater across the commercial HTI membranes which are applied to prevent NH_4HCO_3 diffusion while allow water to pass. The separation of product water and the recovery of NH_4HCO_3 draw solutions can be carried out through thermal processes. Upon moderate heating of around 60 °C, NH_4HCO_3 decomposes to ammonia and carbon dioxide gases and separates from the product water.

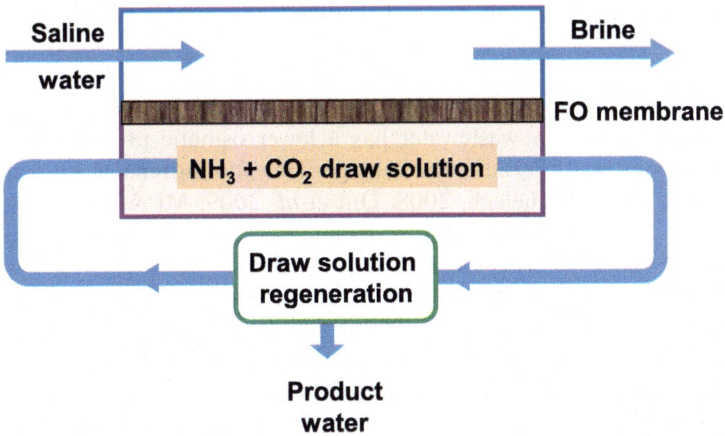

Figure 5.1.4 Schematic drawing of the novel ammonia–carbon dioxide FO process.

The overall energy required in the FO desalination and NH_4HCO_3 draw solution recovery has been predicted by using a chemical process modeling software (HYSYS) (McGinnis & Elimelech, 2007). Both thermal and electrical energy needed were calculated when distillation columns were used to separate draw solutes from the product water. The majority of energy consumption was the thermal process for NH_4HCO_3 recovery. The heat requirement with a single distillation was more than 75 $kWh \cdot m^{-3}$. In contrast, the electrical power requirement of a single FO process in desalination was less than 0.25 $kWh \cdot m^{-3}$. Therefore, although NH_4HCO_3 solution can produce a high osmotic pressure and result in a good water flux, its recovery is still an energy-intensive process. Unless sources of waste heat are available at low costs, the NH_4HCO_3 based seawater desalination may not be practical. In addition, a significant ICP phenomenon has been observed which leads to a significant water flux decline (McCutcheon *et al.* 2006; McGinnis & Elimelech, 2007; Liu *et al.* 2011). Furthermore, the quality of product water may not meet the drinking water standard (USEPA, 2009) when real seawater is used as the feed solution because of the presence of nonvolatile organic species from NH_4HCO_3 and higher boron concentration in the FO permeate (Danasamy, 2009).

To avoid the problems encountered in NH_4HCO_3 based FO desalination, MNPs capped with polyacrylic acid were also tested as draw solutes for seawater desalination (Ling & Chung, 2011a). The highly concentrated MNPs solution is able to extract water from model seawater and the diluted MNPs can be regenerated by a UF process. However, the FO performance of recycled MNPs declines gradually because the MNPs with sizes smaller the UF membrane pores leak out during regeneration. It is believed that by employing appropriate UF membranes with smaller pore sizes may retain the FO performance of MNPs. Meanwhile, more studies are required to optimize the regeneration system with better membrane selection.

5.1.5.2 Wastewater reclamation

Compared to seawater, wastewater has a lower osmotic pressure but higher fouling propensity. As FO has the advantage of low fouling tendency (Cornelissen *et al.* 2008; Mi & Elimelech, 2008; Qin *et al.* 2009; Mi & Elimelech, 2010; Tang *et al.* 2010; Liu & Mi, 2012; Valladares Linares *et al.* 2012), wastewater treatment through FO processes has been extensively investigated (Glew, 1965; Cath *et al.* 2005b; Holloway *et al.* 2007; Cornelissen *et al.* 2011; Ge *et al.* 2012b; Su *et al.* 2012a; Su *et al.* 2013). The major consideration in selecting draw solutes used in FO wastewater treatment is focused on their good performance in terms of high water flux and low reverse flux, harmless to the OMBR, sludge quality and bacteria growth when operated in bioreactor (Bowden *et al.* 2012), effective and economic recycle. So far, seawater and RO retentate are the most widely used draw solutes (Cath *et al.* 2005a, 2010; Logan & Elimelech, 2012) due to their low cost and high availability, while $MgCl_2$ is also recommended in view of its high osmotic efficiency even though it has a higher material cost than NaCl (Achilli *et al.* 2010). Hybrid systems of FO-MD (Cath *et al.* 2005a, 2010; Su *et al.* 2013), FO-RO (Choi *et al.* 2009; Choi *et al.* 2011) and FO-NF (Tan & Ng, 2010; Zhao *et al.* 2012a) have been used to recycle $MgCl_2$; however, the FO-RO integrated system may not be economical as compared with the other two systems.

PAA-Na salts (Ge *et al.* 2012b) have demonstrated their potential as draw solutes for wastewater treatment because of their merits such as high water flux, minimal reverse flux and easy recovery. However, their high viscosity at ambient conditions weakens their FO applications. By using the FO-MD hybrid system at a moderate high temperature (~60 °C), not only does it overcome the high viscosity constraint of PAA-Na solutions, but also enhance water flux without inducing reverse flux (Figure 5.1.5). The availability of low-cost waste heat is essential for the practicality of integrated FO-MD systems. Most recently, sucrose has also been used in wastewater treatment (Herron *et al.* 1994) where sucrose has a much lower reverse flux in FO than NaCl and $MgCl_2$ yet with a comparable water flux. Since sucrose is one of the most abundant saccharide in nature, fully biogradable and nontoxic, and has the advantages of high availability, low cost, and simplicity of production (Colorado State University), its prospect as a draw solute for FO wastewater treatment may be brighter. Hence, a sucrose based FO process for wastewater treatment is worthy of more studies.

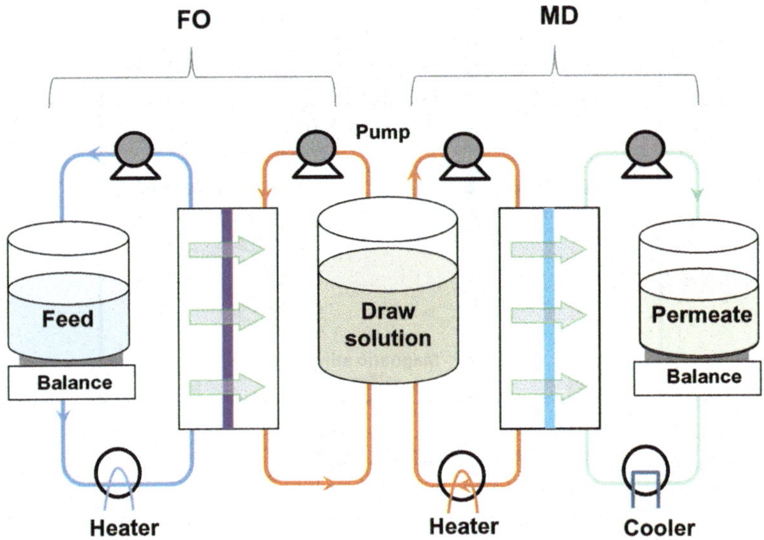

Figure 5.1.5 Schematic diagram of the polyelectrolyte promoted FO-MD hybrid system for wastewater treatment.

5.1.5.3 Protein enrichment

The production of proteins with high productivity has been considered as technically and economically challenging (Ghosh, 2003; van Reis & Zydney, 2007). Since most pharmaceutical products are labile and heat sensitive, an athermal separation approach such as FO is preferable in protein enrichment. The studies of protein enrichment via FO integrated membrane processes have been conducted through various approaches (Yang *et al.* 2009b; Ling & Chung, 2011b; Wang *et al.* 2011). Since protein is highly sensitive to surroundings, they may denature irreversibly when exposed in the environment with inorganic salt presence. Hence, MNPs, saccharide and polyelectrolytes of PAA-Na may be good candidates as draw solutes due to their insignificant reverse fluxes in FO processes. Ling *et al.* has conceptually demonstrated a dual-stage FO system for protein enrichment (Ling & Chung, 2011b). As illustrated in Figure 5.1.6, hydrophilic MNPs were employed as the draw solute to dehydrate the protein solution in the up-stage FO, while RO retentate to regenerate the MNPs solution in the down-stage FO. In this way, a continuous protein enrichment process was established. It has been confirmed that protein enriched by the dual-stage FO system is kept intact as proven by both gel track and CD spectra analyses. Despite these benefits, the dehydration rate in their FO systems is too low. Significant improvements are needed to be practical. Therefore, protein-friendly draw solutes with much higher osmotic pressures are in great demand.

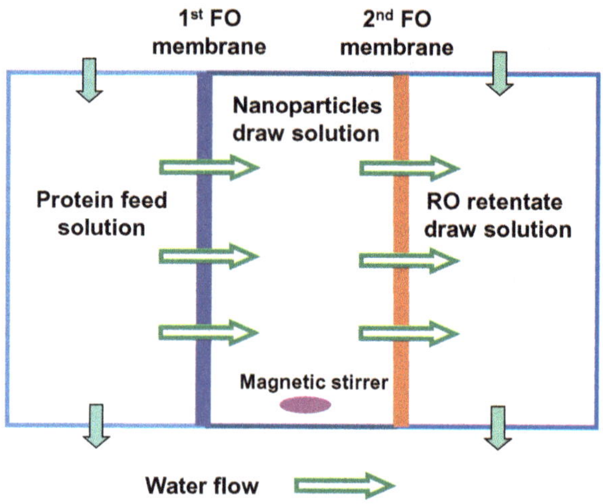

Figure 5.1.6 Schematic diagram of the laboratory-scale dual FO system for protein enrichment.

5.1.5.4 Power regeneration

Power regeneration through membrane process has been investigated in the past years (Pattle, 1954; Loeb *et al.* 1976; Lee *et al.* 1981; Loeb, 1998; Seppälä & Lampinen, 1999; Loeb, 2001; McGinnis *et al.* 2007; Post *et al.* 2007; Gerstandt *et al.* 2008; McGinnis & Elimelech, 2008; Achilli *et al.* 2009a; Thorsen & Holt, 2009; Yip *et al.* 2011; Zhang *et al.* 2013), in which osmotic power was harvested by exploiting the osmotic pressure difference of two solutions with different concentrations. Since freshwater would naturally diffuse to the seawater compartment in FO, the pressure built up in the seawater compartment can generate power by depressurizing the diluted seawater through a hydroturbine, as illustrated in Figure 5.1.7. The concept of osmotic power generation had been proposed and given trials, but most of early works were discontinued due to the lack of effective membranes (Levenspiel & de Nevers, 1974; Norman, 1974; Loeb & Norman, 1975; Loeb, 1976; Lee *et al.* 1981; Curcio & Drioli, 2005). With recent advancements in both FO membranes and draw solutes, attempts have been revived (Loeb, 2001; Gerstandt *et al.* 2008; Skilhagen *et al.* 2008; Achilli *et al.* 2009a; Thorsen & Holt, 2009; Zhang *et al.* 2013). Technically, some aforementioned draw solutes might be useful as draw solutes to generate power, but economically only seawater and RO retentate are the most practical draw solutes.

In fact, the highly concentrated RO retentate is produced through high-pressure and high-energy RO processes. The reuse of RO retentate for osmotic power generation could be somewhat considered as to release its energy. In addition, the integration of osmotic power generation and RO plants can make seawater desalination less energy dependent and more sustainable. It also eliminates environmental issues caused by the disposal of waste RO retentate. Although the PRO process may be technically feasible for power generation, it is not economically viable. The PRO performance is still subjected to the unavailability of suitable FO membranes and draw solutions. Exploration to use synthetic draw solutes for osmotic power generation must aim at applications in remote areas far from seawater or concentrated brine but with plenty provision of waste heat; otherwise, they may not be practical.

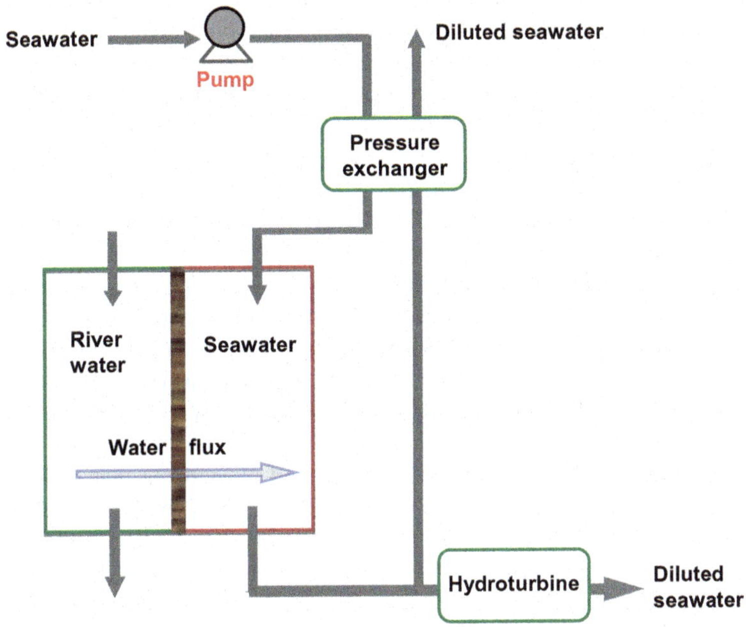

Figure 5.1.7 Schematic drawing of a PRO power plant.

5.1.6 CHALLENGES AND PROSPECTS FOR THE FUTURE

As reviewed above, FO has demonstrated its advantages and potentials in the applications for power regeneration, irrigation, protein and pharmaceutical concentration, water reuse, OMBR and desalination. However, it is still facing many challenges. Breakthroughs on molecular designs of high performance FO membranes and appropriate draw solutions are urgently needed. Since the effects of membranes and draw solutes on FO performance are closely correlated each other, the design strategies must be considered from both aspects. To mitigate ICP, the FO membranes must have characteristics of high permeability and hydrophilicity with a small structure parameter, while the preferred draw solutes must have a high Van't Hoff factor and diffusion coefficient, reasonable ion or molecule size and low viscosity (Zydney, 1997; Cath *et al.* 2006; Gray *et al.* 2006; Zhao & Zou, 2011b).

A balance between osmotic pressure and reverse flux should be considered in the selection of draw solutes. A small and fully dissociated draw solute may generate a high water flux yet with a high reverse flux simultaneously that reduces the overall osmotic driving force, while a large and fully dissociated draw solute may have a low osmotic pressure but a negligible reverse flux that not only maintains the effective osmotic driving force but also compensates the originally low osmotic pressure. In addition, the best choice of draw solutes is also dependent on the progress of FO membranes. FO membranes with ideal semipermeable characteristics may prefer small and fully dissociated draw solutes.

Given the advantages and limitations of each draw solution reported in the literature, none of them is perfect for FO processes. However, draw solutes such as $MgCl_2$, NH_4HCO_3, polyelectrolytes, sucrose, MNPs, organic salts and RO brine have demonstrated their potentials for some specific applications. Obviously, the requirements of selecting a suitable draw solution are subjected to the purpose of FO and the subsequent applications. Developing an ideal draw solution for seawater desalination to get potable water may be still a long course. Firstly, unlike other FO processes, the draw solution used in FO seawater desalination must have an osmotic pressure higher than seawater (~27 atm) in addition to the requirements of material nontoxicity and low cost. Secondly, easy separation between the product water and draw solute is required so that the energy consumption is affordable. Finally, the reverse flux of draw solutes must be minimal in order to meet the drinking water standard without additional complicated posttreatments.

Recent publications have reported complicated phenomena that associate the reverse flux of draw solutes with fouling (Mi & Elimelech, 2008; Lay *et al.* 2010; Lee *et al.* 2010; Tang *et al.* 2010; Jung *et al.* 2011; Hausman & Escobar, 2013; She *et al.* 2013). Fouling is affected by the draw solute's chemistry properties, diffusion rate to the feed and interactions with feed foulants. To alleviate this problem, in addition to designing FO membranes with a much higher flux and salt rejection (Wang *et al.* 2010b; Widjojo *et al.* 2011; Han *et al.* 2012; Wang *et al.* 2012), draw solutions with a much lower reverse flux are preferable. Multivalent ionic solutions are considered to meet this requirement for their low diffusion coefficients. However, an aggravation in fouling may occur due to the interaction between the feed solute and multivalent ions when reverse diffusion takes place (Mi & Elimelech, 2008; Hancock & Cath, 2009; Achilli *et al.* 2010; Mi & Elimelech, 2010; Zou *et al.* 2011; Liu & Mi, 2012). Since multivalent ions tend to have low solution diffusion coefficients because of their large ion sizes (Achilli *et al.* 2010; Zhao & Zou, 2011b), fully dissociated ionic draw solutes with bulky or expanded configurations may be a solution. As demonstrated by Ge *et al.* (2012a), polyelectrolytes of PAA-Na salts with a low molecular weight generate good osmotic pressures but with insignificant reverse fluxes due to their expanded structures (Ge *et al.* 2012a, 2012b). However, more work is needed to understand their diffusion coefficients and ICP effects. In summary, future works should focus on the design of fully dissociated ionic draw solutes with bulky or expanded configurations and investigate their relationship with ICP, reverse flux and membrane fouling in various FO processes. The exploration of novel draw solutes in future should also consider their recycle via well-established low energy processes, such as UF processes. In addition, more efforts should be made to employ solar energy for the regeneration of both conventional and temperature sensitive draw solutes (Schrier, 2012).

5.1.7 SUMMARY

Forward osmosis (FO) has emerged as one of potential technologies to mitigate water and energy shortage. Not only can it produce clean water but also energy by employing draw solutes to induce osmotic gradients across semipermeable membranes as the driving force for water production and power generation. Ideally, the semipermeable membrane performs as a barrier that allows only water to pass through but rejects all others. However, in reality, depending on draw solute's chemistry property and physical structure, the reverse flux of draw solutes may take place across FO membranes which not only results in a lower effective osmotic driving force but also facilitates fouling. In addition, the asymmetric structure of FO membranes and the transport resistance of draw solutes within the FO membranes cause concentration polarization and lower the water flux. Furthermore, the regeneration of draw solutes from diluted draw solutions and the production of clean water might be energy intensive if inappropriate draw solutes and recycle processes are utilized. Therefore, in this work we aim to give a comprehensive review on the progress of draw solution for FO processes. An assessment on the advantages and limitations of the existing draw solutes are made. Various FO integrated processes for water production and draw solute regeneration are exemplified. We also highlight the challenges and future research directions for the molecular design of better draw solutes.

5.1.8 NOMENCLATURE

A water permeability constant
A_m effective membrane surface area (m^2)
C_0 initial salt concentration of the feed (mol L^{-1})
C_t salt concentration in the feed after the operation time interval (mol L^{-1})
D diffusion coefficient of a draw solution
J_s reverse salt leakage (g m^{-2} h^{-1})
J_w product water flux (L m^{-2} h^{-1})
k mass transfer coefficient
K solute resistivity
S membrane structural parameter
t membrane thickness
Δt operation time interval (h)
V_0 initial volume of the feed (L)
V_t volume of the feed after the operation time interval (L)

5.1.8.1 Greek symbols

$\Delta\pi$ osmotic pressure difference (bar)
$\pi_{D,a}$ osmotic pressures of the draw solution at the membrane active layer surface (bar)
$\pi_{D,b}$ osmotic pressure of the draw solution in the bulk (bar)
$\pi_{D,m}$ osmotic pressure of the draw solution at the membrane surface (bar)
$\pi_{D,s}$ osmotic pressures of the draw solution at the membrane support layer surface (bar)
$\pi_{F,a}$ osmotic pressures of the feed solution at the membrane active layer surface (bar)
$\pi_{F,b}$ osmotic pressure of the feed in the bulk (bar)
$\pi_{F,m}$ osmotic pressure of the feed at the membrane surface (bar)
$\pi_{F,s}$ osmotic pressures of the feed solution at the membrane support layer surface (bar)
σ reflection coefficient
τ membrane tortuosity
ε membrane porosity

5.1.9 REFERENCES

Achilli A., Cath T. Y. and Childress A. E. (2009a). Power generation with pressure retarded osmosis: an experimental and theoretical investigation. *Journal of Membrane Science,* **343**, 42–52.

Achilli A., Cath T. Y., Marchand E. A. and Childress A. E. (2009b). The forward osmosis membrane bioreactor: a low fouling alternative to MBR processes. *Desalination,* **239**, 10–21.

Achilli A., Cath T. Y. and Childress A. E. (2010). Selection of inorganic-based draw solutions for forward osmosis applications. *Journal of Membrane Science,* **364**, 233–241.

Adham S., Oppenheimer J., Liu L. and Kumar M. (2007). Dewatering Reverse Osmosis Concentrate from Water Reuse Applications using Forward Osmosis. WateReuse Foundation, Alexandria, VA, USA.

Alturki A., McDonald J., Khan S. J., Hai F. I., Price W. E. and Nghiem L. D. (2012). Performance of a novel osmotic membrane bioreactor (OMBR) system: flux stability and removal of trace organics. *Bioresource Technology,* **113**, 201–206.

Bai H., Liu Z. and Sun D. D. (2011). Highly water soluble and recovered dextran coated Fe_3O_4 magnetic nanoparticles for brackish water desalination. *Separation and Purification Technology,* **81**, 392–399.

Bamaga O., Yokochi A. and Beaudry E. (2009). Application of forward osmosis in pretreatment of seawater for small reverse osmosis desalination units. *Desalination and Water Treatment,* **5**, 183–191.

Bamaga O., Yokochi A., Zabara B. and Babaqi A. (2011). Hybrid FO/RO desalination system: preliminary assessment of osmotic energy recovery and designs of new FO membrane module configurations. *Desalination,* **268**, 163–169.

Batchelder G. W. (1965). Process for the demineralization of water. US Patent 3,171,799.

Bowden K. S., Achilli A. and Childress A. E. (2012). Organic ionic salt draw solutions for osmotic membrane bioreactors. *Bioresource Technology,* **122**, 207–216.

Cai G., Gorey C., Zaky A., Escobar I. and Gruden C. (2011). Thermally responsive membrane-based microbiological sensing component for early detection of membrane biofouling. *Desalination,* **270**, 116–123.

Carmignani G., Sitkiewitz S. and Webley J. W. (2012). Recovery of retrograde soluble solute for forward osmosis water treatment. US Patent 20120267308 A1. Trevi Systems Inc., (Petaluma, CA, US), United States.

Cath T. Y., Adams D. and Childress A. E. (2005a). Membrane contactor processes for wastewater reclamation in space: II. Combined direct osmosis, osmotic distillation, and membrane distillation for treatment of metabolic wastewater. *Journal of Membrane Science,* **257**, 111–119.

Cath T. Y., Gormly S., Beaudry E. G., Flynn M. T., Adams V. D. and Childress A. E. (2005b). Membrane contactor processes for wastewater reclamation in space: Part I. Direct osmotic concentration as pretreatment for reverse osmosis. *Journal of Membrane Science,* **257**, 85–98.

Cath T. Y., Childress A. E. and Elimelech M. (2006). Forward osmosis: principles, applications, and recent developments. *Journal of Membrane Science,* **281**, 70–87.

Cath T. Y., Hancock N. T., Lundin C. D., Hoppe-Jones C. and Drewes J. E. (2010). A multi-barrier osmotic dilution process for simultaneous desalination and purification of impaired water. *Journal of Membrane Science,* **362**, 417–426.

Chekli L., Phuntsho S., Shon H. K., Vigneswaran S., Kandasamy J. and Chanan A. (2012). A review of draw solutes in forward osmosis process and their use in modern applications. *Desalination and Water Treatment,* **43**, 167–184.

Choi J.-S., Kim H., Lee S., Hwang T.-M., Oh H., Yang D. R. and Kim J. H. (2010). Theoretical investigation of hybrid desalination system combining reverse osmosis and forward osmosis. *Desalination and Water Treatment,* **15**, 114–120.

Choi Y.-J., Choi J.-S., Oh H.-J., Lee S., Yang D. R. and Kim J. H. (2009). Toward a combined system of forward osmosis and reverse osmosis for seawater desalination. *Desalination,* **247**, 239–246.

Choi Y.-J., Hwang T.-M., Oh H., Nam S.-H., Lee S., Jeon J.-c., Han S. J. and Chung Y. (2011). Development of a simulation program for the forward osmosis and reverse osmosis process. *Desalination and Water Treatment,* **33**, 273–282.

Chon K., KyongShon H. and Cho J. (2012). Membrane bioreactor and nanofiltration hybrid system for reclamation of municipal wastewater: removal of nutrients, organic matter and micropollutants. *Bioresource Technology,* **122**, 181–188.

Chung T.-S., Li X., Ong R. C., Ge Q., Wang H. and Han G. (2012a). Emerging forward osmosis (FO) technologies and challenges ahead for clean water and clean energy applications. *Current Opinion in Chemical Engineering,* **1**, 246–257.

Chung T.-S., Zhang S., Wang K. Y., Su J. and Ling M. M. (2012b). Forward osmosis processes: yesterday, today and tomorrow. *Desalination,* **287**, 78–81.

Colorado State University. http://www.ext.colostate.edu/pubs/foodnut/09301.html

Cornelissen E. R., Harmsen D., de Korte K. F., Ruiken C. J., Qin J.-J., Oo H. and Wessels L. P. (2008). Membrane fouling and process performance of forward osmosis membranes on activated sludge. *Journal of Membrane Science,* **319**, 158–168.

Cornelissen E., Harmsen D., Beerendonk E., Qin J., Oo H., De Korte K. and Kappelhof J. (2011). The innovative osmotic membrane bioreactor (OMBR) for reuse of wastewater. *Water Science & Technology,* **63**, 1557–1565.

Crow J. (2012). Keeping the tap on. *Chemistry World,* **9**, 44–47.

Curcio E. and Drioli E. (2005). Membrane distillation and related operations – a review. *Separation and Purification Reviews,* **34**, 35–86.

Danasamy G. (2009). Sustainability of Seawater Desalination Technology – the Study of Forward Osmosis as a New Alternative. MSc thesis, Imperial College, London.

Elimelech M. and Phillip W. A. (2011). The future of seawater desalination: energy, technology, and the environment. *Science,* **333**, 712–717.

Escobar I. C. (2010). Conclusion: a summary of challenges still facing desalination and water reuse. *Sustainability Science and Engineering,* **2**, 389–397.

Fang W., Wang R., Chou S., Setiawan L. and Fane A. G. (2012). Composite forward osmosis hollow fiber membranes: integration of RO-and NF-like selective layers to enhance membrane properties of anti-scaling and anti-internal concentration polarization. *Journal of Membrane Science,* **394**, 140–150.

Frank B. S. (1972). Desalination of sea water. US Patent 3,670,897.

Garcia-Castello E. M., McCutcheon J. R. and Elimelech M. (2009). Performance evaluation of sucrose concentration using forward osmosis. *Journal of Membrane Science,* **338**, 61–66.

Ge Q., Su J., Chung T.-S. and Amy G. (2010). Hydrophilic superparamagnetic nanoparticles: synthesis, characterization, and performance in forward osmosis processes. *Industrial & Engineering Chemistry Research,* **50**, 382–388.

Ge Q., Su J., Amy G. L. and Chung T.-S. (2012a). Exploration of polyelectrolytes as draw solutes in forward osmosis processes. *Water Research,* **46**, 1318–1326.

Ge Q., Wang P., Wan C. and Chung T.-S. (2012b). Polyelectrolyte-Promoted Forward Osmosis-Membrane Distillation (FO-MD) Hybrid Process for Dye Wastewater Treatment. *Environmental Science & Technology,* **46**, 6236–6243.

Gerstandt K., Peinemann K.-V., Skilhagen S. E., Thorsen T. and Holt T. (2008). Membrane processes in energy supply for an osmotic power plant. *Desalination,* **224**, 64–70.

Ghosh R. (2003). Protein Bioseparation Using Ultrafiltration. Imperial College Press, River Edge, NJ.

Glew D. N. (1965). Process for liquid recovery and solution concentration. US Patent 3,216,930.

Gray G. T., McCutcheon J. R. and Elimelech M. (2006). Internal concentration polarization in forward osmosis: role of membrane orientation. *Desalination,* **197**, 1–8.

Greenlee L. F., Lawler D. F., Freeman B. D., Marrot B. and Moulin P. (2009). Reverse osmosis desalination: water sources, technology, and today's challenges. *Water Research,* **43**, 2317–2348.

Han G., Zhang S., Li X., Widjojo N. and Chung T.-S. (2012). Thin film composite forward osmosis membranes based on polydopamine modified polysulfone substrates with enhancements in both water flux and salt rejection. *Chemical Engineering Science,* **80**, 219–231.

Hancock N. T. and Cath T. Y. (2009). Solute coupled diffusion in osmotically driven membrane processes. *Environmental Science & Technology,* **43**, 6769–6775.

Hausman R. and Escobar I. C. (2013). A comparison of silver- and copper-charged polypropylene feed spacers for biofouling control. *Journal of Applied Polymer Science,* **128**, 1706–1714.

Herbig S., Cardinal J., Korsmeyer R. and Smith K. (1995). Asymmetric-membrane tablet coatings for osmotic drug delivery. *Journal of Controlled Release,* **35**, 127–136.

Herron J. R., Beaudry E. G., Jochums C. E. and Medina L. E. (1994). Osmotic concentration apparatus and method for direct osmotic concentration of fruit juices. US Patent 5,281,430.

Holloway R. W., Childress A. E., Dennett K. E. and Cath T. Y. (2007). Forward osmosis for concentration of anaerobic digester centrate. *Water Research,* **41**, 4005–4014.

Hough W. T. (1970). Process for extracting solvent from a solution. US Patent 3,532,621.

http://en.wikipedia.org/wiki/Osmotic_pressure.

Iyer S. (2012). Systems and methods for forward osmosis fluid purification using cloud point extraction. US Patent 20120267297 A1.

Jiao B., Cassano A. and Drioli E. (2004). Recent advances on membrane processes for the concentration of fruit juices: a review. *Journal of Food Engineering,* **63**, 303–324.

Jung D. H., Lee J., Lee Y. G., Park M., Lee S., Yang D. R. and Kim J. H. (2011). Simulation of forward osmosis membrane process: effect of membrane orientation and flow direction of feed and draw solutions. *Desalination,* **277**, 83–91.

Kessler J. and Moody C. (1976). Drinking water from sea water by forward osmosis. *Desalination,* **18**, 297–306.

Kim T.-W., Kim Y., Yun C., Jang H., Kim W. and Park S. (2012). Systematic approach for draw solute selection and optimal system design for forward osmosis desalination. *Desalination,* **284**, 253–260.

King C. W. and Webber M. E. (2008). Water intensity of transportation. *Environmental Science & Technology,* **42**, 7866–7872.

Kravath R. E. and Davis J. A. (1975). Desalination of sea water by direct osmosis. *Desalination,* **16**, 151–155.

Lay W. C. L., Chong T. H., Tang C. Y., Fane A. G., Zhang J. and Liu Y. (2010). Fouling propensity of forward osmosis: investigation of the slower flux decline phenomenon. *Water Science and Technology,* **61**(4), 927–936.

Lee K. L., Baker R. W. and Lonsdale H. K. (1981). Membranes for power generation by pressure-retarded osmosis. *Journal of Membrane Science,* **8**, 141–171.

Lee K. P., Arnot T. C. and Mattia D. (2011). A review of reverse osmosis membrane materials for desalination – Development to date and future potential. *Journal of Membrane Science,* **370**, 1–22.

Lee S., Boo C., Elimelech M. and Hong S. (2010). Comparison of fouling behavior in forward osmosis (FO) and reverse osmosis (RO). *Journal of Membrane Science,* **365**, 34–39.

Levenspiel O. and de Nevers N. (1974). The osmotic pump. *Science,* **183**, 157–160.

Li D., Zhang X., Yao J., Simon G. P. and Wang H. (2011a). Stimuli-responsive polymer hydrogels as a new class of draw agent for forward osmosis desalination. *Chemical Communications,* **47**, 1710–1712.

Li D., Zhang X., Yao J., Zeng Y., Simon G. P. and Wang H. (2011b). Composite polymer hydrogels as draw agents in forward osmosis and solar dewatering. *Soft Matter,* **7**, 10048–10056.

Lin Y.-K. and Ho H.-O. (2003). Investigations on the drug releasing mechanism from an asymmetric membrane-coated capsule with an in situ formed delivery orifice. *Journal of Controlled Release,* **89**, 57–69.

Ling M. M. and Chung T.-S. (2011a). Desalination process using super hydrophilic nanoparticles via forward osmosis integrated with ultrafiltration regeneration. *Desalination,* **278**, 194–202.

Ling M. M. and Chung T.-S. (2011b). Novel dual-stage FO system for sustainable protein enrichment using nanoparticles as intermediate draw solutes. *Journal of Membrane Science,* **372**, 201–209.

Ling M. M. and Chung T.-S. (2012). Surface-dissociated nanoparticle draw solutions in forward osmosis and the regeneration in an integrated electric field and nanofiltration system. *Industrial & Engineering Chemistry Research,* **51**, 15463–15471.

Ling M. M., Wang K. Y. and Chung T.-S. (2010). Highly water-soluble magnetic nanoparticles as novel draw solutes in forward osmosis for water reuse. *Industrial & Engineering Chemistry Research,* **49**, 5869–5876.

Ling M. M., Chung T. S. and Lu X. (2011). Facile synthesis of thermosensitive magnetic nanoparticles as "smart" draw solutes in forward osmosis. *Chemical Communications,* **47**, 10788–10790.

Liu Y. and Mi B. (2012). Combined fouling of forward osmosis membranes: synergistic foulant interaction and direct observation of fouling layer formation. *Journal of Membrane Science,* **407–408**, 136–144.

Liu Z., Bai H., Lee J. and Sun D. D. (2011). A low-energy forward osmosis process to produce drinking water. *Energy & Environmental Science,* **4**, 2582–2585.

Loeb S. (1976). Production of energy from concentrated brines by pressure-retarded osmosis: I. Preliminary technical and economic correlations. *Journal of Membrane Science,* **1**, 49–63.

Loeb S. (1998). Energy production at the Dead Sea by pressure-retarded osmosis: challenge or chimera? *Desalination,* **120**, 247–262.

Loeb S. (2001). One hundred and thirty benign and renewable megawatts from Great Salt Lake? The possibilities of hydroelectric power by pressure-retarded osmosis. *Desalination,* **141**, 85–91.

Loeb S. and Norman R. S. (1975). Osmotic power plants. *Science,* **189**, 654–655.

Loeb S., Van Hessen F. and Shahaf D. (1976). Production of energy from concentrated brines by pressure-retarded osmosis: II. Experimental results and projected energy costs. *Journal of Membrane Science,* **1**, 249–269.

Loeb S., Titelman L., Korngold E. and Freiman J. (1997). Effect of porous support fabric on osmosis through a Loeb-Sourirajan type asymmetric membrane. *Journal of Membrane Science,* **129**, 243–249.

Logan B. E. and Elimelech M. (2012). Membrane-based processes for sustainable power generation using water. *Nature,* **488**, 313–319.

Martinetti C. R., Childress A. E. and Cath T. Y. (2009). High recovery of concentrated RO brines using forward osmosis and membrane distillation. *Journal of Membrane Science,* **331**, 31–39.

McCormick P., Pellegrino J., Mantovani F. and Sarti G. (2008). Water, salt, and ethanol diffusion through membranes for water recovery by forward (direct) osmosis processes. *Journal of Membrane Science,* **325**, 467–478.

McCutcheon J. and Elimelech M. (2006). Influence of concentrative and dilutive internal concentration polarization on flux behavior in forward osmosis. *Journal of Membrane, Science* **284**, 237–247.

McCutcheon J. R. and Elimelech M. (2007). Modeling water flux in forward osmosis: implications for improved membrane design. *AIChE Journal,* **53**, 1736–1744.

McCutcheon J. R., McGinnis R. L. and Elimelech M. (2005). A novel ammonia – carbon dioxide forward (direct) osmosis desalination process. *Desalination,* **174**, 1–11.

McCutcheon J. R., McGinnis R. L. and Elimelech M. (2006). Desalination by ammonia–carbon dioxide forward osmosis: influence of draw and feed solution concentrations on process performance. *Journal of Membrane Science,* **278**, 114–123.

McGinnis R. L. (2002). Osmotic desalinization process. US Patent 6391205 B1.

McGinnis R. L. and Elimelech M. (2007). Energy requirements of ammonia–carbon dioxide forward osmosis desalination. *Desalination,* **207**, 370–382.

McGinnis R. L. and Elimelech M. (2008). Global challenges in energy and water supply: the promise of engineered osmosis. *Environmental Science & Technology,* **42**, 8625–8629.

McGinnis R. L., McCutcheon J. R. and Elimelech M. (2007). A novel ammonia–carbon dioxide osmotic heat engine for power generation. *Journal of Membrane Science*, **305**, 13–19.

Mehta G. D. and Loeb S. (1979). Internal polarization in the porous substructure of a semipermeable membrane under pressure-retarded osmosis. *Journal of Membrane, Science*, **4**, 261–265.

Mi B. and Elimelech M. (2008). Chemical and physical aspects of organic fouling of forward osmosis membranes. *Journal of Membrane Science,* **320**, 292–302.

Mi B. and Elimelech M. (2010). Organic fouling of forward osmosis membranes: fouling reversibility and cleaning without chemical reagents. *Journal of Membrane Science,* **348**, 337–345.

Moßhammer M. R., Stintzing F. C. and Carle R. (2006). Evaluation of different methods for the production of juice concentrates and fruit powders from cactus pear. *Innovative Food Science & Emerging Technologies,* **7**, 275–287.

Montgomery M. A. and Elimelech M. (2007). Water and sanitation in developing countries: including health in the equation. *Environmental Science & Technology,* **41**, 17–24.

Morão A. M., Nunes J. C., Sousa F., de Amorim M. T. P., Escobar I. C. and Queiroz J. A. (2011). Ultrafiltration of supercoiled plasmid DNA: modeling and application. *Journal of Membrane Science,* **378**, 280–289.

Nayak C. A. and Rastogi N. K. (2010). Forward osmosis for the concentration of anthocyanin from Garcinia indica Choisy. *Separation and Purification Technology,* **71**, 144–151.

Neff R. A. (1964). Solvent extractor. US Patent 3,130,156.

Noh M., Mok Y., Lee S., Kim H., Lee S. H., Jin G.-w., Seo J.-H., Koo H., Park T. H. and Lee Y. (2012). Novel lower critical solution temperature phase transition materials effectively control osmosis by mild temperature changes. *Chemical Communications,* **48**, 3845–3847.

Norman R. S. (1974). Water salination: a source of energy. *Science,* **186**, 350–352.

Pattle R. (1954). Production of electric power by mixing fresh and salt water in the hydroelectric pile. *Nature,* 174–660.

Petrotos K. B. and Lazarides H. N. (2001). Osmotic concentration of liquid foods. *Journal of Food Engineering,* **49**, 201–206.

Phillip W. A., Yong J. S. and Elimelech M. (2010). Reverse draw solute permeation in forward osmosis: modeling and experiments. *Environmental Science & Technology,* **44**, 5170–5176.

Phuntsho S., Shon H. K., Hong S., Lee S. and Vigneswaran S. (2011). A novel low energy fertilizer driven forward osmosis desalination for direct fertigation: evaluating the performance of fertilizer draw solutions. *Journal of Membrane Science,* **375**, 172–181.

Phuntsho S., Shon H. K., Majeed T., El Saliby I., Vigneswaran S., Kandasamy J., Hong S. and Lee S. (2012). Blended fertilizers as draw solutions for fertilizer-drawn forward osmosis desalination. *Environmental Science & Technology,* **46**, 4567–4575.

Post J. W., Veerman J., Hamelers H. V. M., Euverink G. J. W., Metz S. J., Nymeijer K. and Buisman C. J. N. (2007). Salinity-gradient power: evaluation of pressure-retarded osmosis and reverse electrodialysis. *Journal of Membrane Science,* **288**, 218–230.

Qin J.-J., Liberman B. and Kekre K. A. (2009). Direct osmosis for reverse osmosis fouling control: principles, applications and recent developments. *The Open Chemical Engineering Journal*, **3**, 8–16.

Santus G. and Baker R. W. (1995). Osmotic drug delivery: a review of the patent literature. *Journal of Controlled Release*, **35**, 1–21.

Schrier J. (2012). Ethanol concentration by forward osmosis with solar-regenerated draw solution. *Solar Energy*, **86**, 1351–1358.

Seppälä A. and Lampinen M. J. (1999). Thermodynamic optimizing of pressure-retarded osmosis power generation systems. *Journal of Membrane Science*, **161**, 115–138.

Shannon M. A., Bohn P. W., Elimelech M., Georgiadis J. G., Marinas B. J. and Mayes A. M. (2008). Science and technology for water purification in the coming decades. *Nature*, **452**, 301–310.

Shao L., Low B. T., Chung T.-S. and Greenberg A. R. (2009). Polymeric membranes for the hydrogen economy: contemporary approaches and prospects for the future. *Journal of Membrane Science*, **327**, 18–31.

She Q., Wong Y. K. W., Zhao S. and Tang C. Y. (2013). Organic fouling in pressure retarded osmosis: experiments, mechanisms and implications. *Journal of Membrane Science*, **428**, 181–189.

Sivertsen E., Holt T., Thelin W. and Brekke G. (2012). Modelling mass transport in hollow fibre membranes used for pressure retarded osmosis. *Journal of Membrane Science*, **417**, 69–79.

Skilhagen S. E., Dugstad J. E. and Aaberg R. J. (2008). Osmotic power – power production based on the osmotic pressure difference between waters with varying salt gradients. *Desalination*, **220**, 476–482.

Stache K. (1989). Apparatus for transforming sea water, brackish water, polluted water or the like into a nutrious drink by means of osmosis. US Patent 4,879,030.

Stone M. L., Wilson A. D., Harrup M. K. and Stewart F. F. (2013). An initial study of hexavalent phosphazene salts as draw solutes in forward osmosis. *Desalination*, **312**, 130–136.

Su J. and Chung T.-S. (2011). Sublayer structure and reflection coefficient and their effects on concentration polarization and membrane performance in FO processes. *Journal of Membrane Science*, **376**, 214–224.

Su J., Yang Q., Teo J. F. and Chung T.-S. (2010a). Cellulose acetate nanofiltration hollow fiber membranes for forward osmosis processes. *Journal of Membrane Science*, **355**, 36–44.

Su J., Zhang S., Chen H., Chen H., Jean Y. and Chung T.-S. (2010b). Effects of annealing on the microstructure and performance of cellulose acetate membranes for pressure-retarded osmosis processes. *Journal of Membrane Science*, **364**, 344–353.

Su J., Chung T.-S., Helmer B. J. and de Wit J. S. (2012a). Enhanced double-skinned FO membranes with inner dense layer for wastewater treatment and macromolecule recycle using Sucrose as draw solute. *Journal of Membrane Science*, **396**, 92–100.

Su J., Zhang S., Ling M. and Chung T.-S. (2012b). Forward osmosis: an emerging technology for sustainable supply of clean water. *Clean Technologies and Environmental Policy*, **14**(4), 507–511.

Su J., Ong R. C., Wang P., Chung T. S., Helmer B. J. and Wit J. S. (2013). Advanced FO membranes from newly synthesized CAP polymer for wastewater reclamation through an integrated FO-MD hybrid system. *AIChE Journal*, **59**, 1245–1254.

Subramani A., Badruzzaman M., Oppenheimer J. and Jacangelo J. G. (2011). Energy minimization strategies and renewable energy utilization for desalination: a review. *Water Research,* **45**, 1907–1920.

Tan C. H. and Ng H. Y. (2008). Modified models to predict flux behavior in forward osmosis in consideration of external and internal concentration polarizations. *Journal of Membrane Science,* **324**, 209–219.

Tan C. H. and Ng H. Y. (2010). A novel hybrid forward osmosis-nanofiltration (FO-NF) process for seawater desalination: draw solution selection and system configuration. *Desalination and Water Treatment,* **13**, 356–361.

Tang C. Y., She Q., Lay W. C. L., Wang R. and Fane A. G. (2010). Coupled effects of internal concentration polarization and fouling on flux behavior of forward osmosis membranes during humic acid filtration. *Journal of Membrane Science,* **354**, 123–133.

Thombre A., Cardinal J., DeNoto A., Herbig S. and Smith K. (1999). Asymmetric membrane capsules for osmotic drug delivery: I. Development of a manufacturing process. *Journal of Controlled Release,* **57**, 55–64.

Thorsen T. and Holt T. (2009). The potential for power production from salinity gradients by pressure retarded osmosis. *Journal of Membrane Science,* **335**, 103–110.

USEPAE. (2009). 816-F-09-0004, National Primary Drinking Water Regulations. http://www.epa.gov/safewater/consumer/pdf/mcl.pdf

Valladares Linares R., Yangali-Quintanilla V., Li Z. and Amy G. (2012). NOM and TEP fouling of a forward osmosis (FO) membrane: foulant identification and cleaning. *Journal of Membrane Science,* **421–422**, 217–224.

van Reis R. and Zydney A. (2007). Bioprocess membrane technology. *Journal of Membrane Science,* **297**, 16–50.

Wang C.-Y., Ho H.-O., Lin L.-H., Lin Y.-K. and Sheu M.-T. (2005). Asymmetric membrane capsules for delivery of poorly water-soluble drugs by osmotic effects. *International Journal of Pharmaceutics,* **297**, 89–97.

Wang H., Chung T. S., Tong Y. W., Jeyaseelan K., Armugam A., Chen Z., Hong M. and Meier W. (2012). Highly permeable and selective pore-spanning biomimetic membrane embedded with aquaporin Z. *Small,* **8**, 1185–1190.

Wang K. Y., Yang Q., Chung T.-S. and Rajagopalan R. (2009). Enhanced forward osmosis from chemically modified polybenzimidazole (PBI) nanofiltration hollow fiber membranes with a thin wall. *Chemical Engineering Science,* **64**, 1577–1584.

Wang K. Y., Ong R. C. and Chung T.-S. (2010a). Double-skinned forward osmosis membranes for reducing internal concentration polarization within the porous sublayer. *Industrial & Engineering Chemistry Research,* **49**, 4824–4831.

Wang R., Shi L., Tang C. Y., Chou S., Qiu C. and Fane A. G. (2010b). Characterization of novel forward osmosis hollow fiber membranes. *Journal of Membrane Science,* **355**, 158–167.

Wang K. Y., Teoh M. M., Nugroho A. and Chung T.-S. (2011). Integrated forward osmosis–membrane distillation (FO–MD) hybrid system for the concentration of protein solutions. *Chemical Engineering Science,* **66**, 2421–2430.

Warne B., Buscall R., Mayes E., Oriard T. and Norris I. (2008). Water purification method. GB Patent 2,464,956.

Widjojo N., Chung T.-S., Weber M., Maletzko C. and Warzelhan V. (2011). The role of sulphonated polymer and macrovoid-free structure in the support layer for thin-film composite (TFC) forward osmosis (FO) membranes. *Journal of Membrane Science,* **383**, 214–223.

Yaeli J. (1992). Method and apparatus for processing liquid solutions of suspensions particularly useful in the desalination of saline water. US Patent 5098575 A.

Yang Q., Wang K. Y. and Chung T.-S. (2009a). Dual-layer hollow fibers with enhanced flux as novel forward osmosis membranes for water production. *Environmental Science & Technology,* **43**, 2800–2805.

Yang Q., Wang K. Y. and Chung T.-S. (2009b). A novel dual-layer forward osmosis membrane for protein enrichment and concentration. *Separation and Purification Technology,* **69**, 269–274.

Yangali-Quintanilla V., Li Z., Valladares R., Li Q. and Amy G. (2011). Indirect desalination of Red Sea water with forward osmosis and low pressure reverse osmosis for water reuse. *Desalination,* **280**, 160–166.

Yap W. J., Zhang J., Lay W. C., Cao B., Fane A. G. and Liu Y. (2012). State of the art of osmotic membrane bioreactors for water reclamation. *Bioresource Technology,* **122**, 217–222.

Yen S. K., Mehnas Haja N. F., Su M., Wang K. Y. and Chung T.-S. (2010). Study of draw solutes using 2-methylimidazole-based compounds in forward osmosis. *Journal of Membrane Science,* **364**, 242–252.

Yip N. Y., Tiraferri A., Phillip W. A., Schiffman J. D., Hoover L. A., Kim Y. C. and Elimelech M. (2011). Thin-film composite pressure retarded osmosis membranes for sustainable power generation from salinity gradients. *Environmental Science & Technology,* **45**, 4360–4369.

Yong J. S., Phillip W. A. and Elimelech M. (2012). Coupled reverse draw solute permeation and water flux in forward osmosis with neutral draw solutes. *Journal of Membrane Science,* **392–393**, 9–17.

Zhang S., Wang K. Y., Chung T.-S., Chen H., Jean Y. C. and Amy G. (2010). Well-constructed cellulose acetate membranes for forward osmosis: minimized internal concentration polarization with an ultra-thin selective layer. *Journal of Membrane Science,* **360**, 522–535.

Zhang S., Wang K. Y., Chung T.-S., Jean Y. and Chen H. (2011). Molecular design of the cellulose ester-based forward osmosis membranes for desalination. *Chemical Engineering Science,* **66**, 2008–2018.

Zhang S., Fu F. and Chung T.-S. (2013). Substrate modifications and alcohol treatment on thin film composite membranes for osmotic power. *Chemical Engineering Science,* **87**, 40–50.

Zhao S. and Zou L. (2011a). Effects of working temperature on separation performance, membrane scaling and cleaning in forward osmosis desalination. *Desalination,* **278**, 157–164.

Zhao S. and Zou L. (2011b). Relating solution physicochemical properties to internal concentration polarization in forward osmosis. *Journal of Membrane Science,* **379**, 459–467.

Zou S., Gu Y., Xiao D. and Tang C. Y. (2011). The role of physical and chemical parameters on forward osmosis membrane fouling during algae separation. *Journal of Membrane Science,* **366**, 356–362.

Zhao S., Zou L. and Mulcahy D. (2012a). Brackish water desalination by a hybrid forward osmosis–nanofiltration system using divalent draw solute. *Desalination,* **284**, 175–181.

Zhao S., Zou L., Tang C. Y. and Mulcahy D. (2012b). Recent developments in forward osmosis: opportunities and challenges. *Journal of Membrane Science,* **396**, 1–21.

Zydney A. L. (1997). Stagnant film model for concentration polarization in membrane systems. *Journal of Membrane Science,* **130**, 275–281.

Chapter 5.2

Cellulose acetate membrane: minimized internal concentration polarization*

5.2.1 INTRODUCTION

Water shortage is a serious problem facing the world nowadays due to overpopulation and environmental pollution (Semiat, 2008). As new technologies continue to emerge to reduce the cost of desalination of seawater, more and more countries turn to membrane desalination to address their water scarcity (Kranhold). Many membrane processes have been utilized, such as ultrafiltration (UF), microfiltration (MF), nanofiltration (NF), membrane distillation (MD) and reverse osmosis (RO). In recent years, forward osmosis (FO) has gained much attention as an alternative desalination and water recycle process. Driven by an osmotic pressure gradient, FO does not require significant energy input. In addition, it may offer the advantages of high rejection of a wide range of ion contaminants and lower membrane-fouling propensities than traditional pressure-driven membrane processes (Mi & Elimelech, 2010).

While the osmotic phenomenon was discovered long time ago, the concept of FO had not been proposed until 1970's (Kessler & Moody, 1976). By means of a concentrated draw solution running against a dilute feed solution on the other side of a semi-permeable membrane, water is drawn automatically from the feed to the draw side due to the osmotic pressure gradient generated by the concentration difference between the two streams. Along with the fast development in polymeric membranes for separation processes in the past decade, FO has gained vast popularity among researchers in the areas of food processing (Petrotos & Lazarides, 2001; Jiao *et al.* 2004), pharmaceutical concentration (Yang *et al.* 2009c), power generation (Seppälä & Lampinen, 1999; Loeb, 2001; McGinnis *et al.* 2007), portable water reuse (Cath *et al.* 2005), and desalination (McCutcheon *et al.* 2005; Semiat, 2008).

---

* This chapter is based, with permission from the copyright holder, on the previously published paper. (Zhang S., Wang K. Y., Chung T.-S., Chen H., Jean Y. C. and Amy G. (2010). Well-constructed cellulose acetate membranes for forward osmosis: minimized internal concentration polarization with an ultra-thin selective layer. *Journal of Membrane Science*, **360**, 522–535. doi:10.1016/j.memsci.2010.05.056)

The development of appropriate membranes specifically for FO, however, remains one of the major challenges for researchers. Currently used membranes for FO include conventional RO membranes made from thin-film polymerization on a polysulfone layer supported by nonwoven fabrics, HTI (Hydration Technologies Inc., OR, USA) FO membranes made of cellulose triacetate coated on polyester mesh, both of which have flat sheet configurations, and hollow fiber membranes developed recently by our group (Wang *et al.* 2007; Wang *et al.* 2009; Yang *et al.* 2009b). All these membranes have a non-selective and low-porosity support layer rendering them suffer from severe internal concentration polarization (ICP) which greatly reduces the effective driving force (Loeb *et al.* 1997; McCutcheon & Elimelech, 2006,2007) and results in a significant loss in water flux. With these membranes, the water flux in the FO mode, where dilutive ICP occurs since the porous support faces the draw solution, is much lower than that in the pressure retarded osmosis (PRO) mode where the draw solution runs against the selective layer. A porous support layer, nevertheless, is necessary so as to sustain desirable mechanical strength of the membrane. New explorations and developments in phase inversion and membrane formation processes, therefore, are urgently needed to fabricate membranes with a thin and fully porous support layer that can minimize ICP effect, while at the same time maintaining a thin and dense selective layer for sufficient water flux and salt rejection.

Ever since Loeb and Sourirajan (1963) invented the phase inversion process to fabricate asymmetric polymeric membranes more than 40 years ago, cellulose acetate (CA) has been a popular material for various separation applications, especially for RO (Manjikian, 1967; Pilon *et al.* 1971; Sourirajan *et al.* 1985). It possesses unique advantages such as relative hydrophilicity which favors high water flux and low fouling propensity, good mechanical strength and wide availability. Meanwhile, in a world where environmental protection becomes more and more important nowadays, it is also notable that CA is a green polymer as it derives from acetyl substitution of cellulose, which is one of the most common natural organic compounds on earth. Moreover, the membrane formation process for CA has served for long as a model for the study of phase inversion mechanism from the day it was invented (Wijmans *et al.* 1983; Yilmaz & McHugh, 1988; Tsay & McHugh, 1990). In addition, recent data show that CA has a less fouling propensity than the conventional polyamide in FO processes (Mi & Elimelech, 2010). Hence, it is worthwhile to investigate further into the phase inversion of CA polymers to explore new progress that might have been overlooked in the past and yet can be good supplement to the traditional theories, and also to search for better design of membrane structure for FO processes for the ultimate purpose of seawater desalination.

So far, nearly all existing studies about phase inversion focus on the formation of a dense selective layer on the top of asymmetric membranes (Wijmans *et al.* 1983; Yilmaz & McHugh, 1988; Tsay & McHugh, 1990). This dense top layer is formed due to the fast evaporation and outflow of the solvent which is often acetone in the case of CA, when the as-cast membrane is immersed in a coagulant bath, which is normally water. It is usually considered that the selectivity of a membrane is determined by the pore size and pore size distribution of this dense layer. Much less attention has been paid to the formation mechanism of the bottom layer. It is either neglected or considered to be the same as the middle porous part. Limited discussions have been reported in literatures, which noticed that the bottom surface might be more hydrophilic than the top due to hydrophilic-hydrophilic interaction of the polymers with the glass plate during membrane casting (Wang *et al.* 2008; Jiang & Chung, 2010). It is of great interest, however, to ask whether the interaction with the casting substrate of different chemistry would affect the pore structure of the bottom layer and how substantial the influence can be. Meanwhile, the formation of a fully porous layer in the middle is another issue that deserves attention. By careful choosing of the solvent and experimental conditions, an interconnected, open-cell porous structure can be achieved. Little resistance towards solute diffusion would then be encountered within this structure, which is one of the most important properties highly needed for FO membranes.

In this paper, CA is employed as the polymeric material for the fabrication of superior membranes that are structurally feasible for FO desalination. The phase inversion process of CA is studied by the introduction of different casting substrates and coagulant baths. The specific objectives of this paper are to reveal the role of substrate chemistry on the structure of bottom layer and to design membranes that have good rejection, desirable water flux and minimized ICP for FO process.

5.2.2 EXPERIMENTAL

5.2.2.1 Materials

CA-398–10 supplied by Eastman Chemical Company was studied in this work. The basic physicochemical properties of CA-398–10 obtained from the company are listed in Table 5.2.1. Polymers were dried in 80°C vacuum oven prior to use. N-Methyl-2-pyrolidone (NMP, 99.5%, Merck) and acetone (99.5%, TEDIA) were used as the solvents. Magnesium chloride (98%, Alfa Aesar) and sodium chloride (99.5%, Merck) were employed to prepare feed and draw solutions. Solutions characteristics, such as osmotic pressure, density and dynamic viscosity were calculated using OLI software (OLI systems, Inc.). Solute diffusivity used is from reference (Lobo, 1993). To do the fouling test, aluminum oxide nanoparticles (<50 nm, Sigma Aldrich) were agitatedly stirred in deionised (DI) water at a concentration of $5\ g \cdot L^{-1}$ for 48 h and then sonicated for 2 h. The particle size of the obtained colloidal solution was found by dynamic light scattering measurements (Zetasizer Nano System, Malvern, United Kingdom) to be approximately 200 nm.

Table 5.2.1 The physicochemical properties of the CA-398–10 polymer and substrates.

Substance	Content (%)		Density ($g \cdot cm^{-3}$)	δsp* ($cal^{1/2} \cdot cm^{-3/2}$)	Water Contact Angle (°)	M_n	Intrinsic Viscosity ($dL \cdot g^{-1}$)
	Hydroxyl	acetyl					
CA-398–10	3.5	39.8	1.31	12.7	71.5 ± 2.4	40,000	1.32
Teflon plate	–	–	–	–	107.2 ± 1.4	–	–
Glass plate	–	–	–	–	43.9 ± 1.0	–	–

*Solubility parameter, calculated according to reference (Barton, 1991).

5.2.2.2 Membrane preparation

Polymer solutions were prepared by dissolving CA-398–10 (22.5 wt%) in a mixed solvent of acetone (5 wt%) and NMP (72.5 wt%). Solutions were then cast on a glass plate with a 50 μm casting knife, followed by immediate immersion in a tap water bath at 25 °C (unless otherwise specified). After removal of residual solvents, membranes were ready for use or underwent heat treatment prior to use. To investigate into the effect of different substrates on membrane morphology, another two sets of membrane casting conditions were tried: (1) CA-398–10 solution was dispensed and cast on a Teflon plate and then the whole plate was immersed in tap water; (2) CA-398–10 solution was cast on a glass plate, followed by immediate immersion in a NMP/water (v:v = 40:60, 80:20 and 90:10, respectively) mixed bath for 3 seconds, and then placed in a tap water bath.

5.2.2.3 Forward osmosis and fouling tests

FO experiments were conducted as described in reference (Wang *et al.* 2010). The permeation cell has one water channel with 9.0 cm in length, 1.0 cm in width and 1.0 cm in height on each side of the membrane. The volume flow rate was $0.2\ L \cdot min^{-1}$ (corresponding to the linear velocity $3.3\ cm \cdot s^{-1}$) for both feed and draw solutions which flow co-currently through the cell in the cross-flow mode. The temperature was maintained at 22 ± 0.5 °C. NaCl and $MgCl_2$ solutions of different concentrations were prepared as the draw solution. The water permeation flux (J_v, $L \cdot m^{-2} \cdot h^{-1}$, abbreviated as LMH) was calculated as below:

$$J_s = \frac{\Delta V}{A \Delta_t} \tag{5.2.1}$$

where ΔV (L) is the permeation water collected over a predetermined time Δ_t (h) in the duration of tests; A is the effective membrane surface area (m^2).

When deionized (DI) water was used as the feed, conductivity was recorded and salt leakage, the salt back-diffusion from the draw solution to the feed, J_s in

$g \cdot m^{-2} \cdot h^{-1}$ (abbreviated as gMH), was therefore determined from the increase in conductivity of the feed:

$$J_s = \frac{\Delta C_t V_t}{A \Delta_t} \tag{5.2.2}$$

where C_t and V_t are the salt concentration and volume of the feed at the end of tests, respectively.

Fouling tests were done in a similar way as forward osmosis, with 2 M NaCl as the draw solution which faces the dense selective layer, and 5 $g \cdot L^{-1}$ aluminum oxide colloidal solution as the feed.

5.2.2.4 Pure water permeability, salt rejection and salt permeability tests

The pure water permeability A (PWP) and salt rejection of different CA membranes were tested at 22 ± 0.5 °C with their denser layers facing the feed. PWP measurements were conducted with Milli-Q DI water and a trans-membrane pressure of 5.0 bar. For salt rejection, 200 ppm NaCl (or $MgCl_2$) solution was used and the trans-membrane pressure was 5.0 bar. The concentrations of NaCl in the feed (C_f) and permeate (C_p) were determined by conductivity measurements. Salt rejection, R was calculated according to Equation (5.2.3):

$$R = \left(1 - \frac{C_p}{C_f}\right) \times 100\% \tag{5.2.3}$$

To get the salt (NaCl) permeability B of the membranes, water permeability and salt rejection were tested under different hydraulic pressures. B was then determined by a linear fitting based on Equation (5.2.4):

$$\frac{1-R}{R} = \frac{1}{(\Delta P - \Delta \pi)A} B \tag{5.2.4}$$

where ΔP is the hydraulic pressure applied and $\Delta \pi$ is the osmotic pressure difference between the feed and permeate.

5.2.2.5 Pore size and pore size distribution

The pore size and pore size distribution of CA membranes were characterized by solute transport experiments according to the procedure described in references (Wang *et al.* 2004; Yang *et al.* 2009a). Generally, PEG, PEO and PVP polymers, and small molecules including raffinose, sucrose, glucose and glycerol with different molecular weights were employed and the solute rejection of the membranes were tested against them under a hydraulic pressure of 5.0 bar. The solute diameter d_s

is related to the molecular weight M of the polymer as described in the following equations:

For PEG:

$$d_s = 33.46 \times 10^{-12} \times M^{0.557} \tag{5.2.5}$$

For PEO:

$$d_s = 20.88 \times 10^{-12} \times M^{0.587} \tag{5.2.6}$$

For PVP:

$$d_s = 16.8 \times 10^{-12} \times M^{0.593} \tag{5.2.7}$$

For small molecules:

$$\log \frac{d_s}{2} = -1.4962 + 0.4654 \log MW \tag{5.2.8}$$

When the solute rejection R is plotted versus d_s on a log-normal probability paper, a straight line is yielded and the mean effective pore size μ_p (diameter) is found at $R = 50\%$, and the geometric standard deviation σ_p is found as the ratio of d_s at $R = 84.13\%$ and 50%. Then the pore size distribution of the membrane can be expressed as following:

$$\frac{dR(d_p)}{dd_p} = \frac{1}{d_p \ln \sigma_p \sqrt{2\pi}} \exp\left[-\frac{(\ln d_p - \ln \mu_p)^2}{2(\ln \sigma_p)^2} \right] \tag{5.2.9}$$

5.2.2.6 Water contact angle

Water contact angles of the casting substrates and membranes dried by a freeze dryer (ModulyoD, Thermo Electron Corporation, USA) were measured with Milli-Q DI water using a Contact Angle Geniometer (Rame Hart, USA) at 22 ± 0.5 °C. Ten measurements were made and results were averaged for each substance.

5.2.2.7 Porosity *P*

For the measurement of porosity, wet membranes were taken out from the water bath followed by careful and quick removal of the excess water on the surface by tissue paper. The membrane were then weighted (m_1, g), freeze dried overnight, and re-weighted (m_2, g). The water content is thus calculated as $m_1 - m_2$, and the dry weight of the membrane is m_2. Since the densities of both water (ρ_w, 1.00 g·cm^{-3}) and CA (ρ_p, 1.31 g·cm^{-3}) are known, their volumes can be calculated separately, and the overall porosity P (%) is then obtained:

$$P = \frac{\frac{m_1 - m_2}{\rho_w}}{\frac{m_1 - m_2}{\rho_w} + \frac{m_2}{\rho_p}} \times 100 \tag{5.2.10}$$

5.2.2.8 Field emission scanning electronic microscopy (FESEM)

Membranes were dried in a freeze-dryer, fractured in liquid nitrogen and sputtered with platinum by a Jeol JFC-1100E Ion Sputtering device for morphology study. The membrane morphology was observed using a FESEM (JEOL JSM-6700).

5.2.2.9 Atomic force microscope (AFM)

The surface roughness was studied by AFM (Nanoscope IIIa, Digital Instrument, USA). A tapping mode was operated at room temperature, and images in the range of 500 nm × 500 nm were obtained. The mean roughness (*Ra*) was used for the membrane surface characterization.

5.2.2.10 Positron annihilation spectroscopy (PAS)

The membranes cast on glass plates and phase transited in water were further studied by Doppler broadening energy spectra (DBES) and positron annihilation lifetime spectroscopy (PALS) using PAS coupled with a slow positron beam to investigate the depth profile of free volume and its distribution. When the top layer is oriented toward the beam and studied, it is termed as the top mode; otherwise it is the bottom mode. The DBES spectra were recorded using an HP Ge detector at a counting rate of approximately 4800 cps, and the total number of counts for each spectrum was 1.0 million. The DBES spectra were characterized by S parameters, which were measured as a function of positron implantation energy from 100 eV to 30 keV at room temperature. The incident position energy can be expressed in terms of depth, as given in the following equation:

$$Z(E_+) = \frac{40}{\rho} E_+^{1.6} \tag{5.2.11}$$

where Z is the depth in nm, ρ is the density of the polymer material in $g \cdot cm^{-3}$, and E_+ is the incident positron energy in keV (Chen *et al.* 2007; Jean *et al.* 2008). The S parameter is defined as the ratio of integrated counts between 510.3 and 511.7 keV, which represents the low momentum part of positron-electron annihilation radiation, to the total counts after the background is deducted. Therefore, it is sensitive to the changes of the positron and positronium states resulted from microstructural changes. When the positrons and Ps are confined in a free volume with a finite size, the *S* parameter is a measure of the momentum broadening according to the uncertainty principle. A larger free volume leads to a larger *S* parameter value. The *S* parameter has been successfully used to detect the free-volume depth profile in polymeric systems (Lo *et al.* 2009; Tung *et al.* 2009). Meanwhile, the *R* parameter which provides information about large holes (nm to μm) where o-Ps undergoes 3γ annihilation, is also reported in this work. The *R* parameter is defined as the ratio of the total counts from the valley region with an energy width between 364.2 and 496.2 keV (from 3γ annihilation) to those from the 511 keV peak region with a width between 504.35 and 517.65 (from 2γ annihilation). In the PALS experiments, different incident positron energy was performed at a counting rate of 100–500 counts per second and each spectrum contains 2 million counts. The data were fitted into three lifetimes using the PATFIT program and also into continuous lifetime distributions using MELT. The o-Ps lifetime τ_3 decreases when it is trapped in molecular "holes" and annihilated with electrons in the walls of the holes. It is reported to be in the order of 1–5 ns and is used to calculate the mean free volume radius (Jean, 1990).

5.2.3 RESULTS AND DISCUSSION

5.2.3.1 Morphology influenced by different substrates and phase inversion conditions

Figure 5.2.1 shows the morphology of the CA membrane cast on a glass plate and immediately immersed in water. Its cross-section has a double dense-layer structure consisting of two relatively dense morphologies in both top and bottom skins and a fully porous middle layer. The top relatively dense structure may be attributed to rapid acetone evaporation during the cast and subsequent solvent outflow to the surrounding water when immersing the nascent membrane into the coagulation bath. As a result, the polymer concentration at the top skin increases rapidly and forms a relatively dense outer skin. The fully porous open-cell may arrive from the fact that water diffuses in faster than solvents (acetone and NMP) outflow as shown in Table 5.2.2 (Wilke & Chang, 1955) and the CA-398–10 solution tolerates much water as illustrated in Figure 5.2.2. Consequently, a delayed demixing dominates in the membrane bulk and results in a fully porous inner structure (Chung *et al.* 1997; Wang *et al.* 2010).

Interestingly, a very thin dense bottom layer (measured to be 95 ± 7 nm) is also observed in Figure 5.2.1, which is confirmed by the surface morphology shown in Figure 5.2.3 where both top and bottom surfaces are smooth and dense without any visible pores under a magnification of 10 K. The phase inversion process of the bottom layer, therefore, must be different from the middle part. It can be seen from Table 5.2.1 that the CA polymer used is hydrophilic. It preserves a hydroxyl content of 3.8%, and hence has a relatively high solubility parameter of 12.7 cal$^{1/2}$ cm$^{-3/2}$ (Barton, 1991) as well as a low water contact angle of 71.5 ± 2.4°. It is hypothesized that, during phase inversion, CA molecules near the bottom surface are in direct contact with the glass plate. As the glass is hydrophilic, hydrophilic CA molecules would adhere upon the glass surface and aggregate to form the bottom skin during the membrane formation process.

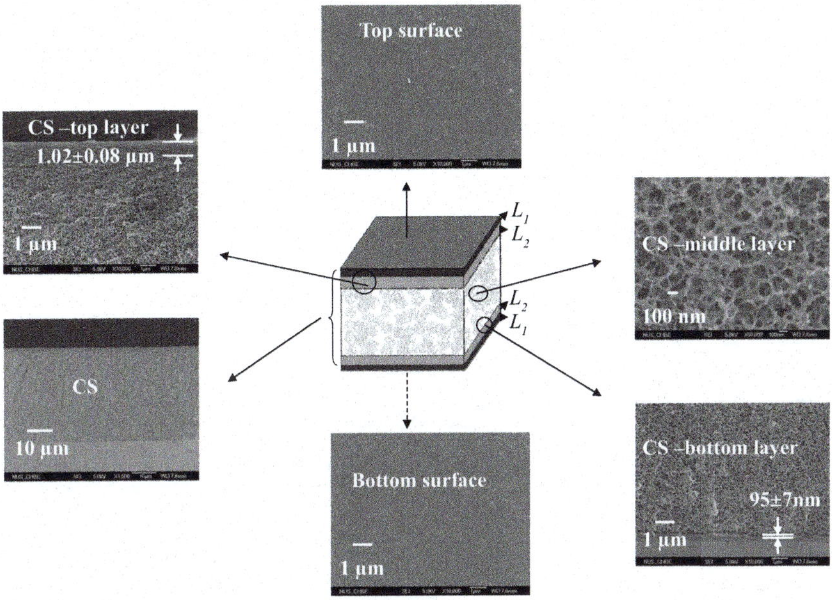

Figure 5.2.1 Schematic diagram and FESEM images of the nascent CA membrane cast on glass plate and phase transited in water.

Table 5.2.2 The physicochemical properties of solvents and non-solvents.

Solvent	NMP	Acetone	Water	NMP/Water (v:v)		
				40/60	80/20	90/10
Density (g · cm^{-3}) (20 °C, 1 atm)	1.028	0.792	0.998	–		
Viscosity (cP) (20 °C)	1.7	0.32	0.89	–		
$D_{w\text{-}s} \times 10^6$ (cm^2 · s^{-1}) (20 °C)*	18	88.6	–	–		
$D_{s\text{-}w} \times 10^6$ (cm^2 · s^{-1}) (20 °C)**	8.9	11.7	–	–		
$D_{s\text{-}w}/D_{w\text{-}s}$	0.494	0.132	–	–		
δsp (cal$^{1/2}$ · cm$^{-3/2}$)	11.8	9.77	23.5	22.20	18.49	14.88

*The diffusion coefficient of water in almost pure solvent.

**The diffusion coefficient of solvent in almost pure water.

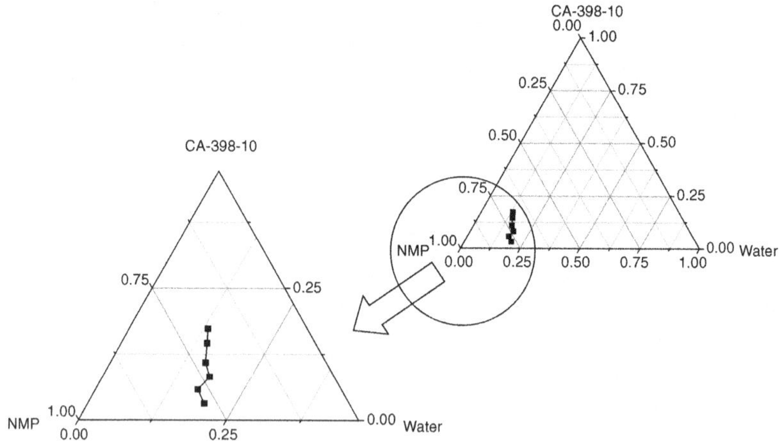

Figure 5.2.2 The phase diagram of CA-398–10.

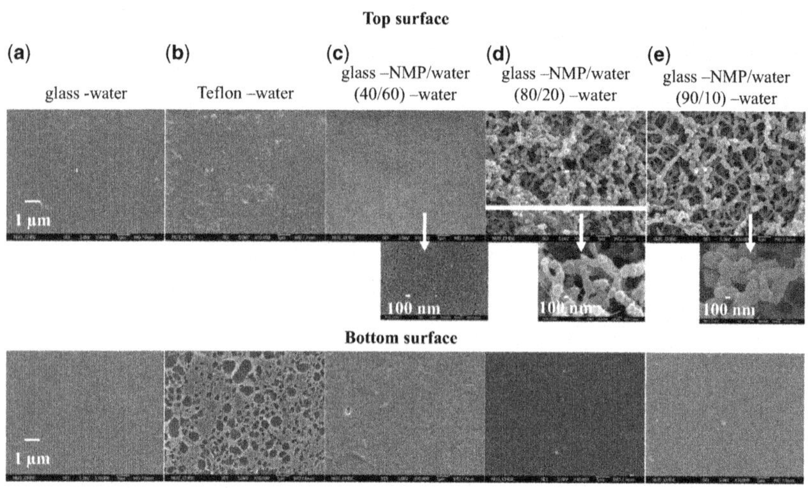

Figure 5.2.3 Surface morphology of fresh CA membranes prepared under different substrates-coagulant combinations: (a) glass – water, (b) Teflon – water, (c) glass – NMP/water (40/60) – water, (d) glass – NMP/water (80/20) – water and (e) glass – NMP/water (90/10) – water.

Several experiments were conducted to confirm our hypothesis and investigate the effect of substrate hydrophobicity on bottom surface morphology using the same CA solution as described in Section 5.2.2.2. As displayed in Figures 5.2.3 and 5.2.4, no significant morphological change can be observed at the top surface and cross section when a Teflon plate is employed as the casting substrate, whereas the bottom surface becomes rather porous. It implies that the formation mechanism near the bottom surface has been greatly changed. Since Teflon is fairly hydrophobic, during phase inversion, hydrophilic CA molecules tend to stay away from the plate, interact closely with the solvent and coagulant water, and finally undergo delay demixing which results in a porous structure at the bottom. Clearly, the hydrophobicity of the casting substrate plays an important role influencing the membrane morphology of the bottom surface.

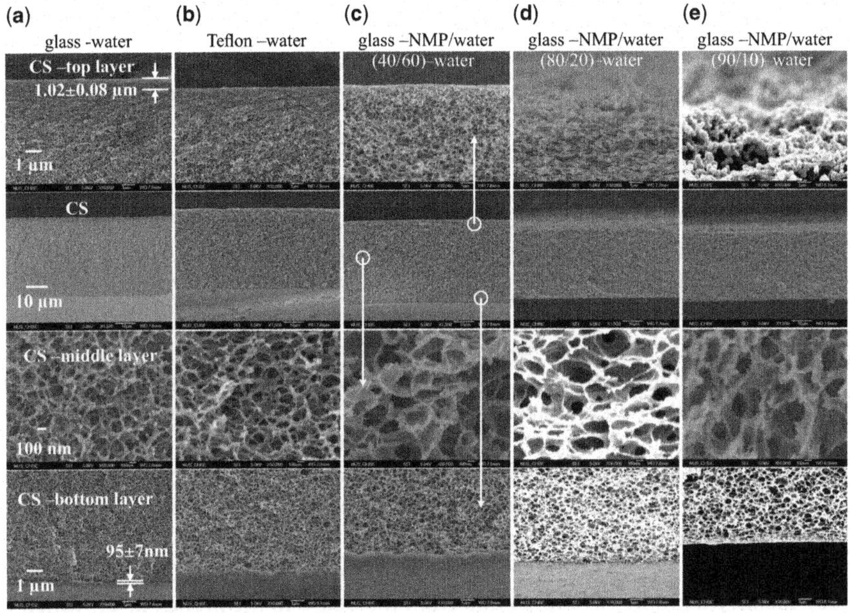

Figure 5.2.4 Cross-section morphology of fresh CA membranes prepared under different substrates and different coagulants: (a) glass – water, (b) Teflon – water, (c) glass – NMP/water (40/60) – water, (d) glass – NMP/water (80/20) – water and (e) glass – NMP/water (90/10) – water.

In addition, when immersing the as-cast membrane on a glass plate into various NMP/water mixed coagulant baths for 3 seconds prior to complete phase inversion in water, the top surface of CA membranes becomes more porous with an increase in NMP content while the bottom surfaces still maintain smooth and dense. The increase in porosity of the top surface is due to the delayed demixing mechanism, while the bottom layer stays intact due to the short dipping period and relatively long phase inversion process for a CA polymer solution to precipitate in a NMP/water mixture. At a magnification of 10 K, Figure 5.2.3 shows no notable distinctions in the bottom surface of the samples a, c, d and e where samples were cast on glass plates but experienced different coagulants chemistry (i.e., water or NMP/water mixtures). The asymmetric CA membranes phase inversed in a bath of high NMP/water ratio and then in water may have great potential for forward osmosis application because of less transport resistance as discussed in the later sessions.

5.2.3.2 Morphology characterized by PALS

Figure 5.2.5 shows the *S* parameter versus positron incident energy or the corresponding mean depth of the Figure 5.2.1 sample (i.e., cast on a glass plate and immediately immersed in water). It can be observed that near the surface *S* increases dramatically as the positron energy increases, which is known to be due to the back diffusion of positronium in polymeric materials in the range of around 1–20 nm. Meanwhile, a maximum of *S* is noted before it decreases till a plateau, which is an indication of a multilayer structure in the membrane (Chen *et al.* 2007; Jean *et al.* 2008; Tung *et al.* 2009). The computer program VEPFIT is employed to simulate the layer structure. For both the top and bottom modes, three-sub-layer models fit the *S* parameters well and give stable results which are shown in Table 5.2.3. For the top mode, a sub-layer with a thickness of 852 ± 480 nm is identified as the dense skin layer. The transition sub-layer is found to have a much larger length of more than 2 μm, which is in between the top dense layer and the bulk porous support below it. The third sub-layer is indentified as the porous support. Similar results are obtained with the bottom mode. The thickness of the dense layer is fitted to be 98 ± 67 nm, much thinner than the top. An intermediate sub-layer of around 2 μm as well as a porous support is also found. The simulated results agree well with visual observations from FESEM images, and thus confirm the existence a dense layer at the bottom; moreover, they identify a progressive transition from the dense to the porous layer, which is an understandable product of phase inversion and yet has not been pointed out by other characterization techniques.

The three-sub-layer structure and the relative thickness of each sub-layer are even more evident in the depth profile of *R* parameters (also shown in Figure 5.2.5). *R* is a measure of the relative amount of 3γ annihilation that could only be attributed to o-Ps in vacuum or large pores. In the top mode, the fast decrease of *R* near the surface is not well understood yet but may be due to the back diffusion of positroniums (Tung *et al.* 2009). Then, a small-scale plateau with the smallest *R* values appears till around 800 nm, which corresponds to the dense skin layer. There is a gradual increase of *R* from about 800 nm to 2 μm, which is identified as the transition layer. Later, a new plateau appears showing the contributions from 3γ annihilation of o-Ps in large pores (μm). Similar phenomena are also presented in the bottom mode. Therefore, the double dense-layer structure of the membrane cast on glass plate and phase transited in water is again evidently proved.

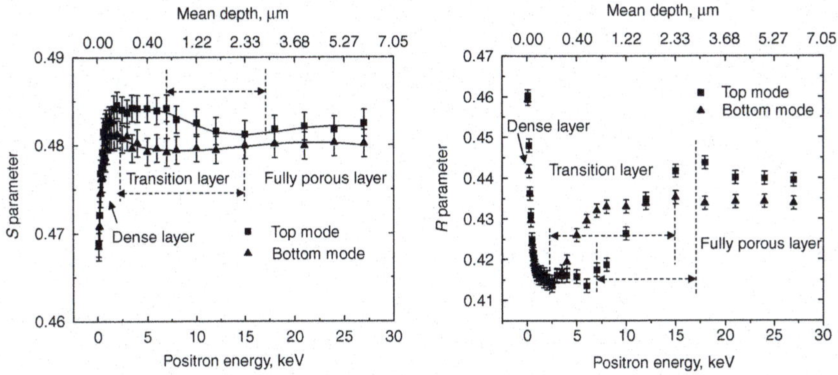

Figure 5.2.5 *S* (left) and *R* (right) parameters versus position incident energy (or depth) in fresh CA membranes cast on glass plate and phase transited in water. The lines for *S* parameters are fitted results from VEPFIT analysis in three-sub-layer model. Top mode refers to the characterization of the top layer of the membrane, and bottom mode refers to the characterization of the bottom layer of the membrane.

Table 5.2.3 *S* parameters and thickness of the sub-layers of the nascent CA membranes prepared on glass plate and immersed in water.

Mode*	S_1	S_2	S_3	L_1	L_2**
Top	0.4841 ± 0.0003	0.4783 ± 0.0016	0.4830 ± 0.0012	852 ± 530	2778 ± 1792
Bottom	0.4824 ± 0.0006	0.4793 ± 0.0004	0.4813 ± 0.0009	98 ± 67	2210 ± 1583

*Top mode refers to the characterization of the top layer of the membrane facing the air, and bottom mode refers to the characterization of the bottom layer of the membrane facing the glass plate.

**L_1 and L_2 are the boundary lengths of the dense and transition sub-layers, respectively. A schematic diagram of the layer and sub-layer structure is illustrated in Figure 5.2.1.

5.2.3.3 Pore size and pore size distribution

Even though both the top and bottom layers of the CA membrane cast on a glass plate and phase inversion in water have dense structures, it is, however, unclear which layer is denser to determine the membrane selectivity. The pore size distribution of the top layer can be approximately characterized by the membrane cast on a Teflon plate because it has the same phase inversion conditions at the top surface but with a fully porous bottom layer. Table 5.2.4 shows that the mean effective pore size of the nascent membrane cast on the Teflon plate is about 12.5 nm, while Figure 5.2.6 displays its relatively wide pore size distribution. Since its molecular weight cut-off (MWCO) is 257.2 kDa, much larger than salt ions, the top layer does not have sufficient selectivity between water and salts for FO application. Meanwhile, the pore size of the double dense-layer membrane is also characterized, which turns out to be 0.65 nm as shown in Table 5.2.4 and Figure 5.2.6. The value is more than one-magnitude lower than that of the top layer. Therefore, the conclusion can be safely made that the bottom layer formed due to hydrophilic-hydrophilic interaction of the polymer with the glass plate is much denser than the top layer which is usually considered to be the sole selective layer in traditional phase inversion theories.

Table 5.2.4 Mean effective pore size (μ_p), geometric standard deviation (σ_p) and cut-off (MWCO) of different CA membranes.

Substrate – Coagulant	Annealing	μ_p (nm)	σ_p	MWCO (Da)
Teflon plate – water	no	12.5	2.08	257,200
Teflon plate – water	yes	9.6	1.36	87,300
Glass plate – water	no	0.65	1.46	428

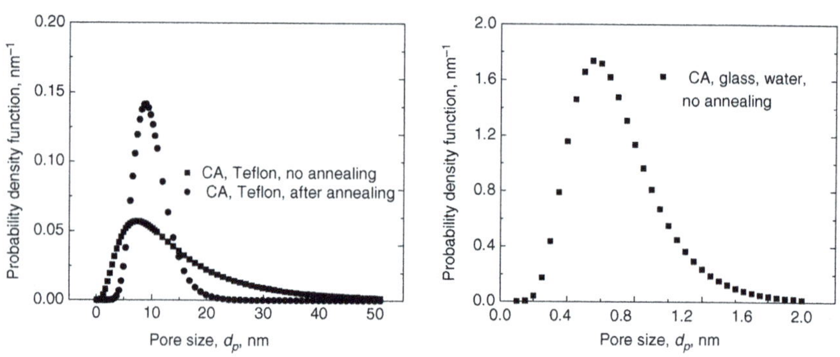

Figure 5.2.6 The probability density function curves of different CA membranes. (Left) membranes cast on Teflon plate before and after heat annealing in 90 °C water bath for 15 min. (Right) nascent membranes cast on glass plate and phase transited in water.

5.2.3.4 PWP, NaCl rejection and FO performance of different membranes

The PWP and NaCl rejections of the nascent CA membranes cast under different conditions are compared in Table 5.2.5. The membrane cast on a Teflon plate has a large PWP of 139.8 LMH · bar^{-1}, with no rejection towards NaCl or $MgCl_2$ at all. Whereas for the membrane cast on a glass plate and immersed directly in water, the PWP is 1.15 LMH · bar^{-1} and the rejections for NaCl and $MgCl_2$ are 51.2% and 79.8%, respectively. These results are consistent with the previous discussion and support the top layer to be denser than the bottom as follows.

Table 5.2.5 Pure water permeability and salt rejection of CA membranes prepared under different phase inversion conditions before and after heat treatment. The hydraulic pressure applied is 5 bar, and for salt rejection test, 200 ppm NaCl or $MgCl_2$ is used as the feed. All membranes are tested with denser layer facing the feed.

Substrate	Coagulant	Without Heat Treatment			Annealed at 90 °C for 15 min		
		PWP ($LMH \cdot bar^{-1}$)	NaCl Rejection (%)	$MgCl_2$ Rejection (%)	PWP ($LMH \cdot bar^{-1}$)	NaCl Rejection (%)	$MgCl_2$ Rejection (%)
Glass plate	Water	1.15 ± 0.10	51.2 ± 1.1	81.5 ± 1.6	0.17 ± 0.01	92.0 ± 1.0	99.0 ± 0.7
Glass plate	NMP/water 90/10	1.11 ± 0.13	57.3 ± 2.1	85.4 ± 2.3	0.13 ± 0.02	87.4 ± 1.8	96.7 ± 1.4
Teflon plate	Water	139.8 ± 1.4	0	0	50.0 ± 1.7	0	0

The water transport resistance cross the double dense-layer membrane can be described by the following equation:

$$\frac{1}{A} = \frac{1}{A^{top}} + \frac{1}{A^{mid}} + \frac{1}{A^{bot}} \tag{5.2.12}$$

where A is the PWP of the double dense-layer membrane, A^{top}, A^{mid} and A^{bot} are the PWP of the top, middle and bottom layers, respectively. Since the middle porous layer has almost no resistance, the term $1/A^{mid}$ can be ignored. According to the CA membrane cast on a Teflon plate, the PWP of the top layer is 139.8 LMH · bar^{-1}, which is several orders larger than that of the double dense-layer membrane, the term $1/A^{top}$ can be also neglected. Therefore, the PWP of the double dense-layer membrane can be regarded as that of the bottom layer. Additionally, the PWP and salt rejection of the membrane immersed in a 90:10 (v:v) NMP/water mixed coagulant bath (and hence has only a single dense layer at the bottom) are very close to that of the double dense-layer one, further proving that the bottom layer is predominant in the resistance and selectivity of the double dense-layer membrane.

However, for the membranes to be applied in FO processes, a higher salt rejection must be achieved. Heat treatment of CA membranes has already been widely practiced for several decades to densify the membrane structure and enhance the rejection in RO processes for desalination (Kopeček & Sourirajan, 1969; Wang *et al.* 1994; Su *et al.* 2010). Annealing as-cast membranes in a 90 °C water bath for 15 min is chosen in this work to enhance their rejection towards NaCl. It is noticed from Figure 5.2.6 that the pore size distribution of the top layer of the membrane cast on a Teflon plate becomes much sharper after annealing. In addition to shrinking pores to smaller ones, heat treatment has reconstructed the whole dense layer structure through polymer chain relaxation and rearrangement to make the pore size distribution narrower in a macroscopic view. In addition, the mean effective pore size drops to 9.6 nm, as shown in Table 5.2.4 and the MWCO decreases to 87.3 kDa. This implies that this type of membrane after annealing is still not dense enough for FO. It is indeed shown in Table 5.2.5 that the NaCl rejection of this membrane remains zero after annealing.

Figure 5.2.7 shows the free volume radius and distribution of the annealed bottom dense layer of the double dense-layer membrane characterized by PALS using positron energy from 0.5 keV to 2 keV where positrons are confined in the dense layer. The curve is quite similar to that of S parameters with an increasing trend at low energy levels owing to surface back diffusion. Stabilized results are obtained from 1.5 keV (a mean depth of 58.4 nm) onwards and Table 5.2.6 summarizes the details at 1.5 keV. The mean free volume radius of the bottom layer is 2.977 Å with a narrow lifetime distribution as illustrated in Figure 5.2.7. There have been diverse reports on the hydrated radii of Na and Cl ions, which give quite different results generally in the range from 1.5 Å to 3.5 Å (Nightingale, 1959; Collins, 1995; Kiriukhin & Collins, 2002; Zhou *et al.* 2002). Since the annealed double dense-layer membrane exhibits a PWP of 0.17 LMH·bar^{-1} and a high rejection of 92.0% towards NaCl, as shown in Table 5.2.5, it can be postulated that at least one of the ions should have a hydrated radius in the vicinity of (or higher than) 2.977 Å. Similarly, the membrane immersed in a 90:10 (v:v) NMP/water mixed bath exhibits interesting results because it has a PWP of 0.13 LMH·bar^{-1} and a NaCl rejection of 87.4% after annealing.

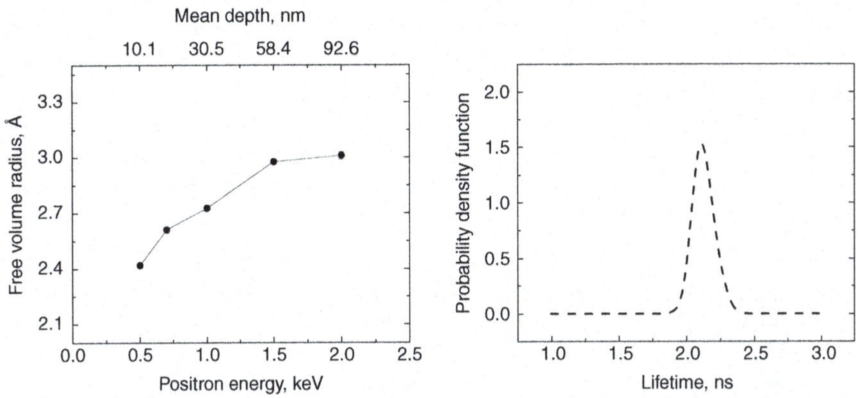

Figure 5.2.7 The mean free volume radius under different positron energy and o-Ps lifetime distributions of CA membranes cast on glass plate, phase transited in water and heat annealed in 90 °C water bath for 15 min.

Table 5.2.6 Free volume obtained from positron lifetime for the bottom layer (bottom mode) of the CA membranes cast on a glass plate, phase inversion in water and annealed in a 90°C water bath. The data are obtained at 1.5 keV positron energy (mean depth 58 nm).

Membrane	τ_3 (ns)	I_3 (%)	Free Volume Radius (Å)	Free Volume Fraction (%)
CA	2.136 ± 0.020	25.97 ± 0.34	2.977 ± 0.014	5.16 ± 0.14

Table 5.2.7 compares FO performance of annealed CA membranes prepared under different phase inversion conditions using a draw solution of 2 M $MgCl_2$. For the membrane cast on a Teflon plate, there is no surprise that it has a very high reverse salt flux. Yet it is surprising to notice that although the water flux is quite small due to the existence of relatively large pores in the dense layer, the flux in the PRO mode (its top layer facing the draw solution) is greater than that in the FO mode. This phenomenon arises from severe ICP in the porous layer in the FO mode and thus this is a dramatic reduction in the effective driving force. For the double dense-layer membrane, the reverse salt flux is as low as around 1 gMH,

implying that this membrane has great potential for seawater desalination. The water flux in the FO mode, nevertheless, is lower than that in the PRO mode (in this case the denser bottom layer faces the draw solution) as a result of ICP. However, the difference is not as significant as compared with other commercial membranes reported (McCutcheon *et al.* 2005; McCutcheon & Elimelech, 2006), probably owing to the fully porous structure of the middle supporting layer. It has been pointed out that a prospective FO membrane must have a supporting layer as porous as possible to mitigate the ICP effect (McCutcheon & Elimelech, 2007). The newly developed membrane proves to carry the properties required for a good FO membrane. Moreover, the FO performance of the membrane immersed in a NMP/water mixed bath also looks promising due to the fully porous top layer which decreases the resistance to salt diffusion. Hence, both membranes are further studied in the following session to explore their potential for seawater desalination via FO and to reveal the internal structure properties in their porous layers.

Table 5.2.7 FO performance of 90°C heat-annealed CA membranes prepared under different phase inversion conditions. Draw solution: 2 M $MgCl_2$, feed solution: DI water, linear flow rate: 3.3 cm · s^{-1}.

Substrate	Coagulant	Water Flux (LMH)		Salt Reverse Flux (gMH)	
		FO	PRO*	FO	PRO
Glass plate	Water	10.3 ± 0.3	17.3 ± 0.4	0.8 ± 0.1	1.2 ± 0.2
Glass plate	NMP/water 90/10	9.8 ± 0.7	14.7 ± 1.0	1.2 ± 0.3	2.2 ± 0.4
Teflon plate	water	7.9 ± 0.5	9.5 ± 0.4	824.8 ± 15.0	698.6 ± 14.8

*PRO refers to such membrane orientation that the denser layer faces the draw solution, while for FO vice versa. Therefore, for the membrane cast on a Teflon plate, the PRO mode means the top layer facing the draw solution, which is different from all the other membranes where the bottom layer faces the draw solution in the PRO mode.

5.2.3.5 Modeling of FO performance and structural parameter S_t

In order to compare FO performance with the prediction from existing theories, efforts were given to obtain basic transport properties of annealed membranes. Figure 5.2.8 shows the water flux and salt rejection of the double dense-layer membrane tested under different hydraulic pressures. A linear fitting of $(1\text{-}R)/R$ versus $1/(\Delta P\text{-}\Delta\pi)A$ is performed and the slope obtained is the NaCl permeability B of the membrane, which is found to be 0.07 LMH. Figure 5.2.9 illustrates the FO performance of the membrane using NaCl as draw solutes and DI water as the feed. The salt reverse flux remains lower than 5 gMH in all the NaCl concentrations tested, and the water flux in the PRO mode shows a better performance than that in the FO mode, which is attributed to the severer ICP of the latter, as to be discussed in the following.

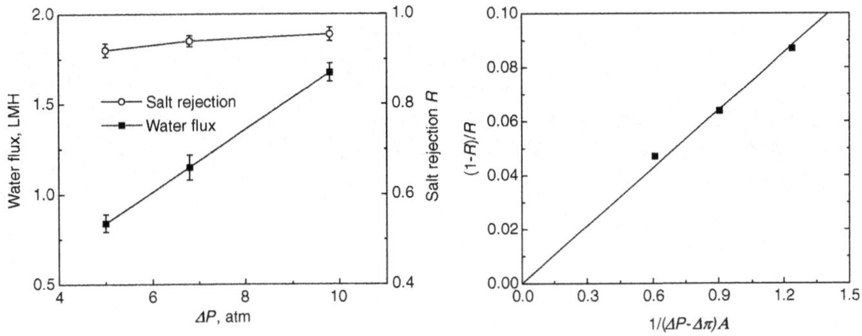

Figure 5.2.8 (Left) water flux and salt rejection of the 90 °C annealed CA membranes cast on glass plates and phase inversion in water, tested under different hydraulic pressures using DI water or 200 ppm NaCl. (Right) Linear fitting of (1-*R*)/*R* vs. 1/(ΔP-$\Delta\pi$)*A*, the slope is the salt permeability *B* (LMH).

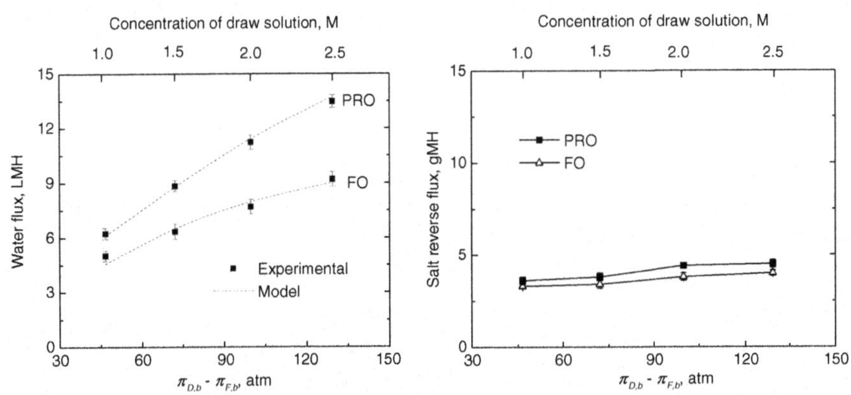

Figure 5.2.9 FO performance of the 90 °C annealed CA double dense-layer membrane. Draw solute: NaCl, feed: DI water, linear velocity: 3.3 cm·s^{-1}. (left) water flux and (right) salt reverse flux.

Since concentration polarization is a serious problem in osmotic-driven processes. The concentration polarization can happen externally on the dense side of the membrane and internally inside the porous structure. Dilutive ECP occurs when the draw solution is placed against the selective layer in the PRO mode, and the osmotic pressure at the membrane surface $\pi_{D,m}$ can be obtained with the following equation:

$$\pi_{D,m} = \pi_{D,b} \exp\left(-\frac{J_w}{k} \right) \tag{5.2.13}$$

where J_w is the experimental water flux, k is the mass transfer coefficient, and $\pi_{D,b}$ is the bulk osmotic pressure of the draw solution. On the contrary, when the feed solution is facing against the selective layer in the FO mode, concentrative ECP would happen which gives the osmotic pressure at the membrane surface $\pi_{F,m}$ by:

$$\pi_{F,m} = \pi_{F,b} \exp\left(\frac{J_w}{k} \right) \tag{5.2.14}$$

where $\pi_{F,b}$ is the bulk feed osmotic pressure. The mass transfer coefficient, k, can be calculated by the following equation (McCutcheon & Elimelech, 2006):

$$k = \frac{ShD}{d_h} \tag{5.2.15}$$

where D is the solute diffusion coefficient, d_h is the hydraulic diameter, and Sh is the Sherwood number of a laminar flow in a rectangular channel given by:

$$Sh = 1.85\left(Re\,Sc\frac{d_h}{L} \right)^{0.33} \tag{5.2.16}$$

Here, *Re* is the Reynolds number, *Sc* is the Schmidt number, and *L* is the length of the channel. Lee *et al.* (1981) have developed the following expression to model the flux under ICP in the PRO application:

$$J_w = \frac{1}{K} \ln \frac{A\pi_{D,m} - J_w + B}{A\pi_{F,b} + B} \tag{5.2.17}$$

Here, *K* is the solute resistivity for diffusion within the porous layer given by:

$$K = \frac{S_t}{D} \tag{5.2.18}$$

where S_t is the structural parameter of the membrane defined by tortuosity, τ, porous layer thickness, χ and porosity, ε, as shown in Equation (5.2.17). S_t has the dimension of length and appears as the apparent thickness of the porous layer.

$$S_t = \frac{\tau\chi}{\varepsilon} \tag{5.2.19}$$

Similarly, in the FO mode, concentrative ECP dominates at the feed side and dilutive ICP happens at the draw side. The expression to model the flux under ICP has been derived as follows (Lee *et al.* 1981):

$$J_w = \frac{1}{K}\ln\frac{A\pi_{D,b} + B}{A\pi_{F,m} + J_w + B} \tag{5.2.20}$$

Figures 5.2.9 and 5.2.10 shows comparisons between experimental data using DI water and seawater (3.5 wt% NaCl solution) as the feeds and the model predictions based on the double dense-layer membrane, respectively. The models fit experimental results closely. The value of S_t turns out to be 54 μm, approximately 1.5 times larger than the average thickness of the membrane of around 35 μm obtained from FESEM images. Therefore, the value of τ/ε is about 1.5, indicating a rather porous internal structure of the membrane. The overall porosity of the membrane, actually, has been measured to be 75% ± 0.4%. Hence, the tortuosity τ is around 1.2.

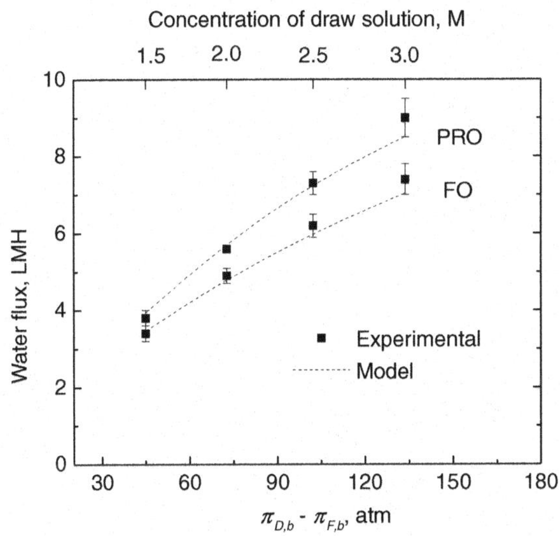

Figure 5.2.10 FO performance of the 90 °C heat annealed CA double dense-layer membrane. Draw solute: NaCl, feed: seawater (3.5 wt% NaCl), linear velocity: 3.3 cm · s^{-1}.

Similar experiments have also been done on the single dense-layer membrane which was first immersed in a NMP/water mixture and then water. The NaCl permeability B is found from Figure 5.2.11 to be 0.09 LMH, resulting in a slightly higher salt reverse flux than the double dense-layer membrane. If we neglect the lower water flux and higher salt leakage brought by the inhomogeneity during membrane formation process, the single dense-layer membrane is even more prosperous for FO, since the structural parameter S_t modeled from Figures 5.2.12 and 5.2.13 is only 51 μm. A comparison of our membrane structure with other membranes published in literature is listed in Table 5.2.8 (Gray *et al.* 2006; McCutcheon & Elimelech, 2006; Achilli *et al.* 2009; Thorsen & Holt, 2009). Obviously, both of our membranes possess much lower S values than others. The lower the S_t value, the less severe the ICP. Therefore, much less ICP would be encountered with our membrane during FO processes, and the effective driving force would be preserved to the highest extent.

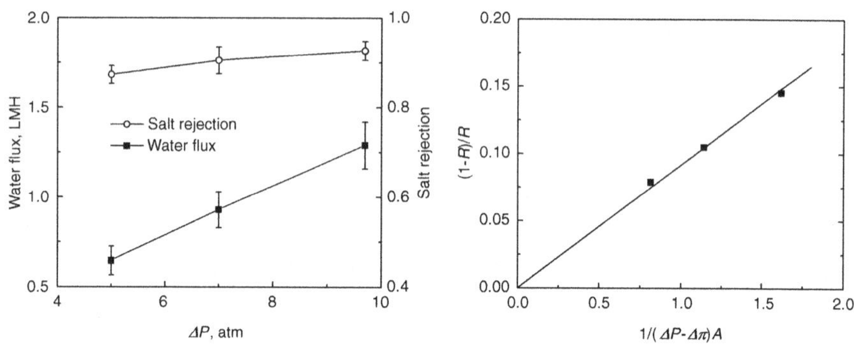

Figure 5.2.11 (Left) water flux and salt rejection of the 90 °C heat-annealed CA membranes cast on glass plate, phase transited firstly in 90:10 (v:v) NMP/water and then in water, tested under hydraulic pressure and DI water or 200 ppm NaCl. (Right) Linear fitting of (1-R)/R vs. 1/(ΔP-$\Delta\pi$)A, the slope is salt permeability B (LMH).

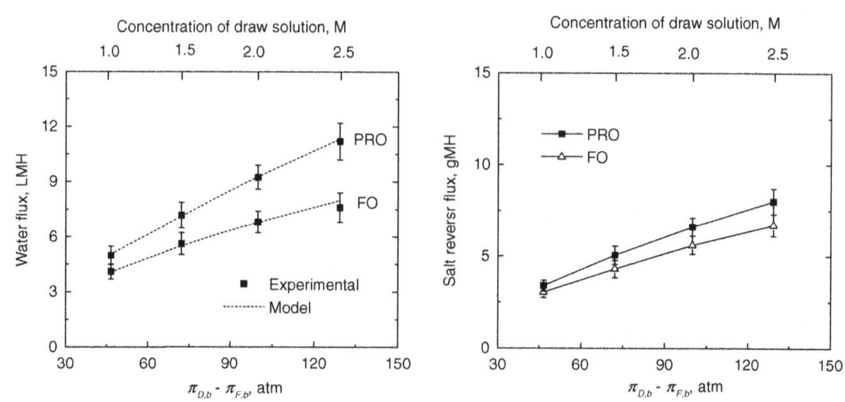

Figure 5.2.12 FO performance of the 90 °C annealed CA single dense-layer membrane. Draw solute: NaCl, feed: DI water, linear velocity: 3.3 cm · s^{-1}. (left) water flux and (right) salt reverse flux.

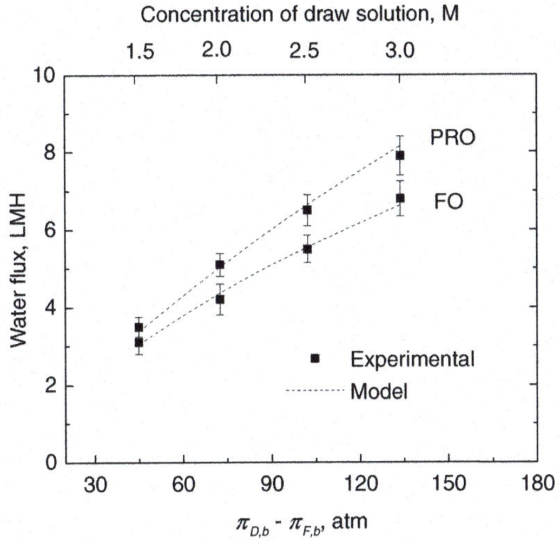

Figure 5.2.13 FO performance of the 90 °C annealed CA single dense-layer membrane. Draw solute: NaCl, feed: seawater (3.5 wt% NaCl), linear velocity: 3.3 cm · s^{-1}.

Table 5.2.8 Intrinsic separation properties and structural parameters of different FO membranes.

Membrane*	Water Permeability *A* (LMH · bar^{-1})	Salt Permeability *B* (LMH)	Membrane Thickness (μm)	Structural Parameter S_t (μm)	Reference
Double-dense-layer membrane	0.17	0.07	35	54	This work
Single-dense-layer membrane	0.13	0.09	34	51	This work
HTI membrane	–	–	<50	329	(McCutcheon & Elimelech, 2006)
HTI membrane	1.13	0.46	<50	575	(Gray *et al.* 2006)
TFC membrane	2.56	0.40	–	670	(Thorsen & Holt, 2009)
CA membrane	0.72	0.22	~250	2500	(Thorsen & Holt, 2009)
HTI membrane	0.67	0.40	–	678	(Achilli *et al.* 2009)

*The double- and single-dense-layer membranes refer to the 90 °C annealed CA membranes cast on glass plates as discussed in the text.

5.2.3.6 Single vs. double dense-layer structure in the FO-MBR integrated system

With small structural parameters and good salt rejections, the above tested membranes can be expected to have high performance in the novel FO-MBR integrated system for waste water treatment. A good membrane configuration for FO-MBR should be that the selective layer faces draw solution to sustain the highest effective driving force, and the feed runs against the porous support. Since the major components of the feed in a MBR process include bio-macromolecules and colloidal particles whose size is in the range from nanometer to micrometer or even higher (Ho & Zydney, 2006; Le-Clech *et al.* 2006), most of them would therefore be rejected by the dense selective layer of the membrane which has a high rejection even for NaCl. However, if the pore size of the porous layer that is in direct contact with the feed is larger than the feed components, pore blockage would happen within this layer leading to significant fouling which is irreversible. If the pore size is smaller than the feed components, on the contrary, fouling might still happen due to the cake formation at the membrane surface, which is, nevertheless, reversible and can be easily removed during maintenance cleaning (Le-Clech *et al.* 2006). The double dense-layer membrane, with a relative dense layer at the porous support side facing the feed, is therefore expected to be subjected to less and more reversible fouling during the FO-MBR process, when compared with the membrane prepared in a NMP/water mixed bath (referred to single dense-layer membrane below) whose pore sizes facing the feed may reach to as high as several hundred nanometers. Moreover, surface roughness also plays a role. Higher roughness usually leads to more severe irreversible fouling (He *et al.* 2005; Le-Clech *et al.* 2006). It is clearly shown in Figure 5.2.14 that the top surface (facing the feed) of the single dense-layer membrane is several hundred times rougher than that of the double one, which also predicts less fouling in the double dense-layer membrane.

Figure 5.2.15 shows the fouling test results with the two membranes in a modeling FO-MBR system. The colloidal solution of aluminum oxide nanoparticles whose actual size is around 200 nm is employed as the feed. Water flux is normalized to the initial value when the feed is DI water. It is seen that more significant fouling occurs with the single-dense-layer membrane, whose flux drops to about 70 percent of its initial value after several hours, whereas for the double dense-layer one, only a slight decrease (less than 10 percent) in water flux is observed. In addition, after washing with DI water for 1 h, the re-tested water flux almost goes back to the initial value for the double dense-layer membrane, while for the single dense-layer one, it only returns to 80 percent. The results show that with double dense-layer configuration, milder and more reversible fouling may be expected in the FO-MBR system when large particles are predominant in the feed, and it is hence more energy-saving as the cleaning frequency can be reduced. It is yet to be examined in the future how small molecules in the feed would affect the performance and how the membrane configuration can be optimized so that minimum fouling would be met in the system.

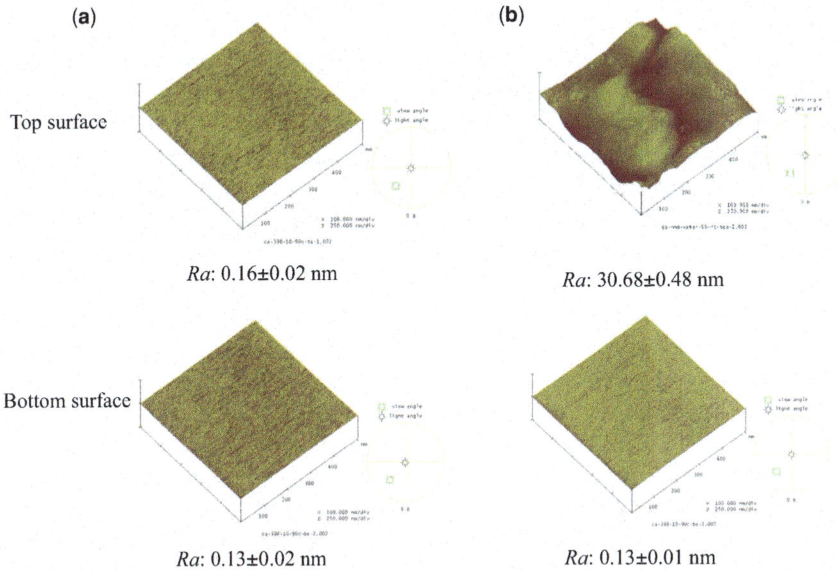

Figure 5.2.14 AFM images and mean roughness of 90 °C heat-annealed CA membranes cast on glass plate and immersed in (a) water and (b) NMP/water (90/10) and then water.

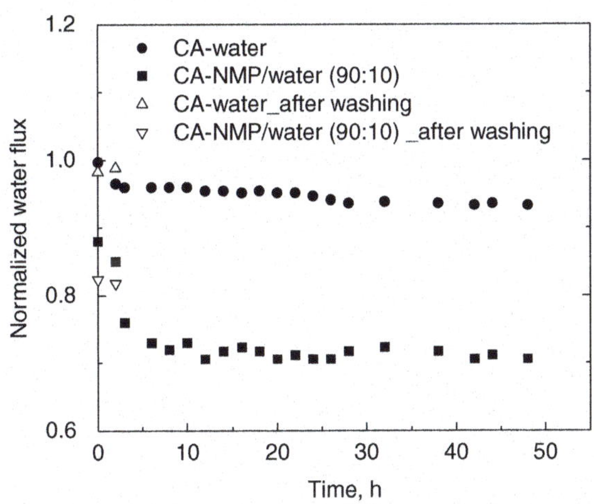

Figure 5.2.15 Normalized water flux in PRO mode in fouling test against time of CA membranes cast on glass plate and immersed in different coagulant baths, with 2 M NaCl as the draw solution and Al_2O_3 colloidal solution (or DI water, after washing) as the feed.

5.2.4 SUMMARY

The design and engineering of membrane structure that produces high water flux, low salt leakage and minimized internal concentration polarization (ICP) in forward osmosis (FO) processes have been explored in this work. The fundamentals of phase inversion of cellulose acetate (CA) regarding the formation of an ultra-thin selective layer at the bottom interface of polymer and casting substrate were investigated by using substrates with different hydrophilicity. An in-depth understanding of membrane structure and pore size distribution has been elucidated with Field Emission Electronic Microscopy (FESEM) and Positron Annihilation Lifetime Spectroscopy (PALS). A double dense-layer structure is formed when glass plate is used as the casting substrate and water as the coagulant. The thickness of the ultra-thin bottom layer resulted from hydrophilic-hydrophilic interaction is identified to be around 95 nm, while a fully porous, open-cell structure is formed in the middle support layer due to spinodal decomposition. Consequently, the membrane shows low salt leakage with mitigated ICP in the FO process for seawater desalination. The structural parameter (S_t) of the membrane is analyzed by modeling water flux using the theory that considers both external concentration polarization (ECP) and ICP, and the S_t value of the double dense-layer membrane is much smaller than those reported in literatures. Furthermore, the effects of an intermediate immersion into a solvent/water mixed bath prior to complete immersion in water on membrane formation have been studied. The resultant membranes may have a single dense layer with an even lower S_t value. A comparison of fouling behavior in a simple FO-membrane bioreactor (MBR) system is evaluated for these two types of membranes. The double dense-layer membrane shows a less fouling propensity. This study may help pave the way to improve the membrane design for new-generation FO membranes.

5.2.5 REFERENCES

Achilli A., Cath T. Y. and Childress A. E. (2009). Power generation with pressure retarded osmosis: an experimental and theoretical investigation. *Journal of Membrane Science*, **343**, 42–52.

Barton A. F. (1991). CRC Handbook of Solubility Parameters and Other Cohesion Parameters. CRC Press, Boca Raton, FL.

Cath T. Y., Adams D. and Childress A. E. (2005). Membrane contactor processes for wastewater reclamation in space: II. Combined direct osmosis, osmotic distillation, and membrane distillation for treatment of metabolic wastewater. *Journal of Membrane Science*, **257**, 111–119.

Chen H., Hung W.-S., Lo C.-H., Huang S.-H., Cheng M.-L., Liu G., Lee K.-R., Lai J.-Y., Sun Y.-M. and Hu C.-C. (2007). Free-volume depth profile of polymeric membranes studied by positron annihilation spectroscopy: layer structure from interfacial polymerization. *Macromolecules*, **40**, 7542–7557.

Chung T. S., Teoh S. K. and Hu X. (1997). Formation of ultrathin high-performance polyethersulfone hollow-fiber membranes. *Journal of Membrane Science*, **133**, 161–175.

Collins K. D. (1995). Sticky ions in biological systems. *Proceedings of the National Academy of Sciences*, **92**, 5553–5557.

Gray G. T., McCutcheon J. R. and Elimelech M. (2006). Internal concentration polarization in forward osmosis: role of membrane orientation. *Desalination*, **197**, 1–8.

He Y., Xu P., Li C. and Zhang B. (2005). High-concentration food wastewater treatment by an anaerobic membrane bioreactor. *Water Research*, **39**, 4110–4118.

Ho C. C. and Zydney A. L. (2006). Overview of fouling phenomena and modeling approaches for membrane bioreactors. *Separation Science and Technology*, **41**, 1231–1251.

Jean Y. (1990). Positron annihilation spectroscopy for chemical analysis: a novel probe for microstructural analysis of polymers. *Microchemical Journal*, **42**, 72–102.

Jean Y., Hung W.-S., Lo C.-H., Chen H., Liu G., Chakka L., Cheng M.-L., Nanda D., Tung K.-L. and Huang S.-H. (2008). Applications of positron annihilation spectroscopy to polymeric membranes. *Desalination*, **234**, 89–98.

Jiang L. Y. and Chung T. S. (2010). Homogeneous polyimide/cyclodextrin composite membranes for pervaporation dehydration of isopropanol. *Journal of Membrane Science*, **346**, 45–58.

Jiao B., Cassano A. and Drioli E. (2004). Recent advances on membrane processes for the concentration of fruit juices: a review. *Journal of Food Engineering*, **63**, 303–324.

Kessler J. and Moody C. (1976). Drinking water from sea water by forward osmosis. *Desalination*, **18**, 297–306.

Kiriukhin M. Y. and Collins K. D. (2002). Dynamic hydration numbers for biologically important ions. *Biophysical Chemistry*, **99**, 155–168.

Kopeček J. and Sourirajan S. (1969). Structure of porous cellulose acetate membranes and a method for improving their performance in reverse osmosis. *Journal of Applied Polymer Science*, **13**, 637–657.

Kranhold K. http://online.wsj.com/article/SB120053698876396483.html?mod=googlenews_wsj.

Le-Clech P., Chen V. and Fane T. A. G. (2006). Fouling in membrane bioreactors used in wastewater treatment. *Journal of Membrane Science*, **284**, 17–53.

Lee K. L., Baker R. W. and Lonsdale H. K. (1981). Membranes for power generation by pressure-retarded osmosis. *Journal of Membrane Science*, **8**, 141–171.

Lo C.-H., Huang J.-K., Hung W.-S., Huang S.-H., De Guzman M., Rouessac V., Li C.-L., Hu C.-C., Lee K.-R. and Lai J.-Y. (2009). Investigation on the variation in the fine structure of plasma-polymerized composite membrane by positron annihilation spectroscopy. *Journal of Membrane Science*, **337**, 297–303.

Lobo V. (1993). Mutual diffusion coefficients in aqueous electrolyte solutions (technical report). *Pure and Applied Chemistry*, **65**, 2613–2640.

Loeb S. (2001). One hundred and thirty benign and renewable megawatts from Great Salt Lake? The possibilities of hydroelectric power by pressure-retarded osmosis. *Desalination*, **141**, 85–91.

Loeb S. and Sourirajan S. (1963). Sea water demineralization by means of an osmotic membrane. *Advances in Chemistry Series*, **38**, 117.

Loeb S., Titelman L., Korngold E. and Freiman J. (1997). Effect of porous support fabric on osmosis through a Loeb-Sourirajan type asymmetric membrane. *Journal of Membrane Science*, **129**, 243–249.

Manjikian S. (1967). Desalination membranes from organic casting solutions. *Industrial & Engineering Chemistry Product Research and Development*, **6**, 23–32.

McCutcheon J. and Elimelech M. (2006). Influence of concentrative and dilutive internal concentration polarization on flux behavior in forward osmosis. *Journal of Membrane Science*, **284**, 237–247.

McCutcheon J. R. and Elimelech M. (2007). Modeling water flux in forward osmosis: implications for improved membrane design. *AIChE Journal*, **53**, 1736–1744.

McCutcheon J. R., McGinnis R. L. and Elimelech M. (2005). A novel ammonia – carbon dioxide forward (direct) osmosis desalination process. *Desalination*, **174**, 1–11.

McGinnis R. L., McCutcheon J. R. and Elimelech M. (2007). A novel ammonia–carbon dioxide osmotic heat engine for power generation. *Journal of Membrane Science*, **305**, 13–19.

Mi B. and Elimelech M. (2010). Gypsum scaling and cleaning in forward osmosis: measurements and mechanisms. *Environmental Science & Technology*, **44**, 2022–2028.

Nightingale E. R. J. (1959). Phenomenological theory of ion solvation. Effective radii of hydrated ions. *The Journal of Physical Chemistry*, **63**, 1381–1387.

Petrotos K. B. and Lazarides H. N. (2001). Osmotic concentration of liquid foods. *Journal of Food Engineering*, **49**, 201–206.

Pilon R., Kunst B. and Sourirajan S. (1971). Studies on the development of improved reverse osmosis membranes from cellulose acetate–acetone–formamide casting solutions. *Journal of Applied Polymer Science*, **15**, 1317–1334.

Semiat R. (2008). Energy issues in desalination processes. *Environmental Science & Technology*, **42**, 8193–8201.

Seppälä A. and Lampinen M. J. (1999). Thermodynamic optimizing of pressure-retarded osmosis power generation systems. *Journal of Membrane Science*, **161**, 115–138.

Sourirajan S., Matsuura T., Sourirajan S., Sourirajan S. and Chemist I. (1985). Reverse Osmosis-Ultrafiltration Process Principles. National Research Council, Canada Ottawa.

Su J., Yang Q., Teo J. F. and Chung T.-S. (2010). Cellulose acetate nanofiltration hollow fiber membranes for forward osmosis processes. *Journal of Membrane Science*, **355**, 36–44.

Thorsen T. and Holt T. (2009). The potential for power production from salinity gradients by pressure retarded osmosis. *Journal of Membrane Science*, **335**, 103–110.

Tsay C. and McHugh A. (1990). Mass transfer modeling of asymmetric membrane formation by phase inversion. *Journal of Polymer Science Part B: Polymer Physics*, **28**, 1327–1365.

Tung K.-L., Jean Y.-C., Nanda D., Lee K.-R., Hung W.-S., Lo C.-H. and Lai J.-Y. (2009). Characterization of multilayer nanofiltration membranes using positron annihilation spectroscopy. *Journal of Membrane Science*, **343**, 147–156.

Wang K. Y., Matsuura T., Chung T.-S. and Guo W. F. (2004). The effects of flow angle and shear rate within the spinneret on the separation performance of poly (ethersulfone) (PES) ultrafiltration hollow fiber membranes. *Journal of Membrane Science*, **240**, 67–79.

Wang K. Y., Chung T.-S. and Qin J.-J. (2007). Polybenzimidazole (PBI) nanofiltration hollow fiber membranes applied in forward osmosis process. *Journal of Membrane Science*, **300**, 6–12.

Wang K. Y., Yang Q., Chung T.-S. and Rajagopalan R. (2009). Enhanced forward osmosis from chemically modified polybenzimidazole (PBI) nanofiltration hollow fiber membranes with a thin wall. *Chemical Engineering Science*, **64**, 1577–1584.

Wang K. Y., Ong R. C. and Chung T.-S. (2010). Double-skinned forward osmosis membranes for reducing internal concentration polarization within the porous sublayer. *Industrial & Engineering Chemistry Research*, **49**, 4824–4831.

Wang Y., Lau W. W. and Sourirajan S. (1994). Effects of pretreatments on morphology and performance of cellulose acetate (formamide type) membranes. *Desalination*, **95**, 155–169.

Wang Y., Jiang L., Matsuura T., Chung T. S. and Goh S. H. (2008). Investigation of the fundamental differences between polyamide-imide (PAI) and polyetherimide (PEI) membranes for isopropanol dehydration via pervaporation. *Journal of Membrane Science*, **318**, 217–226.

Wijmans J., Baaij J. and Smolders C. (1983). The mechanism of formation of microporous or skinned membranes produced by immersion precipitation. *Journal of Membrane Science*, **14**, 263–274.

Wilke C. and Chang P. (1955). Correlation of diffusion coefficients in dilute solutions. *AIChE Journal*, **1**, 264–270.

Yang Q., Chung T.-S. and Weber M. (2009a). Microscopic behavior of polyvinylpyrrolidone hydrophilizing agents on phase inversion polyethersulfone hollow fiber membranes for hemofiltration. *Journal of Membrane Science*, **326**, 322–331.

Yang Q., Wang K. Y. and Chung T.-S. (2009b). Dual-layer hollow fibers with enhanced flux as novel forward osmosis membranes for water production. *Environmental Science & Technology*, **43**, 2800–2805.

Yang Q., Wang K. Y. and Chung T.-S. (2009c). A novel dual-layer forward osmosis membrane for protein enrichment and concentration. *Separation and Purification Technology*, **69**, 269–274.

Yilmaz L. and McHugh A. (1988). Modeling of asymmetric membrane formation. II. The effects of surface boundary conditions. *Journal of Applied Polymer Science*, **35**, 1967–1979.

Zhou J., Lu X., Wang Y. and Shi J. (2002). Molecular dynamics study on ionic hydration. *Fluid Phase Equilibria*, **194**, 257–270.

Part 6
Modeling

Chapter 6.1

Modeling water flux*

6.1.1 INTRODUCTION

Recent developments in osmotic processes have prompted a flurry of research activity in the field of forward osmosis. Forward osmosis (FO) has recently been studied in the areas of food processing (Petrotos *et al.* 1998; Petrotos & Lazarides, 2001; Jiao *et al.* 2004), electricity production (Lee *et al.* 1981; Loeb, 1998, 2002a, b), and desalination (McCutcheon *et al.* 2005; Gray *et al.* 2006; McCutcheon & Elimelech, 2006b). These processes all rely on the use of a concentrated "draw" solution to drive the transport of water from a dilute feed solution across a semi-permeable membrane. In each case, process performance is evaluated based on solvent (water) recovery, permeate water flux, and solute rejection. Solute rejection is a membrane specific property and will not be discussed in detail in this paper. Recovery and permeate water flux are more pertinent to evaluating FO processes, especially in comparison to pressure driven membrane processes.

Most current membrane processes, such as reverse osmosis (RO), are pressure driven such that permeate water flux and recovery are controlled by the hydraulic pressure applied to the feed water. Concentration polarization, which results in a region of increased solute concentration at the rejecting surface of the membrane, also limits water flux and recovery. In FO, the primary factor controlling water flux and recovery is the transmembrane osmotic pressure. However, concentration polarization, now occurring on both sides of the membrane, also plays a prominent role in reducing the effective transmembrane osmotic pressure. The coupled effect of two polarized boundary layers has significant consequences on FO performance, especially when asymmetric membranes are considered.

---

* This chapter is based, with permission from the copyright holder, on the previously published paper. (Mccutcheon J. R. and Elimelech M. (2007). Modeling water flux in forward osmosis: implications for improved membrane design. *AIChE Journal*, **53**(7), 1736–1744. doi:10.1002/aic.11197)

Most membranes today are designed for pressure driven membrane processes and have an asymmetric structure which consists of a thin selective active layer supported by thick layers of porous polymer and fabric support. It has been reported that these asymmetric polymeric membranes perform poorly when used for osmotic processes due to the boundary layers that occur within these supporting layers (Lee *et al.* 1981; McCutcheon *et al.* 2005; Gray *et al.* 2006; McCutcheon & Elimelech, 2006b). This phenomenon is called internal concentration polarization (ICP). Similar in concept to the concentration polarization discussed above, internal concentration polarization refers to the development of a boundary layer within the protective confines of the membrane support structure.

For asymmetric membranes, two types of ICP exist depending on the membrane orientation (McCutcheon & Elimelech, 2006a). In one orientation, referred to as the pressure retarded osmosis (PRO) mode, the feed solution is directed against the support layer. This orientation is commonly used for PRO applications when osmotic pressure gradients are used to generate electricity. The other orientation, where the draw solution is directed against the support layer, is the FO mode. This is the primary orientation considered for water treatment processes, such as FO desalination. In either case, ICP occurs within the porous support where, unlike concentration polarization in pressure driven membrane processes, it cannot be mitigated by hydrodynamics, such as turbulence, and hence drastically reduces the osmotic driving force.

Several previous investigations (Kravath & Davis, 1975; Kessler & Moody, 1976; Moody & Kessler, 1976; Lee *et al.* 1981) found that osmosis tests using RO membranes yielded poor water flux performance. These results were attributed primarily to ICP while assuming that external boundary layers on either side of the membrane to be negligible. This was a reasonable assumption, being that the water fluxes were often low and the solutions were well mixed. In more recent studies (McCutcheon *et al.* 2005; Gray *et al.* 2006; McCutcheon & Elimelech, 2006a; McCutcheon & Elimelech, 2006b), permeate fluxes have shown marked improvement and external concentration polarization (ECP) can no longer be ignored. Understanding the relative impacts of ICP and ECP on osmotic driving force would help to improve performance of osmotic processes by optimizing system parameters and membrane characteristics. Using models to predict water flux for both the PRO and FO modes under the influence of both ECP and ICP can help achieve this, though only once have existing models for ICP and ECP been combined. In this previous study, McCutcheon and Elimelech (2006a) showed that a combined model incorporating internal and external concentration polarization accurately described experimental data for both the PRO and FO modes and thus could work as a standalone predictor of flux under a variety of experimental conditions for either orientation.

The purpose of this study is to use this combined model to predict water flux performance of osmosis through asymmetric membranes that are oriented in both the PRO and FO modes under a range of feed and draw solution concentrations and membrane structural properties. The negative impacts on osmotic driving force caused by ICP and ECP will be quantified both for individual and coupled scenarios. Changes in membrane structural characteristics will elucidate performance improvements due to hypothetical changes in membrane design. The ramifications of tailoring membrane structural properties for both PRO and FO applications will be discussed.

6.1.2 GOVERNING EQUATIONS FOR PERMEATE FLUX

Two variations on the 'film theory' model for concentration polarization are considered in this investigation. External concentration polarization is modeled using traditional film theory (Mulder, 1996) which incorporates the use of the mass transfer coefficient. Internal concentration polarization is modeled similarly by considering the resistance to diffusion of solute molecules within the porous support layer. Both symmetric and asymmetric membranes are considered. For the asymmetric membrane, two orientations of the membrane are studied: one with the active layer facing the feed solution (FO mode) and the other with the active layer facing the draw solution (PRO mode) (McCutcheon & Elimelech, 2006a).

6.1.2.1 External concentration polarization

Concentration polarization on the feed side of a membrane is a significant problem in pressure-driven membrane desalination processes. This phenomenon inhibits permeate flow due to an increased osmotic pressure at the membrane active layer interface on the feed side of the membrane. In an osmotic process, this phenomenon occurs on both sides of the membrane, with the effect being dilutive on the permeate side. We refer to these two phenomena collectively as external concentration polarization (ECP). Specifically, this phenomenon on the feed and permeate side will be referred to as concentrative and dilutive ECP, respectively.

In order to predict flux in the presence of ECP, we must determine the effective osmotic driving force at the membrane-solution interface on both the feed and permeate sides of the membrane. This is accomplished by using film theory. For the purposes of this model, we will use a similar method of calculating ECP as provided in McCutcheon and Elimelech (2006a).

For a pressure driven membrane process, such as RO, in the absence of ECP and salt passage, the generalized flux equation is

$$J_w = A\left(\Delta P - \pi_{F,b}\right) \tag{6.1.1}$$

where A is the pure water permeability coefficient, ΔP is the transmembrane pressure, and $\pi_{F,b}$ is the osmotic pressure of the bulk feed solution. Note that we assume complete rejection of the feed solute (i.e., the reflection coefficient $\sigma = 1$). Equation 6.1.1 is valid only when the flux is low or the feed solution is very dilute. If flux becomes higher, the concentration polarization effect becomes significant.

The membrane surface concentration on the feed side becomes larger than that of the bulk as solute is rejected, concentrating the feed solute. We therefore refer to this phenomenon as concentrative ECP. By using film theory, we can modify Equation 6.1.1 to account for concentrative ECP:

$$J_w = A\left(\Delta P - \pi_{F,b}\exp\left(\frac{J_w}{k}\right)\right) \tag{6.1.2}$$

Here, the exponential term is the concentrative ECP modulus (McCutcheon & Elimelech, 2006a) which is a function of water flux and mass transfer coefficient, k. Mass transfer coefficient may be calculated from the appropriate Sherwood number correlations, incorporating viscosity, density, diffusion coefficient, and flow velocity, as explained in our previous investigation (Kessler & Moody, 1976). Even though Equation 6.1.2 presents the water flux implicitly, it can still be solved iteratively to model flux for a given set of experimental conditions as is done later in this study. This equation, as will subsequent equations below, assumes a proportional relationship between concentration and osmotic pressure.

For osmotically driven membrane processes with a non-dilute feed, a similar concentrative ECP will occur. In an osmotically driven membrane process, however, we must also consider the dilutive effect that occurs on the permeate side of the membrane. Dilutive ECP occurs as permeate water flow displaces draw solute at the membrane-draw solution interface, reducing the effective driving force of the draw solution. These two ECP phenomena are coupled for osmotic flow when solute is present on both sides of the membrane and they must be accounted for when modeling flux behavior, as shown below.

The standard flux equation for forward osmosis is given as:

$$J_w = A\left(\pi_{D,b} - \pi_{F,b}\right) \tag{6.1.3}$$

which predicts flux as a function of the difference in bulk osmotic pressures of the draw ($\pi_{D,b}$) and feed ($\pi_{F,b}$) solutions. As above, this equation assumes complete rejection of the feed and draw solutes. This equation does not account for ECP, which may be valid only if the permeate flux is very low. When flux rates are higher, though, the equation must be modified to include both the concentrative and dilutive ECP moduli:

$$J_W = A\left[\pi_{D,b}\exp\left(-\frac{J_w}{k_D}\right) - \pi_{F,b}\exp\left(\frac{J_w}{k_F}\right)\right] \tag{6.1.4}$$

Note that the dilutive effect is indicated by the negative exponential term modifying the draw solution osmotic pressure. We must also consider individual mass transfer coefficients on the feed, k_F, and permeate, k_D, sides of the membrane, though for the model parameters chosen below, $k_F$and k_D are essentially equal.

Equation 6.1.4 represents an implicit model for osmotic flux using a dense *symmetric* membrane. However, no synthetic dense symmetric membranes are in use today for osmotic processes, and therefore, the usefulness of this particular flux model is limited. We must therefore consider the case where the membrane is *asymmetric*, for which internal concentration polarization effects are most significant.

6.1.2.2 Internal concentration polarization

Asymmetric membranes, commonly used in pressure driven membrane processes, use porous layers to mechanically support a thin salt rejecting active layer. In osmotic processes, salt must pervade these porous layers, which do not reject the salt to any appreciable degree, yet still hinder its diffusion, to establish the osmotic driving force across this active layer. When water begins to permeate the membrane, concentration polarization occurs on both sides of this active layer. However, the porous layer provides a protected environment on one side of the active layer where the polarized layer can form without the mitigating effects of crossflow (Gray *et al.* 2006; McCutcheon & Elimelech, 2006b). There are two types of ICP depending on the orientation of the membrane. In the PRO mode, the porous layer is against the feed solution and the feed solute will be concentrated within the membrane. In the FO mode, the porous layer is against the permeate side. The draw solute diffuses into this porous layer but becomes diluted as water permeates the membrane. We refer to these phenomena as concentrative and dilutive ICP, respectively (Gray *et al.* 2006; McCutcheon & Elimelech, 2006a).

Both the structure of the membrane support layer and the solute diffusion coefficient play a significant role in determining the severity of these phenomena and must be considered when modeling flux. These characteristics affect the ability of the solute to diffuse out of (concentrative ICP) and into (dilutive ICP) the porous support layer, and in part control the magnitude of ICP. Lee *et al.* (1981) defined a term signifying the solute resistance to diffusion within the membrane support layer, K:

$$K = \frac{t\tau}{D\varepsilon} \tag{6.1.5}$$

where D is the bulk diffusion coefficient of the solute, and t, τ, and ε are the thickness, tortuosity, and porosity of the support layer, respectively. Note that this term is essentially the inverse mass transfer coefficient within the support layer, where $D_{eff} = D\varepsilon/\tau$. In this case, the thickness of the effective 'boundary layer' is the thickness of the support layer. We maintain the use of the K term due to convention established in previous studies on ICP (Kessler & Moody, 1976; Lee *et al.* 1981; Gray *et al.* 2006). This term can be incorporated into Equation 6.1.3 to describe flux in the presence of internal concentration polarization as done in McCutcheon and Elimelech (2006a). The ICP modulus will modify the bulk osmotic pressure of the solution that is in contact with the porous support layer. In the case of the

PRO mode, Equation 6.1.3 is modified to yield (Loeb *et al.* 1997; McCutcheon & Elimelech, 2006a):

$$J_w = A\left[\pi_{D,b} - \pi_{F,b}\exp\left(J_w K\right)\right] \tag{6.1.6}$$

Note how the ICP modulus has a positive exponential term, indicating the concentrative effect. For moderate and high fluxes, dilutive ECP on the permeate (draw) side is not negligible and hence must be accounted for as well (McCutcheon & Elimelech, 2006a):

$$J_w = A\left[\pi_{D,b}\exp\left(-\frac{J_w}{k_D}\right) - \pi_{F,b}\exp\left(J_w K\right)\right] \tag{6.1.7}$$

All of the terms in Equation 6.1.7 are readily determined through experiments or calculations. By solving Equation 6.1.7 numerically, we can predict the water flux through an asymmetric membrane in the PRO mode. It is important to note that this model assumes that the support layer creates no hydraulic resistance to solvent (water) transport and that solute may freely enter the support structure at the membrane interface such that no concentrative ECP occurs on the support layer. This equation was found to accurately model flux for osmosis through an asymmetric membrane in the PRO mode (McCutcheon & Elimelech, 2006a).

A similar approach can be taken for membranes used in the FO mode. Following the work of Loeb *et al.* (1997) and McCutcheon and Elimelech (2006a), the result is an equation resembling Equation 6.1.7:

$$J_w = A\left[\pi_{D,b}\exp\left(-J_w K\right) - \pi_{F,b}\exp\left(\frac{J_w}{k_F}\right)\right] \tag{6.1.8}$$

Equation 6.1.8 resembles Equation 6.1.7 except that the ICP now occurs on the draw solution side and is dilutive while the ECP occurs on the feed side and is concentrative. Equation 6.1.8 has been shown to accurately model water flux for an asymmetric membrane used in the FO mode (McCutcheon & Elimelech, 2006a). Note that in the FO mode, dilutive ICP is coupled with concentrative ECP. As above, it is assumed that no ECP occurs on the permeate side of the membrane since the support layer is completely permeable to the draw solute.

In each of these cases, the ECP and ICP moduli all contribute negatively to the overall osmotic driving force. The negative contribution of each increases with higher flux, which suggests a self-limiting flux behavior. This implies that increasing osmotic driving force will provide diminishing increases in flux. Below, a quantitative analysis of the negative impact that ICP and ECP have on flux is evaluated for a variety of experimental conditions, draw and feed solution concentrations, and hypothetical symmetric and asymmetric membranes.

6.1.2.3 Model parameters

The models for the PRO and FO modes – Equations 6.1.7 and 6.1.8, respectively – were solved iteratively using Microsoft Excel Solver for a range of feed and draw solution concentrations to determine permeate water flux. Assumed experimental conditions are similar to those presented in McCutcheon and Elimelech (2006a) and are summarized in Table 6.1.1. Draw solution concentrations were limited to a range of 0.05 to 1.5 M NaCl. Feed solution concentrations range from 0 (deionized water) to 1.4 M NaCl. Sodium chloride is used as the model solute for the feed and draw solutions because diffusion coefficients for a wide range of salt concentrations are readily available (Lobo, 1993). Viscosity, density, and osmotic pressure of each solution were calculated for the feed and draw solutions using software from OLI Systems, Inc. (Morris Plains, NJ) and Aspen HYSYS® (Cambridge, MA). For the membrane channel, a crossflow, plate and frame assembly of rectangular geometry with dimensions of 7.7 cm × 2.6 cm × 0.3 cm ($L \times W \times D$) was assumed, similar to our previous studies (McCutcheon & Elimelech, 2006a; McCutcheon & Elimelech, 2006b). The membrane characteristics are based on those determined previously (McCutcheon & Elimelech, 2006b) for a commercially available FO membrane made by Hydration Technologies (Albany, OR). The K values were taken from a previous study using this membrane (McCutcheon & Elimelech, 2006a).

Table 6.1.1 Parameters for forward osmosis flux modeling.

Parameter	Symbol	Units	Value/Range
Bulk draw osmotic pressure[a]	$\pi_{d,b}$	atm (psi)	0 – 70 (0 – 1028.3)
Bulk feed osmotic pressure[a]	$\pi_{f,b}$	atm (psi)	0 – 65.4 (0 – 960.7)
Water permeability coefficient[b]	A	$m \cdot s^{-1} \cdot atm^{-1}$	3.11×10^{-7}
Mass transfer coefficient[c]	k	$m \cdot s^{-1}$	1.74×10^{-5}
Solute resistance to diffusion (PRO mode)[d]	K	$s \cdot m^{-1}$	2.24×10^{5}
Solute resistance to diffusion (FO mode)[d]	K	$s \cdot m^{-1}$	2.67×10^{5}
Temperature		°C	20
Crossflow velocity		$cm \cdot s^{-1}$	21.3

[a]Calculated using Hysis/OLI.
[b]Determined from RO experiments (McCutcheon & Elimelech, 2006b).
[c]Calculated with values from Hysis/OLI, same for feed and draw solution.
[d]Determined by McCutcheon and Elimelech (2006a).

6.1.3 RESULTS AND DISCUSSION

The following results model permeate water flux behavior under a variety of experimental conditions, most of which are provided in Table 6.1.1. These results quantify the deleterious effects of internal and external CP both individually and coupled.

6.1.3.1 Dense symmetric membrane

Concentration polarization in pressure driven membrane processes occurs as convective water flux drags solute up against the dense, rejecting layer of the membrane, thereby increasing the concentration of the feed solute at the membrane surface. This increases the osmotic pressure that must be overcome before water can permeate the membrane. The severity of this concentrative ECP can be modeled based on film theory as presented above in Equation 6.1.2. Shown in Figure 6.1.1, this equation was solved iteratively after accounting for all system parameters (temperature, solution characteristics, and hydrodynamics) and by using the model parameters in Table 6.1.1 to predict the water flux behavior based on the net driving force, $\Delta P - \pi_{F,b}$, for feed concentrations ranging from deionized water to 1 M NaCl. A deionized water feed is indicated by the hydraulic permeability line in Figure 6.1.1. Figure 6.1.1 illustrates, as deviations from this dashed line, that higher net driving forces are needed to achieve the same flux for higher feed concentrations because of the greater prevalence of concentrative ECP. Using Figure 6.1.1, we determine the excess net driving force required to obtain a typical operating flux for RO of 10 gfd (4.71 μm · s^{-1}) for each feed concentration, given in the fourth column of Table 6.1.2.

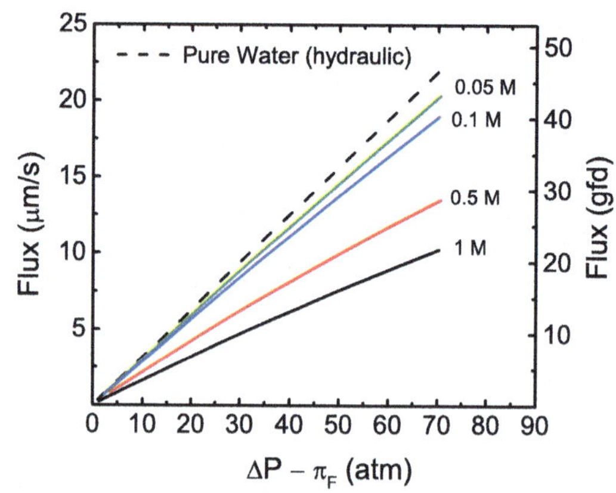

Figure 6.1.1 Effect of concentrative ECP on flux performance based on net driving force in reverse osmosis mode for various NaCl feed solutions. Flux is driven hydraulically and is predicted based on Equation 6.1.2. Conditions: listed in Table 6.1.1.

The excess pressure is indicative of the portion of the hydraulic pressure that is used to compensate for the concentrative ECP. This excess pressure exceeds 50% of the total net driving force, severely limiting flux performance and requiring higher hydraulic pressures for flux maintenance at higher feed water concentrations (or high feed water recoveries).

Table 6.1.2 Performance data from Figure 6.1.1 for a set flux of 10 gfd (4.71 μm·s^{-1}).

Feed Concentration	$\pi_{F,b}$	Required ($\Delta P - \pi_{F,b}$)	Excess Pressure Required for 10 gfd	% of Net Driving Force
M	atm (psi)	atm (psi)	atm (psi)	
0	0.0 (0.0)	14.7 (216.1)	0.0 (0.0)	0.0
0.05	2.3 (33.8)	16 (235.2)	1.3 (14.7)	8.1
0.1	4.7 (69.1)	16.6 (244.0)	1.9 (23.5)	11.4
0.5	23.3 (342.5)	22.6 (332.2)	7.9 (111.7)	35.0
1	46.7 (686.5)	30.3 (445.4)	15.6 (224.9)	51.5

In osmosis across a dense membrane, we must consider both concentrative and dilutive ECP as described in Equation 6.1.4. First, we consider dilutive ECP exclusively. In Figure 6.1.2, the solid line is the modeled osmotically-driven flux for a symmetric membrane under a range of NaCl draw solution concentrations from 0 to 1.5 M NaCl. The deviation from the pure water hydraulic permeability line (the straight dashed line) and this solid line is caused entirely by dilutive ECP on the permeate side of the membrane. When solute is assumed to be present in the feed solution, dilutive ECP on the permeate side of the membrane is coupled with concentrative ECP on the feed side of the membrane. The dotted line in Figure 6.1.2 indicates the coupled effect of concentrative and dilutive ECP at a draw solution concentration of 1.5 M NaCl and a varying feed concentration from 0 to 1.4 M NaCl. As feed concentration is increased, flux will decrease due to a decreased osmotic driving force and concentrative ECP. Simultaneously, however, lower flux will yield a decreased severity of dilutive ECP on the permeate side. The coupling of concentrative and dilutive ECP yields a lower flux than either phenomenon alone, but the effects are not purely additive.

The scenarios described above take into account only external boundary layers associated with dense membranes. However, current generation membranes, especially those used in pressure driven membrane processes, are *asymmetric* with a porous layer which provides mechanical support to a thin, dense active layer. In the following sections, flux is predicted in the presence of ECP along the active layer and internal concentration polarization (ICP) that occurs within this porous layer. Further analysis elucidating the impact of support layer properties on flux performance is presented.

6.1.3.2 Asymmetric membrane in PRO mode

Asymmetric membranes can be oriented in either the PRO or the FO mode, as described above. In the PRO mode, the draw solution is on the active layer while the feed solution is on support layer, as is typical of PRO applications (Lee *et al.*

1981; Loeb *et al.* 1997; McCutcheon & Elimelech, 2006a). To model flux through these membranes, we use Equation 6.1.7, again, solved iteratively as above. If the feed is deionized water or very dilute, concentrative ICP will not occur to a significant extent, and the modeled flux will be the same as given by the solid line in Figure 6.1.2. For a non-dilute feed solution, internal CP will have a significant impact on water flux, the severity of which is determined by the value of the solute resistance to diffusion within the porous layer, *K*.

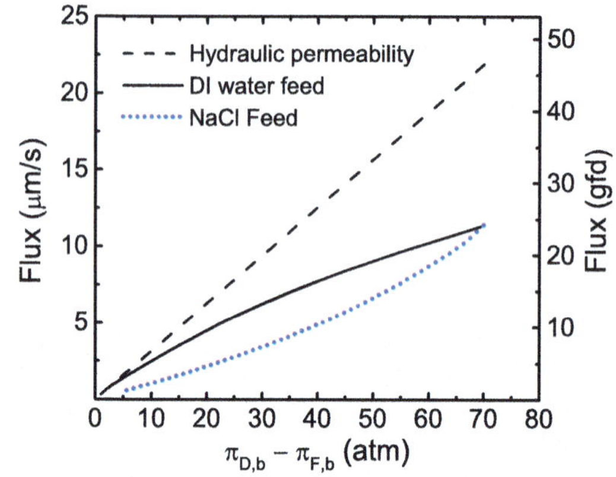

Figure 6.1.2 Effect of dilutive ECP (solid line) and coupled concentrative and dilutive ECP (dotted line) in forward osmosis as determined from Equation (6.1.4). The hydraulic permeability (dashed) line is indicative of pure water flux under hydraulic pressure. Model conditions were as follows. For dilutive ECP only (solid line), feed solution was DI water and draw solution concentration ranged from 0.05 to 1.5 M. For the coupled concentrative and dilutive ICP (dotted line), draw solution was fixed at 1.5 M NaCl and feed solution ranged from 0 to 1.4 M. Other conditions are listed in Table 6.1.1.

Flux simulation in the PRO mode. A simulation using a draw solution of 1.5 M NaCl and a feed solution concentration varied as high as 1.4 M under conditions specified in Table 6.1.1 is shown in Figure 6.1.3 for membranes with a variety of *K* values. The line labeled "*K*" is indicative of flux performance in the PRO mode with the *K* value provided in Table 6.1.1. This value was also used for comparing experimental flux behavior with model predictions in our recent publication (McCutcheon & Elimelech, 2006a). Lines are indicative of the model flux over a range of transmembrane osmotic pressure differences using variable factors of *K* as indicated by the label. These different *K* values are representative of changes in membrane support layer structural characteristics (thickness, tortuosity, and porosity as represented in Equation 6.1.5). For this investigation, it is assumed that *D* will not change since we are only examining NaCl solutions at 20 °C. Diffusion coefficient is also assumed to change negligibly over the concentration range imposed by the model (no greater than 1.5 M NaCl). Higher values of K, typically associated with thicker and denser support layers which have an increased solute diffusion resistance, resulted in lower fluxes for a given transmembrane osmotic pressure. This is due to the buildup of solute within the support layer (concentrative ICP) as the solute diffusion out into the bulk is significantly hindered. As K is lowered, flux increases because solute can diffuse out of the support layer more easily, decreasing the prevalence of concentrative ICP.

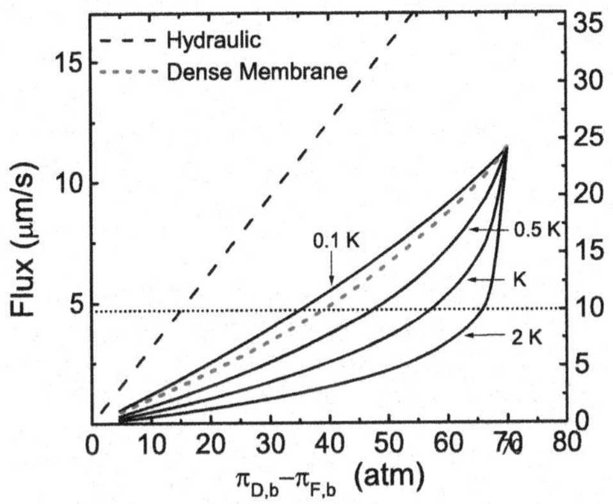

Figure 6.1.3 Flux modeled in the PRO mode (concentrative ICP coupled with dilutive ECP) for membranes with varying solute resistivity, K. Simulations are based on Equation 6.1.7. The hydraulic permeability (dashed) line is indicative of pure water flux under hydraulic pressure. The line labeled "K" is indicative of the performance of a membrane with the PRO value of K given in Table 6.1.1. The other values of K have been changed by the indicated factor. Model assumes that no concentrative ECP occurs at the surface of the porous support layer. The dotted line is the dotted line from Figure 6.1.2 and indicates the performance of a dense symmetric membrane under the same conditions. Conditions used in the simulations: NaCl draw solution concentration of 1.5 M (or $\pi_{D,b} = 70$ atm), NaCl feed solution range from 0 to 1.4 M, and other conditions as summarized in Table 6.1.1.

The value of K can be lowered to zero, hypothetically approaching a dense membrane. However, as K gets small, the model assumption of negligible ECP on the support layer becomes invalid. Superimposed on Figure 6.1.3 is the dotted line from Figure 6.1.2, showing the dense symmetric membrane flux in the presence of concentrative and dilutive ECP. This line represents the theoretical maximum of water flux performance for any membrane with the same water permeability in a forward osmosis process in either PRO or FO mode under these conditions. Hence, this line indicates the actual performance for a membrane where $K = 0$. We therefore conclude that the flux profile labeled by 0.1 K as determined from our model does not actually exist for a real membrane. Hence, while improving flux performance is possible by tailoring membrane structural properties, there is a limit of performance improvement due to ECP.

Performance assessment of the PRO mode. For PRO applications, the salinity of the feed water plays a major role in the performance of the process since this solution is subject to concentrative ICP. Therefore, we can more explicitly quantify the performance for different membranes used in the PRO mode by selecting a desired operating flux of 10 gfd (4.71 $\mu m \cdot s^{-1}$) as indicated by the horizontal dotted line in Figure 6.1.3, and determining the maximum feed concentration that will allow this flux to occur for a fixed draw solution concentration of 1.5 M NaCl. The comparison of the different membranes listed in Figure 6.1.3 at this flux is provided in Table 6.1.3. The required driving force ($\Delta\pi = \pi_{F,b} - \pi_{D,b}$) is given by where the 10 gfd (4.71 $\mu m \cdot s^{-1}$) line intersects each flux profile. The ideal behavior is defined by deionized water being hydraulically driven through the membrane (flux with no external or internal concentration polarization).

Table 6.1.3 Performance data from Figure 6.1.3 (PRO mode) for a flux set at 10 gfd (4.71 $\mu m \cdot s^{-1}$). Draw solution is fixed at 1.5 M NaCl (or 70 atm).

Membrane	Required $\pi_{F,b} - \pi_{D,b}$[a] atm (psi)	Max. $\pi_{F,b}$ atm (psi)	% of Ideal
Ideal[b]	14.7[b] (216.1)	55.3 (812.9)	100
Dense	38.9 (571.8)	31.1 (457.2)	56.2
0.5 *K*	47.4 (696.8)	22.6 (332.2)	40.9 *K*
	57.2 (840.8)	12.8 (188.2)	23.1
2 *K*	66.4 (976.1)	3.6 (52.9)	6.5

[a]Draw solution osmotic pressure is 70 atm.
[b]Based on deionized water feed under hydraulic pressure.

In the presence of coupled concentrative and dilutive ECP, the dense membrane can maintain a flux of 10 gfd with a feed solution having an osmotic pressure 56.2% of a feed solution for a system experiencing no ECP. This indicates that the coupled concentrative and dilutive ECP phenomena have a significant impact on performance of this a membrane in the PRO mode under these conditions at this flux. When concentrative ICP is considered, the maximum feed concentration drops even more precipitously, with the performance at the given value of *K* allowing for just 23.1% of the ideal maximum feed concentration. Doubling that *K* value reduces the concentration to 6.5% of the maximum. These values will change with a different desired flux, since the ECP and ICP moduli are both functions of flux.

6.1.3.3 Asymmetric membrane in FO mode

In the FO mode, the draw solution on the support layer will be subject to dilutive ICP while the feed solution on active layer will be vulnerable to concentrative

ECP. This is typical orientation for FO desalination or other osmotic-driven water treatment membrane processes (Cath *et al.* 2005; Cath *et al.* 2006). To model flux in this mode, we use Equation 6.1.8. If the feed is deionized water or very dilute, concentrative ECP will be negligible. For a non-dilute feed solution, ECP will impact flux performance. Both of these scenarios are discussed individually below.

Simulating flux in the FO mode: deionized water feed. Modeling flux in the FO mode normally incorporates the coupled effects of dilutive ICP and concentrative ECP. We can, however, examine dilutive ICP independently by first considering a deionized water feed. This evaluation is also reasonable for very dilute feed solutions. Figure 6.1.4 shows the flux performance of different membranes using a deionized water feed and draw solutions ranging in concentration from 0.05 M to 1.5 M NaCl. Again, we observe the effect of changing membrane structural properties on flux performance. Lower K values improve flux due to the increased ability of the draw solute to diffuse into the porous support layer to replenish the diluted draw solution. However, as in Figure 6.1.3, there is a limit to flux improvement. As K gets small, the model assumption that no dilutive ECP is occurring on the support layer begins to fail and the dense membrane model is approached (solid line from Figure 6.1.2). Therefore, as above, the flux profile indicating the performance of a membrane with a K value one–tenth of that given in Table 6.1.1 does not exist for a real membrane.

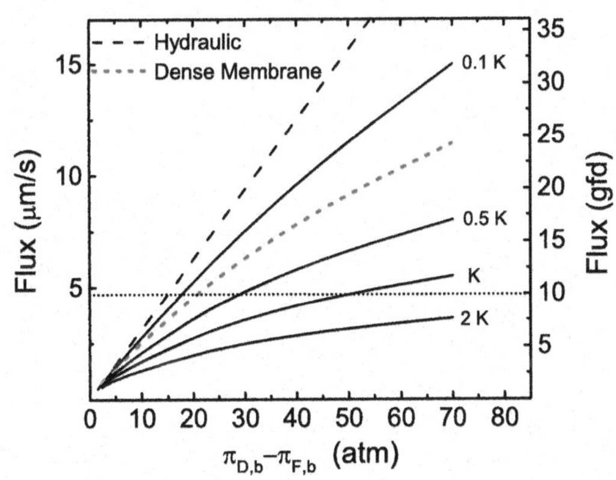

Figure 6.1.4 Flux modeled in the FO mode (dilutive ICP in the absence of concentrative ECP) for membranes with variable solute resistivity (from equation (6.1.8)). Some data taken from McCutcheon and Elimelech (2006a). The line labeled "*K*" is indicative of the performance of a membrane with the FO value of *K* given in Table 6.1.1. The other values of *K* have been changed by the indicated factor. The dotted line is identical to the solid line from Figure 6.1.2 and indicates the performance of a dense membranes under the same conditions. Conditions: NaCl draw solution concentration ranging from 0.05 to 1.5 M, deionized water feed, and other conditions listed in Table 6.1.1.

Performance assessment of the FO mode: deionized water feed. For these experimental conditions, performance is evaluated for each type of membrane by determining the driving force required to obtain a desired water flux set at 10 gfd (4.71 μm · s^{-1}) (horizontal dotted line in Figure 6.1.4). Table 6.1.4 lists the required osmotic driving force (equivalent to the osmotic pressure of the draw solution alone since $\pi_{F,b} = 0$) for each membrane and compares it to the hydraulic driving force required to obtain a permeate flux of 10 gfd (4.71 μm · s^{-1}) with pure water. A dense membrane in the presence of only dilutive ECP requires 1.4 times more driving force (provided by osmotic pressure) than pure water that is hydraulically driven. In the presence of ICP, the required driving forces increases substantially. The driving force must be 3.4 times greater for the membrane described in Table 6.1.1. Increasing K two folds and extrapolating from Figure 6.1.4 indicates that more than 190 atm of osmotic pressure (more than 13 times the ideal driving force) is required to obtain the desired flux of 10 gfd. These results show that dilutive ICP has a considerable diminishing effect on effective driving force.

Table 6.1.4 Performance data from Figure 6.1.4 (FO mode) for a flux of 10 gfd (4.71 μm·s^{-1}).

Membrane	Required $\pi_{F,b} - \pi_{D,b}$[a] atm (psi)	Ratio of Actual: Ideal Driving Force
Ideal	14.7 (216.1)	1
Dense	20.7 (304.3)	1.41
0.5 K	28.9 (424.8)	1.97 K
K	50.3 (424.8)	3.42
2 K	190.5 (2798.4)	12.96

[a]Feed solution osmotic pressure is 0 atm.

Simulating flux in the FO mode: non-dilute feed. When salt is assumed to be in the feed solution at an appreciable concentration, concentrative ECP occurs, further reducing driving force and hence flux. Figure 6.1.5 shows model results, based on Equation 6.1.8, assuming the draw solution is constant at 1.5 M NaCl and the feed solution is varied from 0 to 1.4 M NaCl. As in Figure 6.1.4, the dotted line is indicative of the dense membrane flux behavior (dotted line from Figure 6.1.2). This is the theoretical limit of flux under the specified conditions. As K approaches zero, dilutive ICP no longer solely controls flux behavior since dilutive ECP effects are no longer negligible on the support side.

It should also be noted that in the presence of a non-dilute feed solute, the differences in flux behavior between the different membranes is diminished. As the osmotic pressure of the feed solution is increased, decreased osmotic driving force reduces flux. Lower flux reduces the severity of ICP and hence the impact of K on flux behavior. The presence of concentrative ECP only increases the effective osmotic pressure of the feed solution, intensifying this effect.

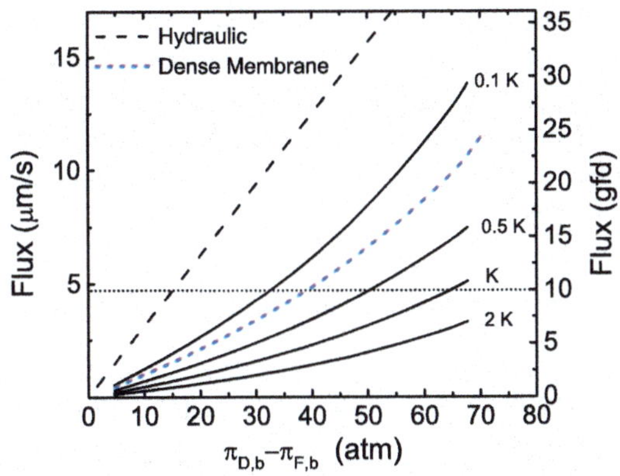

Figure 6.1.5 Flux modeled in the FO mode (dilutive ICP coupled with concentrative ECP) for membranes with variable solute resistivity (from equation (6.1.8)). The line labeled "*K*" is indicative of the performance of a membrane with the FO value of *K* given in Table 6.1.1. The other values of *K* have been changed by the indicated factor. The dotted curved line is identical to the dotted line from Figure 6.1.2 and indicates the performance of a dense membranes under the same conditions. Conditions: NaCl draw solution concentration of 1.5 M, feed concentrations ranging from 0 to 1.4 M NaCl, and other conditions listed in Table 6.1.1.

Performance assessment FO mode: non-dilute feed. Table 6.1.5 quantifies performance at a given permeate flux of 10 gfd or 4.71 $\mu m \cdot s^{-1}$ (see horizontal dotted line in Figure 6.1.5). Since the draw solution concentration is set at 1.5 M NaCl, performance will be quantified by the maximum feed osmotic pressure that can exist at 10 gfd flux for a given membrane. When the fourth column of Table 6.1.5 is compared with the fourth column of Table 6.1.3 (both sets of data taken under the same sets of conditions except that the membranes are oppositely oriented), it is clear that the effects of dilutive ICP are far more significant than the effects of concentrative ICP. At twice the provided value of *K*, even a deionized water feed cannot obtain 10 gfd (4.71 $\mu m \cdot s^{-1}$) under the constraints of the model for the FO mode, which is also apparent in Figure 6.1.4. This is due to the fact that dilutive ICP is impacting the concentrated draw solution, while concentrative ICP impacts the more dilute feed solution. Dilutive ICP therefore has a greater impact on overall driving force.

6.1.3.4 Implications for improved membrane design

Tailoring membranes for optimal performance in osmotic processes is critical for the future success of FO and PRO. In general, membranes used for osmotic processes should have high salt rejection, high water permeability coefficient, and lowest possible *K* value. If a membrane cannot be made without any support layer (i.e., $K = 0$), then the *K* value can be reduced by altering the membrane support layer by making it thinner and more porous (lower t and larger ε in Equation 6.1.5). This is in contrast to RO membrane design, where the support layer has no effect on flux and can be made as thick as needed to provide mechanical strength. By designing new membranes in this fashion, osmotically driven membrane processes, such as PRO and FO, can be made more economically viable for power generation and desalination, respectively.

Table 6.1.5 Performance data from Figure 6.1.5 (FO mode) for a flux of 10 gfd (4.71 $\mu m \cdot s^{-1}$).

Membrane	Required $\pi_{F,b} - \pi_{D,b}$[a] atm (psi)	Max. $\pi_{F,b}$ atm (psi)	% of Ideal
Ideal[b]	14.7[b] (216.1)	55.3 (812.9)	100
Dense	20.7 (304.3)	31.1 (457.2)	56.2
0.5 K	28.9 (424.8)	20.2 (296.9)	36.5 K
	50.3 (739.4)	5.6 (82.3)	10.1
2 K	>70 (>1029)	n.a.	n.a.

[a]Draw solution osmotic pressure is 70 atm.
[b]Based on deionized water feed under hydraulic pressure.

The use of PRO as a means of power generation near river deltas has been suggested for some time (Lee *et al.* 1981; Loeb, 1998). However, feed water salinity significantly impacts the flux performance due to concentrative ICP as shown above. Any salt leakage through the membrane from the draw solution side will also contribute to ICP, further reducing flux. Tailored membranes with a thinner or more porous support layer, while maintaining a very high selectivity (or solute rejection), will reduce the prevalence of ICP and improve water flux, as suggested in Tables 6.1.4 and 6.1.5. Higher flux rates in PRO result in higher energy production per unit area of membrane. If the energy produced per unit area of membrane is increased adequately, capital expenditures for a PRO power plant will be reduced. Being economically competitive with current energy producing technologies, this renewable energy resource could produce electricity where any freshwater river or stream meets a saline water body.

Tailoring membranes for use in the FO process, such as for FO desalination, will improve the overall FO process in two ways. In FO desalination, such as that described by McCutcheon *et al.* (2005), the draw solute is recovered and recycled, yielding fresh water and a reconcentrated draw solution. This recovery process is where the vast majority of the energy required for the overall FO process is consumed (McGinnis & Elimelech, 2007). The concentration of this draw solution must be minimized to reduce the energy used by the solute recovery system. Unfortunately, the prevalence of dilutive ICP forces the use of highly concentrated draw solutions to obtain desired fluxes, as demonstrated by the results of this paper.

By reducing the prevalence of dilutive ICP, more dilute draw solutions can be used, thereby reducing the energy required to recover and reconcentrate the draw solute. This is also suggested in Table 6.1.4, where it is shown that less osmotic pressure is required for obtaining a desired flux for FO membranes tailored to have lower K. The second improvement is the increased recovery possible with tailored membranes. By reducing the prevalence of dilutive ICP, feed waters with higher salinity can be treated with a given draw solute concentration, therefore improving the recovery of the FO desalination system. This is depicted in Table 6.1.5. Increased recoveries will reduce the volume of the brine discharge, which is the single most environmentally harmful byproduct of all desalination processes.

6.1.4 SUMMARY

Osmotically-driven membrane processes, such as forward osmosis and pressure retarded osmosis, operate on the principle of osmotic transport of water across a semi permeable membrane from a dilute feed solution into a concentrated draw solution. The major hindrance to permeate water flux performance is the prevalence of concentration polarization on both sides of the membrane. This paper evaluates the external and internal boundary layers which decrease the effective osmotic driving force. By modeling permeate flux performance, the role that feed and draw concentrations, membrane orientation, and membrane structural properties play in overall permeate flux performance are elucidated and linked to prevalence of external and internal concentration polarization. External concentration polarization is found to play a significant role in the reduction of driving force, though internal concentration polarization has a far more pronounced effect for the chosen system conditions. Reduction of internal concentration polarization by way of membrane modification was found to improve the predicted flux performance significantly, suggesting that alteration of membrane design will lead to improved performance of osmotic membrane processes.

6.1.5 REFERENCES

Cath T. Y., Gormly S., Beaudry E. G., Flynn M. T., Adams V. D. and Childress A. E. (2005). Membrane contactor processes for wastewater reclamation in space Part I. Direct osmotic concentration as pretreatment for reverse osmosis. *Journal of Membrane Science,* **257**, 85–98.

Cath T. Y., Childress A. E. and Elimelech M. (2006). Forward osmosis: principles, applications, and recent developments. *Journal of Membrane Science,* **281**, 70–87.

Gray G. T., McCutcheon J. R. and Elimelech M. (2006). Internal concentration polarization in forward osmosis: role of membrane orientation. *Desalination,* **197**, 1–8.

Jiao B., Cassano A. and Drioli E. (2004). Recent advances on membrane processes for the concentration of fruit juices: a review. *Journal of Food Engineering,* **63**, 303–324.

Kessler J. O. and Moody C. D. (1976). Drinking-water from Sea-Water by forward osmosis. *Desalination,* **18**, 297–306.

Kravath R. E. and Davis J. A. (1975). Desalination of Sea-Water by direct osmosis. *Desalination,* **16**, 151–155.

Lee K. L., Baker R. W. and Lonsdale H. K. (1981). Membranes for power-generation by pressure-retarded osmosis. *Journal of Membrane Science,* **8**, 141–171.

Lobo V. M. M. (1993). Mutual diffusion-coefficients in aqueous-electrolyte solutions (Technical Report). *Pure and Applied Chemistry,* **65**, 2614–2640.

Loeb S. (1998). Energy production at the Dead Sea by pressure-retarded osmosis: challenge or chimera? *Desalination,* **120**, 247–262.

Loeb S. (2002a). Large-scale power production by pressure-retarded osmosis, using river water and sea water passing through spiral modules (vol 143, p. 115 (2002)). *Desalination,* **150**, 205–205.

Loeb S. (2002b). One hundred and thirty benign and renewable megawatts from Great Salt Lake? The possibilities of hydroelectric power by pressure-retarded osmosis with spiral module membranes (vol 141, p. 85 (2001)). *Desalination,* **142**, 207–207.

Loeb S., Titelman L., Korngold E. and Freiman J. (1997). Effect of porous support fabric on osmosis through a Loeb-Sourirajan type asymmetric membrane. *Journal of Membrane Science,* **129**, 243–249.

McCutcheon J. and Elimelech M. (2006a). Influence of concentrative and dilutive internal concentration polarization on flux behavior in forward osmosis. *Journal of Membrane Science,* **284**, 237–247.

McCutcheon J. R. and Elimelech M. (2006b). Desalination by ammonia-carbon dioxide forward osmosis: influence of draw and feed solution concentrations on process performance. *Journal of Membrane Science,* **278**, 114–123.

McCutcheon J. R., McGinnis R. L. and Elimelech M. (2005). A novel ammonia-carbon dioxide forward (direct) osmosis desalination process. *Desalination,* **174**, 1–11.

McGinnis R. L. and Elimelech M. (2007). Energy requirements of ammonia–carbon dioxide forward osmosis desalination. *Desalination*, **207**, 370–382.

Moody C. D. and Kessler J. O. (1976). Forward osmosis extractors. *Desalination,* **18**, 283–295.

Mulder M. (1996). Basic Principles of Membrane Technology, 2nd edn., Kluwer Academic, Dordrecht, Boston.

Petrotos K. B. and Lazarides H. N. (2001). Osmotic concentration of liquid foods. *Journal of Food Engineering,* **49**, 201–206.

Petrotos K. B., Quantick P. and Petropakis H. (1998). A study of the direct osmotic concentration of tomato juice in tubular membrane – module configuration. I. The effect of certain basic process parameters on the process performance. *Journal of Membrane Science,* **150**, 99–110.

Chapter 6.2

Biofouling in FO systems: An experimental and numerical study*

6.2.1 INTRODUCTION

Forward osmosis (FO) is a membrane technology in rapid development, based on the difference in osmotic pressure between two liquids separated by a semi-permeable membrane, with a range of possible water treatment applications. Compared to other membrane filtration processes like reverse osmosis or nanofiltration, in FO almost no external hydraulic pressure is required for the process (Mulder, 1996). The driving force in FO is the osmotic pressure difference, such that water is extracted from the low osmotic pressure solution, also referred to as feed water, into the high osmotic pressure draw solution. Forward osmosis (FO) can play a bridging role in the integration of upstream and downstream water treatment processes, and could reduce energy consumption of the entire desalination or water recovery and reuse processes (Valladares Linares *et al.* 2014b). One of the main advantages of FO compared to conventional membrane processes for desalination and wastewater recovery (i.e., nanofiltration, NF, and reverse osmosis, RO) is the limited amount of external energy required to extract water from the feed solution (Valladares Linares *et al.* 2014b).

Use of FO for water reclamation was demonstrated in applications like oil/gas wastewater (Hickenbottom *et al.* 2013), landfill wastewater treatment and water recycling in space missions (Cath *et al.* 2005; Holloway *et al.* 2007). Bench scale tests demonstrated the potential of FO in municipal wastewater recovery combined with seawater desalination processes (Valladares Linares *et al.* 2013; Werner *et al.* 2013). Other studies evaluated the possibility of using FO systems as a pretreatment in membrane desalination technologies for low-pressure desalination processes

* This chapter is based, with permission from the copyright holder, on the previously published paper. (Szilard S. Bucs, Rodrigo Valladares Linares, Johannes S. Vrouwenvelder, Cristian Picioreanu. (2016). Biofouling in forward osmosis systems: an experimental and numerical study. *Water Research*, **106**, 86–97. doi.org/10.1016/j.watres.2016.09.031)

while simultaneously recovering water from a recycled feed solution (Achilli *et al.* 2014; Boo *et al.* 2013; Hoover *et al.* 2011). For desalination applications of FO, there are two possible approaches, direct and indirect desalination. In a direct FO desalination processes fresh water is directly extracted from seawater or brackish water by using the high salinity water as feed solution and an osmotic reagent as draw solution. An additional process is required for fresh water recovery from the diluted draw solution (Li *et al.* 2013; McCutcheon *et al.* 2005). A full-scale plant commissioned in 2012 is demonstrating the benefits and the viability of direct FO desalination (Modern Water Inc., 2012). In indirect FO desalination, the high salinity water is used as draw solution and an impaired water such as secondary wastewater effluent is used as of feed solution (Valladares Linares *et al.* 2013). The attractiveness of indirect FO desalination beside the low-cost draw solution (seawater or brackish water) relies on the diluted draw solution which can be used as feed water in a subsequent desalination process like reverse osmosis, thus reducing the cost of the entire desalination process (Yangali-Quintanilla *et al.* 2011).

However, as in other membrane filtration processes, FO also has the potential of membrane fouling. In certain FO operation conditions, the membrane can suffer from fouling on one side or on both sides. During indirect FO desalination when both the feed (secondary wastewater effluent) and the draw (seawater) solutions have fouling potential, deposition of unwanted material on the membrane surface can occur. Similar to other membrane filtration processes (e.g., reverse osmosis), accumulation and growth of microorganisms on the membrane surface is highly possible, leading to biofouling (Shannon *et al.* 2008; Vrouwenvelder *et al.* 2008). Only a few studies investigated the impact of biofouling on FO systems (Valladares Linares *et al.* 2014a; Yoon *et al.* 2013; Zou *et al.* 2013). Yoon *et al.* (2013) showed that the occurrence of biofouling in FO is less severe regarding flux decline when comparing to reverse osmosis. Zou *et al.* (2013) and Valladares Linares *et al.* (2014) investigated the impact of different spacer geometries in a cross-flow system on biofouling. Both studies showed an increase in water flux with the spacer inserted in the flow channel. Also noted was that the impact of the same amount of biomass on flux decline was lower when using a thicker spacer (Valladares Linares *et al.* 2014a). However, none of the FO membrane fouling studies investigated the impact of biofouling on both sides of the membrane simultaneously.

An important particularity of FO membrane filtration compared to other membrane systems such as reverse osmosis and/or ultrafiltration is a different occurrence of concentration polarization (CP). In FO systems three kinds of CP are distinguished: concentrative and dilutive external concentration polarization (ECP) and internal concentration polarization (ICP). The concentrative ECP corresponds to that observed in reverse osmosis systems. Unlike in RO systems, the dilutive ECP occurs in the draw solution side of the membrane due to the permeate flux and ICP takes place in the membrane porous support layer. It was shown that both external concentration polarization effects reduce the process driving force (i.e., the osmotic pressure difference), however their impact on water flux is minor. The most important and process- limiting concentration polarization was found to be the internal concentration polarization (McCutcheon & Elimelech, 2006). Consequently, the difference in driving force and its impact on the filtration processes prevent the direct transplantation of research results from different membrane systems, for example from NF and RO to FO.

The novelty of is work consists in the evaluation of biofouling effects on membrane performance in a cross-flow FO system with numerical simulations, also supported by experimental studies. The numerical model was used to study the impact of (i) biofilm properties (thickness, hydraulic permeability, and porosity), biofilm membrane surface coverage, and biofilm location (feed channel, draw channel and both) on FO membrane performance, and (ii) biofouling on external concentration polarization in different FO operation conditions.

6.2.2 EXPERIMENTAL

6.2.2.1 Experimental setup

A lab-scale FO membrane system was used for the experimental evaluation of permeation fluxes without and with biofilm development on the membranes. Two peristaltic pumps (Cole-Parmer) circulated in closed loop the feed and draw solutions at similar cross-flow velocities (8.6 $cm \cdot s^{-1}$). The pumps were connected to two reservoirs with the same volume (1.25 L), one for the feed solution and one for the draw solution. Each reservoir was placed on a digital scale (Mettler Toledo) connected to a computer. The water flux through the membrane was calculated from the change in weight of the reservoirs. All experiments were performed at 20 °C.

Cellulose triacetate (HTI, Scottsdale, AZ) forward osmosis membrane with a thickness of ~50 μm was used for the experiments. In all cases, the membrane active layer faced the feed solution while the support layer was in contact with the draw solution. This configuration was shown to be the most effective to prevent membrane fouling (Mi & Elimelech, 2008; Wang *et al.* 2010).

The membrane was placed into a custom-made flow cell from poly(methyl methacrylate) (PMMA), with an active membrane area of 20 cm^2. The flow cell was designed to allow operation without and with the presence of spacer in the feed and draw channels. The feed and draw channels height were the same and could fit 46 mil (1168 mm) thick spacers. For the studies with spacer, a 46 mil spacer produced by Hydration Technology Innovation (HTI, Scottsdale, AZ) was used on both sides of the membrane. A more detailed description of the flow cell can be found in previous studies (Li *et al.* 2012).

For flux measurements at various osmotic pressure differences sodium chloride (NaCl) solutions were prepared. The feed solution salt concentration was kept constant at 0.58 $kg \cdot m^{-3}$ while the draw solution concentration was varied from 5.8 $kg \cdot m^{-3}$ to 78.8 $kg \cdot m^{-3}$.

The biofouling experiments were run in a repeated batch mode. Each batch cycle was interrupted after 20 h of operation and the feed and draw solutions were replaced by fresh solutions (salt concentrations of 2 $kg \cdot m^{-3}$ in feed and 40 $kg \cdot m^{-3}$ in draw), then the next cycle was started. To ensure biofilm formation in the feed channel, a municipal secondary effluent (Al-Ruwais wastewater treatment plant, Saudi Arabia) was used as feed solution for the first 20 h. The effluent characteristics were: chemical oxygen demand (COD) 25 $mg \cdot L^{-1}$, total organic carbon (TOC) 4.84 $mg \cdot L^{-1}$, pH 7.1, conductivity 2.33 $mS \cdot cm^{-1}$, osmotic pressure 0.38 bar, total suspended solids (TSS) 2.25 $mg \cdot L^{-1}$ and volatile suspended solids (VSS) 2.15 $mg \cdot L^{-1}$. The adenosine triphosphate (ATP) concentration of the effluent was 215 $pg \cdot mL^{-1}$, which corresponds to a bacterial cell concentration of approx. 107 $cells \cdot mL^{-1}$. For the subsequent five FO filtration cycles, the feed solution was a synthetic municipal wastewater as described in (Valladares Linares *et al.* 2014a, 2013), initially with 360 mg $COD \cdot L^{-1}$, 115 mg $TOC \cdot L^{-1}$ and pH 5.66. All the chemicals were purchased from Sigma-Aldrich (Munich, Germany) and had a purity grade higher than 99.5%. All solutions were prepared with deionized (DI) ultra-pure water (Valladares Linares *et al.* 2014a, 2013).

6.2.2.2 Model description

A two-dimensional mathematical model was developed to investigate numerically the impact of biofouling on the performance of forward osmosis processes. The model includes liquid flow coupled with the mass transport of the solute (salt) through the feed and draw channels, through the asymmetric permeable membrane, and through a biofilm present on the FO membrane.

6.2.2.2.1 Model geometry

Without biofilm, the model geometry included three domains: feed and draw channels separated by a membrane support layer (Figure 6.2.1a), while the thin membrane active layer was assimilated with a boundary in this model. The flow channel length (L_X) was chosen tenfold smaller than the length of the flow cell used for the experimental studies. This length was small enough to allow performing computationally efficient numerical simulations, while maintaining a size representative for the FO system (i.e., including several spacer fibers). The heights of the two flow channels (L_Y) and membrane support layer thickness (L_S) were those used in the experimental studies. When the biofilm was present, it constituted a fourth planar domain with thickness L_B. Although not completely realistic, the feed and draw channel spacer fibers were circular and arranged in a zig-zag configuration mimicking a two-dimensional axial slice in a three-dimensional flow channel (Figure 6.2.1b). The diameter (d_M) and distance between the spacer fibers (L_M) corresponded with the 46 mil spacer geometry used for the experimental studies (Table 6.2.1).

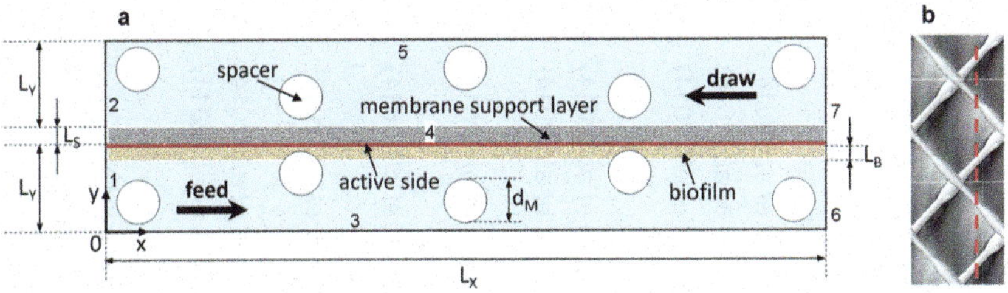

Figure 6.2.1 (a) Illustration of the forward osmosis (FO) model geometry. The model space is divided to three domains: feed and draw channels (blue color, with bold arrows indicating the flow directions) and membrane support layer (woven grey). The membrane active layer (red line) is facing the feed channel side and it represents a boundary. When present, the biofilm (yellow dotted zone) constitutes a fourth domain. The geometry dimensions are explained in Table 6.2.1. Numbers represent the different boundaries of the computational domains, as explained in Table 6.2.2. (b) Spacer fibers are arranged in such a way to mimic an imaginary 2D plane along the red dashed line in a 3D flow channel.

6.2.2.2.2 Fluid flow

The flow in the feed and draw channels was calculated from steady-state laminar Navier-Stokes equations Eq. (6.2.1):

$$\begin{aligned} &\rho(\mathbf{u}\cdot\nabla)\mathbf{u} + \nabla p = \nabla\cdot(\eta\nabla\mathbf{u}) \\ &\nabla\cdot\mathbf{u} = 0 \end{aligned} \tag{6.2.1}$$

where $\mathbf{u} = (u_x, u_y)$ is the vector of local liquid velocity, p is the pressure, ρ and η are the density and dynamic viscosity of the solution at 293 K, respectively. The steady-state laminar flow assumption is justified considering the low Re number used in the experiments (Re ≈ 80, calculated according to Schock & Miquel, 1987), as shown in Bucs *et al.* (2015).

The membrane support layer and biofilm domains were treated as porous media and Brinkman flow was assumed according to Eq. (6.2.2):

$$\begin{aligned} &\frac{\eta}{K}\mathbf{u} + \nabla p = \frac{\eta}{\varepsilon}\nabla^2\mathbf{u} \\ &\nabla\cdot\mathbf{u} = 0 \end{aligned} \tag{6.2.2}$$

where K and ε are the permeability and porosity of membrane support layer (K_{WS} and ε_S) and biofilm (K_B and ε_B).

The boundary conditions used are listed in Table 6.2.2. Fully developed laminar flow profiles with average velocity u_{in} were imposed at the feed and draw channel inlets, while the pressure was set to zero at the outlet boundaries. No-slip conditions were applied to the spacer surface and to the bottom and top

Table 6.2.1 Model parameters and experimental conditions.

Parameter	Symbol	Value	Unit	Source
Model geometry				
Channel length	L_X	10	mm	Chosen
Channel height	L_Y	1.17	mm	46 mil Filmtech™ spacer
Membrane thickness	L_S	50	μm	Valladares Linares *et al.* (2014)
Spacer fiber thickness	d_M	0.58	mm	Valladares Linares *et al.* (2014)
Spacer mesh size	L_M	4.1	mm	Valladares Linares *et al.* (2014)
Experimental conditions				
Average inlet flow velocity	u_{in}	8.6	$cm \cdot s^{-1}$	Valladares Linares *et al.* (2014)
Feed salt concentration inlet or initial (biofouling case)	$C_{F,0}$	2	$kg \cdot m^{-3}$	Valladares Linares *et al.* (2014)
Draw salt concentration inlet or initial (biofouling case)	$C_{D,0}$	40	$kg \cdot m^{-3}$	Valladares Linares *et al.* (2014)
Initial solution volume in feed reservoir	$V_{F,0}$	1.25	L	Valladares Linares *et al.* (2014)
Initial solution volume in draw reservoir	$V_{D,0}$	1.25	L	Valladares Linares *et al.* (2014)
Membrane area	A_M	20	cm^2	Valladares Linares *et al.* (2014)
Temperature	T	293	K	Valladares Linares *et al.* (2014)
Water and solute properties				
Feed water density	ρ_F	999	$kg \cdot m^{-3}$	Craittenden *et al.* (2012)
Draw solution density	ρ_D	1020	$kg \cdot m^{-3}$	McCutcheon and Elimelech (2006)
Water viscosity	η	10^{-3}	$Pa \cdot s$	Crittenden *et al.* (2012)

Salt diffusion coefficient in water	D_W	1.33×10^{-9}	$m^2 \cdot s^{-1}$	McCutcheon *et al.* (2006)
Salt diffusion coefficient in support layer	D_S	0.18×10^{-9}	$m^2 \cdot s^{-1}$	Calculated as L_S/K_S
Membrane properties				
Water permeability	A	0.57	$L \cdot m^{-2} \cdot h^{-1} \cdot bar$	Phillip *et al.* (2010)
Solute permeability	B	0.33	$L \cdot m^{-2} \cdot h^{-1}$	Phillip *et al.* (2010)
Membrane structural parameter	S	360	µm	McCutcheon *et al.* (2006)
Salt resistivity of support layer	K_S	2.7×10^5	$s \cdot m^{-1}$	McCutcheon *et al.* (2006)
Water permeability of support layer	K_{WS}	2×10^{-15}	m^2	Dreszer *et al.* (2013)
Support layer porosity	ε_S	0.3	–	Calculated as $\tau \cdot L_S/S$ (chosen tortuosity $\tau = 2$)
Biofilm properties				
Porosity	ε_B	0.5	–	Radu *et al.* (2010)
Permeability	K_B	10^{-17}	m^2	Dreszer *et al.* (2013)
Salt diffusion coefficient in biofilm	D_B	0.66×10^{-9}	$m^2 \cdot s^{-1}$	Calculated as $\varepsilon_B \cdot D_W$
Biomass specific growth rate	μ	0.81	d^{-1}	Fitted with 46 mil experimental data from Valladares Linares *et al.* (2014)
Initial biofilm thickness	$L_{B,0}$	5	µm	Chosen
Maximum biofilm thickness	$L_{B,E}$	150	µm	Valladares Linares *et al.* (2014)
Biofilm thickness	L_B	150	µm	Valladares Linares *et al.* (2014)

boundaries of the feed and draw channels, respectively. Flow continuity was assumed between the support layer or biofilm surface and the draw or feed channels. A flux of water J_W was imposed on the membrane active layer, proportional with the local osmotic pressure difference ($\Delta\pi$) and the permeability for water (A) Eq. (6.2.3) (Sagiv *et al.* 2014):

$$J_W = A(\pi_{\text{draw}} - \pi_{\text{feed}}) = A \cdot \Delta\pi \tag{6.2.3}$$

Table 6.2.2 Boundary conditions for the FO model geometry presented in Figure 6.2.1a.

Boundary No.	Feed Side		Draw Side	
	Fluid Flow	**Salt Transport**	**Fluid Flow**	**Salt Transport**
1	$u_x = 6u_{in}(1 - y/L_y)y/L_y$ $u_y = 0$	$C = C_F$	–	–
2	–	–	$p = 0$	$-\mathbf{n} \cdot D_w\nabla C = 0$
3 and spacer	$\mathbf{u} = 0$	$-\mathbf{n} \cdot D_w\nabla C = 0$	–	–
4	$u_x = 0$ $u_y = -J_W$	$-\mathbf{n}\cdot\mathbf{J} = J_S$	$u_x = 0$ $u_y = J_W$	$-\mathbf{n}\cdot\mathbf{J} = -J_S$
5 and spacer	–	–	$\mathbf{u} = 0$	$-\mathbf{n} \cdot D_w\nabla C = 0$
6	$p = 0$	$-\mathbf{n} \cdot D_w\nabla C = 0$	–	–
7	–	–	$u_x = -6u_{in}(1 - y/L_y)y/L_y$ $u_y = 0$	$C = C_D$

The osmotic pressures in the draw and feed solution (π_{draw} and π_{feed}) were calculated from Van't Hoff Eq. (6.2.4) (Mulder, 1996, p. 283):

$$\pi = 2C \cdot R \cdot T \tag{6.2.4}$$

function of local salt concentrations C_{draw} and C_{feed} at the membrane surface, universal gas constant R and temperature T (Table 6.2.2).

6.2.2.2.3 *Salt transport*

Sodium chloride was assumed the only soluble compound relevant for this model, transported by convection and diffusion in both of the flow channels Eq. (6.2.5).

$$D\nabla^2 C - \mathbf{u}\nabla C = 0 \tag{6.2.5}$$

where D is the diffusion coefficient and C is the local salt concentration. The salt diffusion coefficient was considered equal in the feed and draw solutions (D_W), whereas lower values were taken in the membrane support layer (D_S) and in the biofilm (D_B) domains (Table 6.2.1).

In the inlet, the solutions contained salt with concentrations C_F (feed channel) and C_D (draw channel). Typical no-diffusion conditions applied at the outlet boundaries for both channels, while the spacer surface, bottom and top boundaries were impermeable walls. In FO processes, draw solute leaks through the membrane into the feed side, as described by Eq. (6.2.6), (Sagiv *et al.* 2014):

$$J_S = B(C_{\text{draw}} - C_{\text{feed}}) \tag{6.2.6}$$

where B is the membrane permeability for salt, and C_{draw} and C_{feed} are the local salt concentrations in the draw and feed channels, respectively. The flux of salt is in opposite direction to the water flux.

6.2.2.2.4 Biofilm development

To study the impact of biofilms formed on the membrane on the FO process performance, three cases were investigated: (i) a continuous planar biofilm with thickness increasing in time (thus roughly simulating biofilm growth); (ii) a continuous planar layer with thickness constant in time; and (iii) a patchy biofilm formed by microbial colonies of various size, but constant in time. In the first two cases, the biofilm of uniform thickness covered the whole membrane area, while in the third case the patchy biofilm covered only 50% of the membrane, but had the same biofilm volume as in the second case. By assuming, for simplicity, unlimited biomass growth at constant specific rate μ and constant biomass concentration in the biofilm, the increase in biofilm thickness L_B in case (i) is represented by Eq. (6.2.7):

$$L_B = L_{B0}\exp(\mu t) \tag{6.2.7}$$

In all cases, the biofilm was represented as a computational domain and flow continuity was applied at the moving biofilm-liquid interface (Figure 6.2.1).

Water and solute transport in biofilms are functions of the biofilm properties and morphology. According to Darcy's law, permeation flux through the biofilm, J_W, is function of the pressure drop over the biofilm, fluid viscosity, biofilm permeability and thickness Eq. (6.2.8) (Dreszer *et al.* 2013):

$$J_W = \frac{K_B}{\eta \cdot L_B}\Delta p \tag{6.2.8}$$

This means that also the hydraulic resistance, R_B defined by Eq. (6.2.9), is time dependent because the biofilm thickness can change in time:

$$R_B = \frac{L_B}{K_B} \tag{6.2.9}$$

6.2.2.2.5 Model solution

The two-dimensional liquid flow and salt transport were solved in COMSOL Multiphysics (v5.1, Comsol Inc., Burlington, MA), on a triangular finite-element mesh with a maximum size of 5 μm, using a stationary solver. In case of biofilm growth, a moving mesh procedure was applied for the time-dependent biofilm-liquid interface, while the liquid flow and salt transport were also solved in time.

6.2.2.2.6 Simulation of batch operation

To simulate the batch process, the model was extended to include the feed solution concentration and draw solution dilution in time, together with the change in feed and draw solution volumes, V_F and V_D Eq. (6.2.10–6.2.13):

$$\frac{dV_F}{dt} = -J_W \cdot A_M \tag{6.2.10}$$

$$\frac{dC_F}{dt} = \left(J_W C_F + J_S\right) \cdot \frac{A_M}{V_F} \tag{6.2.11}$$

$$\frac{dV_D}{dt} = J_W \cdot A_M \tag{6.2.12}$$

$$\frac{dC_D}{dt} = -\left(J_W C_D + J_S\right) \cdot \frac{A_M}{V_D} \tag{6.2.13}$$

where C_F and C_D are the salt concentrations of feed and draw solutions, and A_M is the membrane area. The initial values are listed in Table 6.2.1.

6.2.3 RESULTS AND DISCUSSION

The aim of this study was to evaluate the impact of biofilms on the performance of FO processes. Although concentration polarization is considered in the developed model, a detailed study of its impact on process performance here is out of scope. First, a two-dimensional model was developed for the FO without biofilm presence and the results were compared with experimental data at different osmotic pressures and in a batch operation. Second, the model was extended to include biofilm development in the feed channel and compared with experimental results. Third, the calibrated model was used to investigate the biofilm impact on concentration polarization, and the effects of biofilm properties (i.e., thickness, roughness, porosity, and permeability) on the performance of an FO process. Finally, the influence of biofilm location (i.e., in draw channel, in feed channel or in both) was evaluated.

6.2.3.1 Evaluation of the forward osmosis model

To evaluate the proposed numerical model, simulations were compared with experimental measurements. In total three sets of flux measurements were performed: (i) at different osmotic pressures, (ii) in batch operation to record the flux decline due to solution dilution in time and (iii) with biofilm growth under batch operation.

6.2.3.1.1 Water flux increase with increasing osmotic pressure

Water flux was experimentally measured at various salt concentrations in the feed and draw solutions. For these experimental measurements, no spacer was used in the flow cell and the conditions were set to minimize the possibility of biofilm formation. Consequently, no spacer and no biofilm were considered in this model version. The feed solution salt concentration (NaCl) was kept at 0.58 $kg \cdot m^{-3}$ for all the experimental measurements and the concentration in the draw solution was varied from 5.8 $kg \cdot m^{-3}$ up to 78.8 $kg \cdot m^{-3}$ in steps of 14.6 $kg \cdot m^{-3}$. The experimental setup was operated in batch mode, one hour for each salt concentration, and the average flux in that hour was calculated.

Membrane parameters for the numerical simulations, such as, membrane water permeability ($A = 0.44\ L \cdot m^{-2} \cdot h^{-1} \cdot bar$), membrane solute permeability ($B = 0.26\ L \cdot m^{-2} \cdot h^{-1}$) were taken from Phillip *et al.* 2010. The salt diffusion coefficient in the support layer (D_S), support layer water, and solute permeability (K_{WS}, K_S), were also taken from literature (Dreszer *et al.* 2013; McCutcheon & Elimelech, 2006) (Table 6.2.1). To evaluate the model and the applied parameters, the flux was calculated at the same solute concentrations in the feed and draw solutions as used in the experiments. Comparing the calculated and measured flux increase with increasing osmotic pressure confirmed that, in this case, the model describes the process accurately without further parameter fitting (Figure 6.2.2).

6.2.3.1.2 Flux decline in batch experiments

In a batch FO process, the water flux decreases in time because the draw solution gets diluted, while the feed solution becomes more concentrated in salt. To further test the model in dynamic conditions, the flux decline in a batch process was first experimentally measured. The batch experiment was performed by recirculating 1.25 L of feed and draw solutions through the flow channels with spacers for a period of 1200 minutes (20 h).

For the whole batch period, the model predicts a linear flux decline similar to the experimental results (Figure 6.2.3a). As described by the model Eq. (6.2.8–6.2.11), the flux decline depends on the experimental conditions and membrane properties. For these studies, the 20 cm^2 membrane coupons were cut out from a larger sheet membrane, showing differences in permeability. For FO membranes, a relatively high standard deviation (>30%) in membrane properties is possible, as shown by Phillip *et al.* (2010). While the experimental conditions were set, the values for membrane properties in the numerical model had to be adjusted to describe the measurements (values from Table 6.2.1). In this way, membrane water permeability and the membrane solute permeability were increased by 30 percent in the numerical model compared with the values used in the previous osmotic pressure simulations. By increasing the membrane water permeability, *A*, not only that a higher flux is achieved, but the draw solution is also diluted faster, which results in a more rapid flux decline (Figure 6.2.3a). Higher membrane solute permeability, *B*, also increases the dilution rate of the draw solution, but we found that the water flux decline is more sensitive to the membrane water permeability.

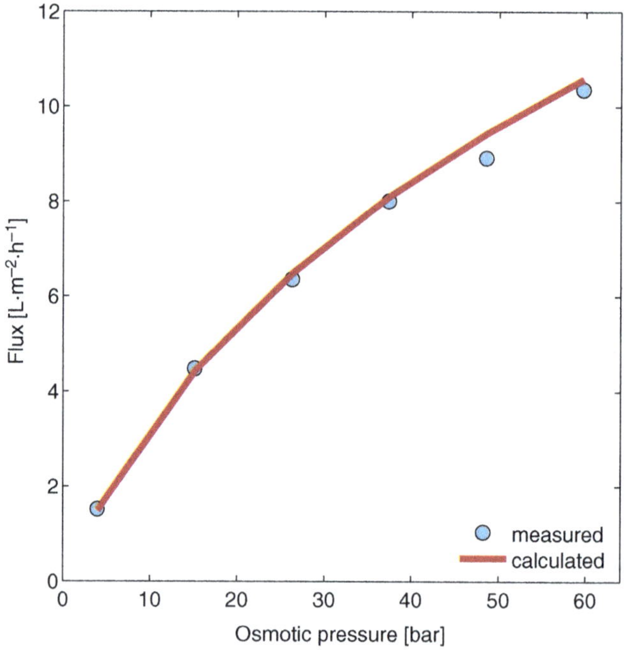

Figure 6.2.2 Experimentally measured (blue circles) and calculated with the developed numerical model (red line) water flux at different osmotic pressure differences, without the presence of spacer or biofilms. Model parameters are presented in Table 6.2.1.

These results were further used to provide consistent initial values and calibrated membrane parameter values for the FO simulations with biofilm.

6.2.3.1.3 Flux decline in repeated batch studies with biofilm formation

Repeated batch experiments were performed under conditions favorable for biofilm development in the feed channel (with spacers). For the draw channel, conditions were kept to avoid biofilm development (i.e., no inoculation and no substrate).

Flux through the membrane, as a performance parameter for the FO process, was monitored during the experimental period. In each cycle, the water flux declined not only because of dilution of the draw solution but also due to biofilm formation. After each cycle (1200 min) the solutions were replaced by fresh solutions, but the biofilm built up in each cycle remained and increasingly affected the flux (Figure 6.2.3b). As the biomass accumulated in the system, the rate of the flux decline within one cycle increased (i.e., steeper flux-time slopes). The faster water flux decline for the later cycles can be explained by the increased resistance of the developed biofilm (Valladares Linares *et al.* 2015).

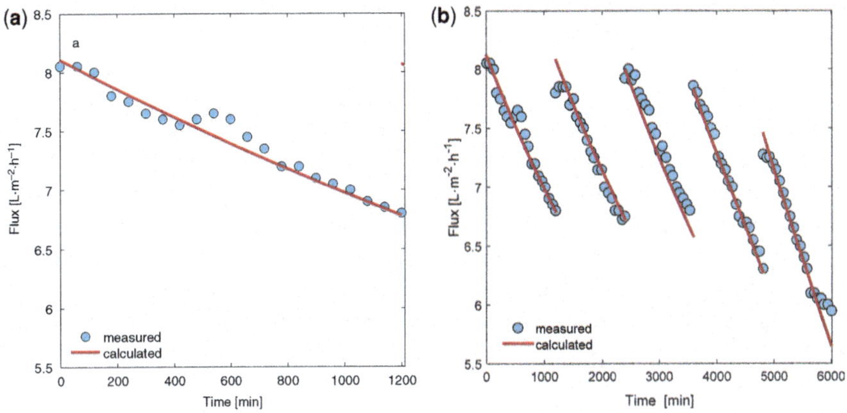

Figure 6.2.3 (a) Measured (blue circles) and calculated (red line) water flux decline over time due to the dilution of draw solution in batch operation, with spacers but without biofilms. (b) Combined impact of dilution and biofilm formation on flux in forward osmosis system operated in repeated batch mode, measured (blue circles) and model (red lines).

Biofilm properties (such as thickness, permeability, porosity and growth rate) may vary as the biofilm develops. Simulations of the repeated batch experiment were run with simple exponential biofilm growth in the feed channel. This ideal scenario for biofilm growth is probably not realistic for long-term biofilm development, but it proves to be representative for the initial phase of biofilm formation when the nutrients are still in excess (Vrouwenvelder *et al.* 2009). In the numerical model the biomass growth rate (μ) and the initial biofilm thickness ($L_{B,0}$) were set such that the maximum biofilm thickness ($L_{B,E}$) reached at the end of the five cycles was 150 µm, similarly to the thickness measured by Valladares-Linares *et al.* (2014). To keep the model simple the biofilm porosity (ε_B) was kept constant, while the water permeability of the biofilm (K_B) had to be decreased with the biofilm age. However, for a real biofilm these parameters (thickness, porosity and water permeability) are changing simultaneously with the biofilm age. For the first two cycles, KB = 10^{-16} m^2, while for last three cycles a smaller value of 10^{-17} m^2 was used in the model to represent the flux measurements (Dreszer *et al.* 2013; Martin *et al.* 2014).

In these conditions, the comparison of measured flux decline with the model results shows that the model can describe well the impact of biofilms on the FO process performance (Figure 6.2.3b).

6.2.3.2 Biofilm effect on FO performance

The biofilm impact on concentration polarization and the effects of biofilm properties and biofilm location on the performance of an FO process were further evaluated with the calibrated model only, no experimental studies were performed.

6.2.3.2.1 Impact of biofilm on water flux and concentration polarization

The effect of spacer presence on water flux was first investigated in a clean FO system. With spacers on both sides of the membrane, a small flux variation was observed along the membrane (±0.15 L·m^{-2}·h^{-1}), while without spacer the flux increased continuously towards the end of the flow channel because of the counter flow operation (Figure 6.2.4a, blue lines). The average water flux was only one percent higher with spacer than without. Visualization of the 2-D solute concentration distribution confirms that the presence of spacer has a small influence on the concentration polarization: the maximum solute concentration at the membrane was 2.35 kg·m^{-3} (in isolated spots only) compared with an inlet concentration of 2 kg·m^{-3} (Figure 6.2.4b). The role of the spacers is to keep the membrane sheets apart and, by enhancing fluid mixing, to reduce concentration polarization at the membrane surface. While in reverse osmosis feed spacers are needed to reduce concentration polarization close to the membrane surface (Radu *et al.* 2010; Subramani *et al.* 2006), it is known that in FO processes the concentration polarization at the membrane surface (external concentration polarization, ECP) does not significantly affect the water flux (Qin *et al.* 2010; Tan & Ng, 2008).

The water flux limiting factor in FO processes is the internal concentration polarization (ICP) within the membrane support layer (Gray *et al.* 2006; McCutcheon & Elimelech, 2006). However, the water flux considerably decreased (~20 percent) when a continuous biofilm on the feed side of the membrane was included in the model. Although the water flux profile along the membrane is similar to the flux profile without biofilm, a sharp flux decrease at the spacer fibers close to the membrane was observed (Figure 6.2.4a, dashed red line). The sharp flux decrease, corresponding to high ECP, is the result of the 150 μm (L_B) thick biofilm that completely fills up the space between the membrane and spacer fibers close to the membrane (Figure 6.2.4c).

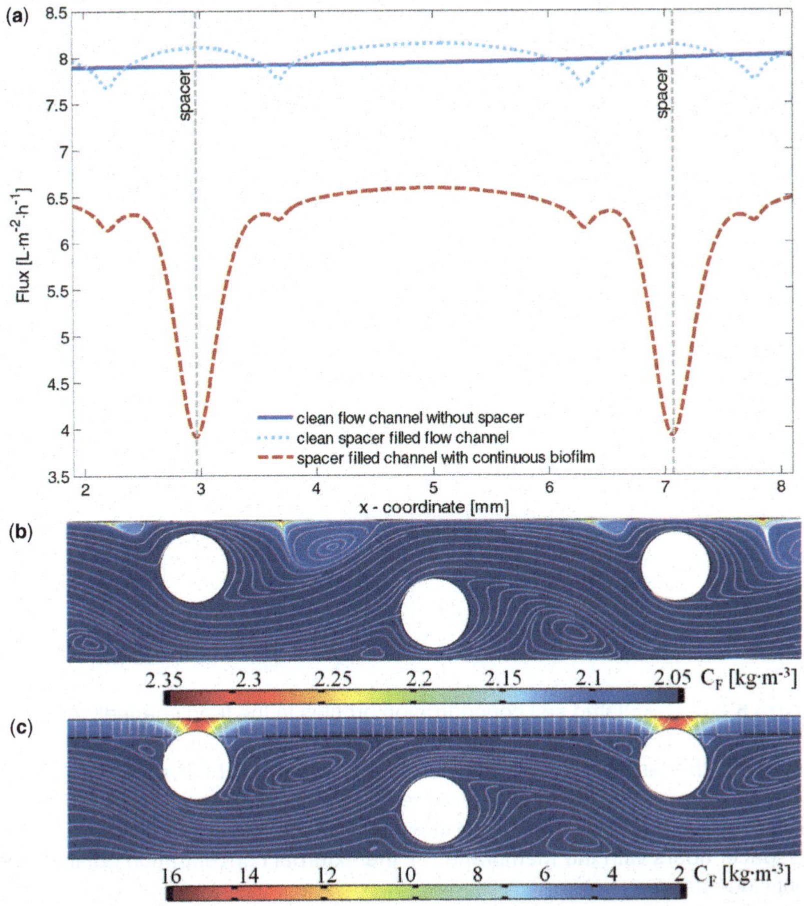

Figure 6.2.4 (a) Calculated water flux along the membrane: without spacer (continuous blue line), with spacer (dotted blue line) and with spacer and biofilm (dashed red line). The vertical dashed gray lines indicate the locations of the spacer fibers close to the membrane surface. (b) Salt concentration field in clean feed channel. (c) Salt concentration field with uniform thickness biofilm on the feed side of the membrane (top figure). The colors in (b) and (c) represent the solute concentration, the white streamlines indicate the water flow (from left to right). The biofilm leads to large concentration polarization and sharp water flux decrease close to the spacer fibers.

As shown in Figure 6.2.5, the presence of spacer and the ECP became important in the presence of biofilms. The osmotic pressure difference (the driving force of the FO process) decreased with 18 percent (an effective $\Delta\pi = 11.7$ bar, compared with 14.3 bar) in the case when the spacer fibers are far from the membrane (Figure 6.2.5a) and with 50 percent (an effective $\Delta\pi = 7.2$ bar, compared with 14.2 bar) when the fibers are close to the membrane and, thus, in contact with the biofilm (Figure 6.2.5b). The biofilm leads to increased ECP because it drastically limits the convective transport and creates a diffusion-dominated mass transfer layer. This is also reflected in the almost linear concentration profiles in the biofilm and support layer (Figure 6.2.5). At higher water fluxes, the increased contribution of convection may lead to nonlinear concentration profiles in the support layer, as reported in earlier studies (Su & Chung, 2011; Suh & Lee, 2013).

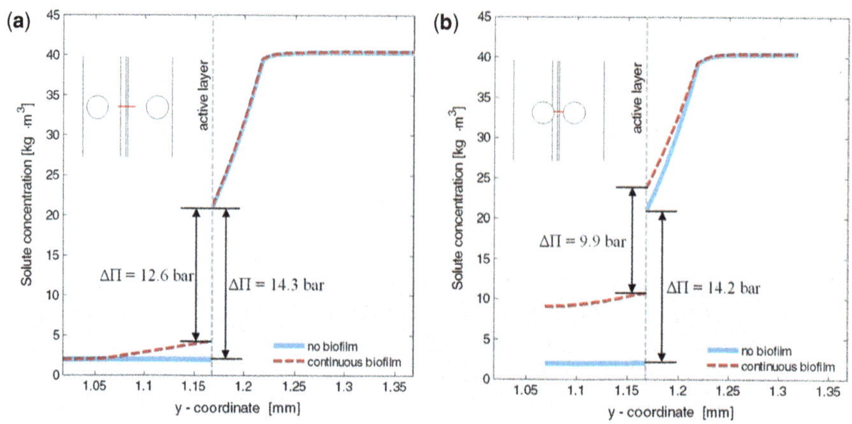

Figure 6.2.5 Calculated salt concentration profiles in the feed channel, membrane support layer and draw channel next to the membrane, without (continuous blue line) and with (dashed red line) a continuous and uniform biofilm layer. (a) Profiles at a location where the spacer fibers are far from the membrane. (b) Profiles at a position where the spacer fibers are close to the membrane. When the biofilm joins the spacer fibers with the membrane (b), the external concentration polarization is highly increased.

In the lab-scale as well as in a full-scale FO system the spacer is in full contact with the membrane surface at several locations. In these locations, the membrane is "shaded", and no water permeation is possible. Areas next to the spacer fibers close to the membrane are favorable locations for biofilm formation (Araújo *et al.* 2012; Baker *et al.* 1995; Vrouwenvelder *et al.* 2009). Furthermore, ECP increases in the biofilm presence and, in the case of a thick and dense biofilm, may even become the limiting factor for the water flux. Similar effects were reported for reverse osmosis systems (Chong *et al.* 2008; Radu *et al.* 2010).

6.2.3.2.2 Effect of biofilm properties

As the biofilm develops on the membrane surface, it becomes part of the separation process. Water and solute transport in biofilms are functions of the biofilm hydraulic permeability K_B and porosity ε_B.

To simulate the impact of biofilms on the FO processes performance, several biofilm parameters had to be defined. However, experimentally measured biofilm hydraulic resistance or permeability values are scarce in the literature (Dreszer *et al.* 2013; Martin *et al.* 2014; McDonogh *et al.* 1994). The reported values for biofilm hydraulic permeability vary between 10^{-15} – 10^{-18} m^2 and most of these were measured for biofilms developed in reverse osmosis or microfiltration systems. To evaluate how the hydraulic permeability influences the permeate production in an FO process, parametric sensitivity simulations were performed with increasing biofilm thickness at various biofilm hydraulic permeability (Figure 6.2.6a).

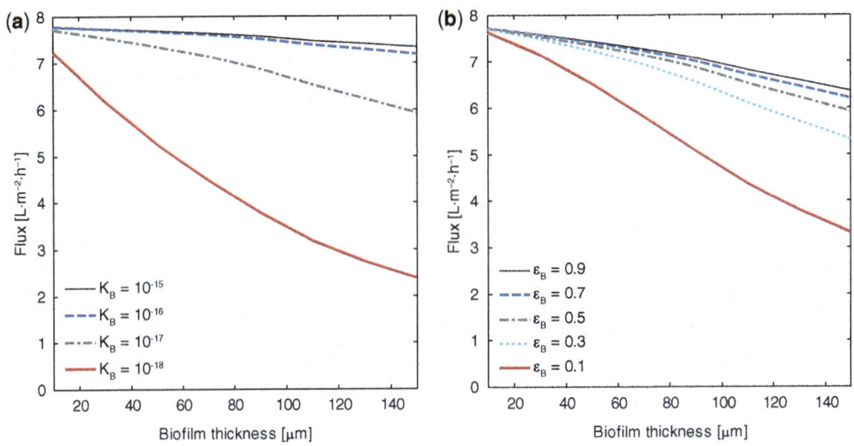

Figure 6.2.6 Calculated flux decline at different biofilm thickness on the membrane feed side, with spacers in the flow channels. (a) Effect of different biofilm permeability (K_B, in m^2). (b) Effect of different biofilm porosities (ε_B).

Results showed significant flux decline even for a relatively thin biofilm ($L_B = 5$ μm) when the biofilm hydraulic permeability was $K_B = 10^{-18}$ m_2 (Figure 6.2.6a, continuous red line). However, a compact biofilm with such a low permeability is unlikely to form in FO systems due to the relatively low water flux and pressure. On the other hand, high hydraulic permeability values such as 10^{-15} and 10^{-16} m^2 had an insignificant impact on flux even with thick (>100 μm) biofilms. Therefore, for the simulations with biofilm development compared with the experimental data, previously presented, the K_B was set to 10^{-16} m^2 for thin biofilm (first two cycles) and changed to 10^{-17} m^2 as the biofilm got thicker. For the rest of the simulations, the hydraulic permeability was kept constant at 10^{-17} m^2, in line with values reported by Martin *et al.* (2014).

Beside hydraulic permeability, the biofilm porosity also affects the permeation flux. Porosity can be defined as the volume fraction of the water phase in the biofilm. In general, as a result of spatial distributions of microbial species, biotic and abiotic components, porosity (along with other biofilm parameters) changes with biofilm age and thickness (Zhang & Bishop, 1994). The impact on flux of different biofilm porosities was evaluated (Figure 6.2.6b). Limited information can be found in the literature about biofilm porosity, with values generally varying from 0.2 to 0.9 (Lewandowski, 2000; Picioreanu *et al.* 2009; Radu *et al.* 2010; Zhang & Bishop, 1994). Simulations with various porosity and biofilm thickness showed that porosity significantly affects the flux only below values of 0.5, which are usually obtained under fast flow conditions with high erosion and detachment rates. In addition, biofilm maturation can also lead to dense cell layers especially in the deeper part of the biofilm, by this decreasing the biofilm porosity. Moreover, another effect of biofilm porosity is related to the decrease of effective salt diffusion coefficient with decreasing porosity, which leads to enhanced concentration polarization and thus less water flux.

The aim of these simulations was to understand how each parameter individually affects the water flux. Nevertheless, all three biofilm parameters (permeability, porosity and thickness) are linked and, in real biofilm systems, all change simultaneously in time. It was shown that in FO (where no applied hydraulic pressure is present) the biofilms tend to grow thicker and more porous and patchy, therefore with less hydraulic resistance (Kwan *et al.* 2015). This is contrary to what was observed for biofilms formed under applied hydraulic pressure in RO systems (Kwan *et al.* 2015; Xie *et al.* 2015). Clearly, measurements obtained in FO conditions are needed to reveal how the biofilm parameters change in time and to supply the numerical models with realistic parameter values.

6.2.3.2.3 Biofilm roughness and membrane surface coverage

In general, an average biofilm thickness is reported in the literature, with no or little information regarding the biofilm distribution along the membrane, due to the difficulty to evaluate the local biofilm thickness under operating conditions. Measurements with dyes coloring the liquid phase in a flow channel (e.g., rhodamine) can give a rough estimate of the average biofilm thickness (Prest *et al.* 2012; Staal *et al.* 2011). With other imaging techniques such as optical coherence tomography (OCT) and confocal laser scanning microscopy (CLSM) the local biofilm thickness can be accurately measured but only on small areas (Dreszer *et al.* 2013; Valladares Linares *et al.* 2015; West *et al.* 2015). Therefore, to evaluate the impact of membrane coverage with biofilm on the FO flux, numerical simulations were performed with the same amount of biofilm in the feed channel (i.e., same biovolume or average thickness) but with different membrane coverage (i.e., different biofilm roughness and membrane coverage).

Two biofilm types were evaluated with the numerical model. A continuous, constant thickness biofilm covering the whole membrane surface, and a patchy, irregular biofilm covering only 50 percent of the total membrane area. Figure 6.2.7a shows the flux profile along the membrane without biofilm and with irregular and continuous biofilm layers. Relative to the flux obtained in the clean channel, the continuous biofilm yielded 20 percent flux decrease. In the case of the patchy biofilm, the flux decrease depends on the local biofilm thickness. Maximum flux is obtained on the membrane areas not covered by biomass, whereas the flux strongly declines under the biomass colonies which lead to more concentration polarization. The biofilm enhanced concentration polarization (BECP) was also observed in the RO process, both experimentally and explained by numerical simulations (Chong *et al.* 2008; Herzberg & Elimelech, 2007; Radu *et al.* 2012, 2010). The average water flux over the whole membrane area was with 7 percent lower for the continuous foulant layer than for the patchy biofilm ($6\ L \cdot m^{-2} \cdot h^{-1}$ compared with $6.4\ L \cdot m^{-2} \cdot h^{-1}$). This effect has been observed also in ultrafiltration (UF) by Martin *et al.* (2014). Their study concluded that the fouling layer resistance depends not only on the thickness or volume of the fouling layer, but also on its spatial distribution. Namely, for two spatial distributions with the same mean thickness, it was found that the more heterogeneous distribution will have the higher permeate flux (Martin *et al.* 2014). Our results confirm that the membrane surface coverage by biofilm is indeed an important parameter to be measured and reported, besides the overall biofilm volume or thickness. However, for FO membranes the impact of biofilm resistance is less pronounced than in UF because the membrane resistance in FO is the dominant one, while UF membranes have a lower hydraulic resistance than the biofilm. Typically microfiltration (MF) and UF membranes resistance are between $10^{-11}\ m^{-1}$ to $10^{-13}\ m^{-1}$, while nanofiltration (NF), RO are $10^{-14}\ m^{-1}$ to $10^{-15}\ m^{-1}$ (Dreszer *et al.* 2013), compared with the biofilm resistance in this study which was between 10^{-16} and $10^{-17}\ m^{-1}$.

6.2.3.2.4 Impact of biofilm location on flux

During indirect water desalination, high salinity water is used as draw solution and impaired-quality water source such as wastewater effluent or urban storm water runoff is used as feed solution (Valladares Linares *et al.* 2014b). Seawater and brackish water have a high potential as draw solutions in indirect FO processes. In such a case, fouling can occur on both feed and draw sides of the membrane. To evaluate the impact of fouling on the different membrane sides, simulations were run with continuous biofilm layers (i) only on the feed side, (ii) only on the draw side and (iii) on both sides of the membrane.

In general fouling develops faster on the membrane support layer as it was shown by Mi and Elimelech (2008) and Wang *et al.* (2010). In our study, the membrane support layer is in contact with the draw solution. However, in this model case the biofilm development was not considered: only the impact on membrane performance of the same amount of biomass on the different membrane sides was evaluated (a constant biofilm thickness of 150 μm). A lower water flux resulted when the biofilm was only on the draw side of the membrane, compared with the biofilm on the feed side ($6.5\ L \cdot m^{-2} \cdot h^{-1}$ vs. $5.9\ L \cdot m^{-2} \cdot h^{-1}$, respectively) (Figure 6.2.8). This is caused by the higher ECP in the draw channel when a biofilm is present compared with the feed channel (Figure 6.2.9). Since the draw solution has a significantly higher solute concentration compared to the feed solution, the impact of dilutive ECP in the presence of biofilm is higher. As expected, the highest flux decline was obtained when biofilms of 150 μm thickness were present on both sides of the membrane (Figure 6.2.8).

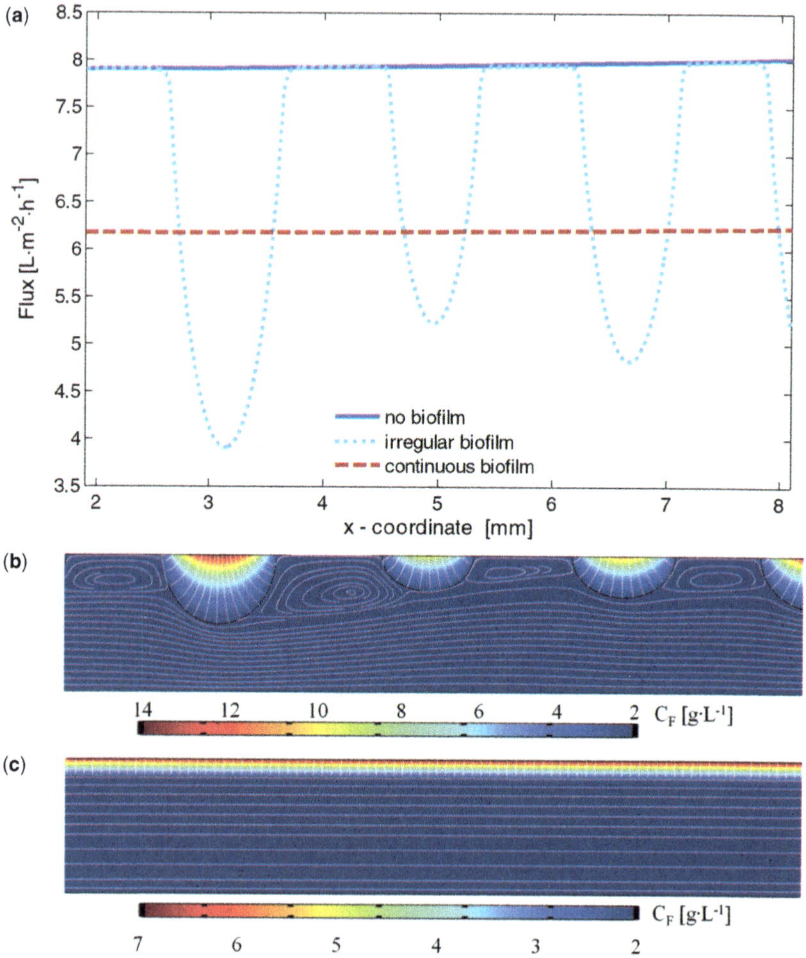

Figure 6.2.7 (a) Calculated water flux along the membrane: without biofilms (continuous blue line), with patchy biofilm (dotted blue line) and with continuous uniform biofilm (dashed red line), on the membrane active side. The average water flux obtained was: 7.9 $L \cdot m^{-2} \cdot h^{-1}$ (no biofilm), 6.4 $L \cdot m^{-2} \cdot h^{-1}$ (patchy biofilm) and 6 $L \cdot m^{-2} \cdot h^{-1}$ (continuous biofilm). (b) Salt concentration field in the patchy biofilm case. (c) Salt concentration field with uniform thickness biofilm. The colors represent the solute concentration, the white streamlines indicate the water flow (from left to right).

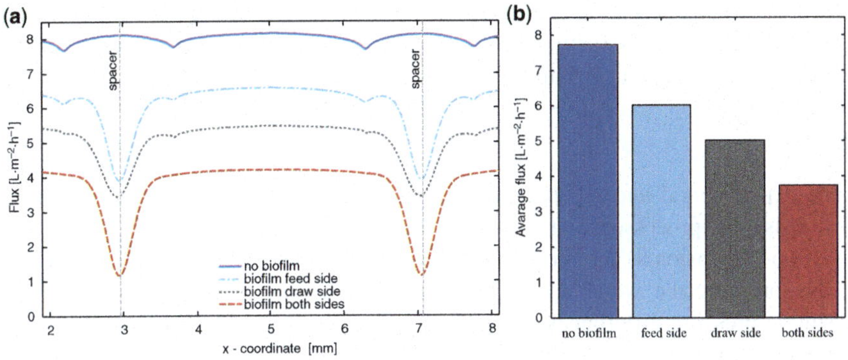

Figure 6.2.8 Calculated impact of monolayer biofilm location on the water flux, with spacers in the flow channels. (a) Calculated flux along the membrane. (b) Average flux over the whole membranem surface. The four considered scenarios were: no biofilm (continuous blue line), biofilm only on the feed side (blue dotted line), biofilm only at the draw side (grey dash-dot line) and biofilm on both sides (red dashed line) of the membrane. Biofilm presence on the draw side of the membrane leads to higher flux decline than the same amount of biomass was present on the feed side.

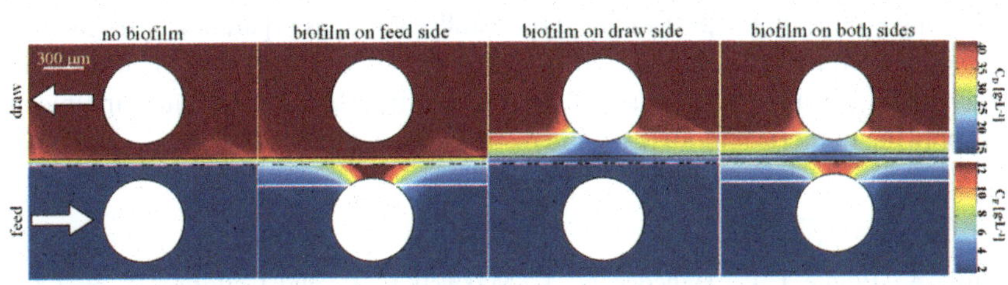

Figure 6.2.9 Calculated salt concentration fields without biofilm and with biofilm on the feed, draw and on both sides of the membrane. The white circles are the spacer fibers in the flow channels, the continuous white line represents the biofilm surface and the white dash-dot line is the membrane active layer. The largest external concentration polarization is induced by the biofilm presence on the draw side of the membrane. Note the different color scales for salt concentration in feed (2–12 kg·m^{-3}) and in draw solution (15–40 kg·m^{-3}).

Based on the simulation results it can be concluded that fouling on the draw side of the membrane has a higher impact on the FO process performance than on the feed side. Foulant deposition on the membrane draw side is possible in indirect desalination with FO when seawater or brackish water is used as draw solution (Valladares Linares *et al.* 2014b). Since the experiments in FO systems mainly focused on fouling accumulation on the membrane feed side (Mi & Elimelech,

2010; Valladares Linares *et al.* 2014a, 2014b; Wang *et al.* 2010), experimental studies are needed to further investigate the impact of biofilm formation in the draw side of the FO membrane.

6.2.4 SUMMARY

The experimental and computational evaluation of biofilm formation in forward osmosis (FO) processes led to the following conclusions.

Two-dimensional numerical simulations combining fluid flow with salt transport describe well the experimental FO flux measurements at (i) various osmotic pressure differences and (ii) in batch operation with and without biofilm on the membrane active layer.

The presence of a biofilm on the membrane significantly affects the external concentration polarization and can become the limiting factor for water permeation in FO systems.

Unlike in RO systems, the same amount of biomass leads to more water flux decline when the biofilm develops on the draw side of the FO membrane compared to the biofilm formed on the feed side, due to higher concentration polarization induced by foulant layer in the draw channel.

Not only biofilm properties such as hydraulic permeability and mean thickness, but also the biofilm porosity and spatial heterogeneity must be considered when evaluating biofouling effects on the performance of membrane-based separation systems. Since FO biofilms tend to grow thicker, and more porous than in RO, measurements obtained in FO conditions are needed to reveal how the biofilm parameters change in time and to supply the numerical models with realistic parameter values.

6.2.5 REFERENCES

Achilli A., Prante J. L., Hancock N. T., Maxwell E. B. and Childress A. E. (2014). Experimental results from RO-PRO: a next generation system for low-energy desalination. *Environmental Science and Technology,* **48**, 6437–6443. doi: 10.1021/es405556s

Araújo P. A., Kruithof J. C., Van Loosdrecht M. C. M. and Vrouwenvelder J. S. (2012). The potential of standard and modified feed spacers for biofouling control. *Journal of Membrane Science,* **403–404**, 58–70. doi: 10.1016/j.memsci.2012.02.015

Baker J., Stephenson T., Dard S. and Cote P. (1995). Characterisation of fouling of nanofiltration membranes used to treat surface waters. *Environmental Technology,* **16**, 977–985. doi: 10.1080/09593331608616335

Boo C., Elimelech M. and Hong S. (2013). Fouling control in a forward osmosis process integrating seawater desalination and wastewater reclamation. *Journal of Membrane Science,* **444**, 148–156. doi: 10.1016/j.memsci.2013.05.004

Bucs S. S., Linares R. V., Marston J. O., Radu A. I., Vrouwenvelder J. S. and Picioreanu C. (2015). Experimental and numerical characterization of the water flow in spacer-filled channels of spiral-wound membranes. *Water Research,* **87**, 299–310. doi: 10.1016/j.watres.2015.09.036

Cath T. Y., Adams D. and Childress A. E. (2005). Membrane contactor processes for wastewater reclamation in space: II. Combined direct osmosis, osmotic distillation, and membrane distillation for treatment of metabolic wastewater. *Journal of Membrane Science,* **257**, 111–119. doi: 10.1016/j.memsci.2004.07.039

Chong T. H., Wong F. and Fane A. (2008). The effect of imposed flux on biofouling in reverse osmosis: role of concentration polarisation and biofilm enhanced osmotic pressure phenomena. *Journal of Membrane Science,* **325**, 840–850. doi: 10.1016/j.memsci.2008.09.011

Crittenden O. C., Rhodes Trussell R., Hand D. W., Howe K. G. and Tchobanoglous G. (2012). MWH's Water Treatment: Principles and Design, 3rd edn., John Wiley & Sons, Inc. doi: 10.1002/9781118131473

Dreszer C., Vrouwenvelder J. S., Paulitsch-Fuchs A. H., Zwijnenburg A., Kruithof J. C. and Flemming H. C. (2013). Hydraulic resistance of biofilms. *Journal of Membrane Science,* **429**, 436–447. doi: 10.1016/j.memsci.2012.11.030

Gray G. T., McCutcheon J. R. and Elimelech M. (2006). Internal concentration polarization in forward osmosis: role of membrane orientation. *Desalination,* **197**, 1–8. doi: 10.1016/j.desal.2006.02.003

Herzberg M. and Elimelech M. (2007). Biofouling of reverse osmosis membranes: role of biofilm-enhanced osmotic pressure. *Journal of Membrane Science,* **295**, 11–20. doi: 10.1016/j.memsci.2007.02.024

Hickenbottom K. L., Hancock N. T., Hutchings N. R., Appleton E. W., Beaudry E. G., Xu P. and Cath T. Y. (2013). Forward osmosis treatment of drilling mud and fracturing wastewater from oil and gas operations. *Desalination,* **312**, 60–66. doi: 10.1016/j.desal.2012.05.037

Holloway R. W., Childress A. E., Dennett K. E. and Cath T. Y. (2007). Forward osmosis for concentration of anaerobic digester centrate. *Water Research,* **41**, 4005–14. doi: 10.1016/j.watres.2007.05.054

Hoover L. A., Phillip W. A., Tiraferri A., Yip N. Y. and Elimelech M. (2011). Forward with osmosis: emerging applications for greater sustainability. *Environmental Science and Technology,* **45**, 9824–9830. doi: 10.1021/es202576h

Kwan S. E., Bar-Zeev E. and Elimelech M. (2015). Biofouling in forward osmosis and reverse osmosis: measurements and mechanisms. *Journal of Membrane Science,* **493**, 703–708. doi: 10.1016/j.memsci.2015.07.027

Lewandowski Z. (2000). Notes on biofilm porosity. *Water Research,* **34**, 2620–2624. doi: 10.1016/S0043-1354(00)00186-X

Li D., Zhang X., Simon G. P. and Wang H. (2013). Forward osmosis desalination using polymer hydrogels as a draw agent: influence of draw agent, feed solution and membrane on process performance. *Water Research,* **47**, 209–215. doi: 10.1016/j.watres.2012.09.049

Li Z. Y., Yangali-Quintanilla V., Valladares-Linares R., Li Q., Zhan T. and Amy G. (2012). Flux patterns and membrane fouling propensity during desalination of seawater by forward osmosis. *Water Research,* **46**, 195–204. doi: 10.1016/j.watres.2011.10.051

Martin K. J., Bolster D., Derlon N., Morgenroth E. and Nerenberg R. (2014). Effect of fouling layer spatial distribution on permeate flux: a theoretical and experimental study. *Journal of Membrane Science,* **471**, 130–137. doi: 10.1016/j.memsci.2014.07.045

McCutcheon J. R. and Elimelech M. (2006). Influence of concentrative and dilutive internal concentration polarization on flux behavior in forward osmosis. *Journal of Membrane Science,* **284**, 237–247. doi: 10.1016/j.memsci.2006.07.049

McCutcheon J. R., McGinnis R. L. and Elimelech M. (2005). A novel ammonia-carbon dioxide forward (direct) osmosis desalination process. *Desalination,* **174**, 1–11. doi: 10.1016/j.desal.2004.11.002

McDonogh R., Schaule G. and Flemming H.-C. (1994). The permeability of biofouling layers on membranes. *Journal of Membrane Science,* **87**, 199–217. doi: 10.1016/0376-7388(93)E0149-E

Mi B. and Elimelech M. (2008). Chemical and physical aspects of organic fouling of forward osmosis membranes. *Journal of Membrane Science,* **320**, 292–302. doi: 10.1016/j.memsci.2008.04.036

Mi B. and Elimelech M. (2010). Organic fouling of forward osmosis membranes: fouling reversibility and cleaning without chemical reagents. *Journal of Membrane Science,* **348**, 337–345. doi: 10.1016/j.memsci.2009.11.021

Modern Water Inc. (2012). Modern Water commissions Al Najdah FO plant. *Membrane Technology,* doi: 10.1016/S0958-2118(12)70202-1

Mulder M. (1996). Basic Principles of Membrane Technology, 2nd edn., Springer, Netherlands. doi: 10.1007/978-94-009-1766-8

Phillip W. A., Yong J. S. and Elimelech M. (2010). Reverse draw solute permeation in forward osmosis: modeling and experiments. *Environmental Science and Technology,* **44**, 5170–5176. doi: 10.1021/es100901n

Picioreanu C., Vrouwenvelder J. S. and van Loosdrecht M. C. M. (2009). Three-dimensional modeling of biofouling and fluid dynamics in feed spacer channels of membrane devices. *Journal of Membrane Science,* **345**, 340–354. doi: 10.1016/j.memsci.2009.09.024

Prest E. I., Staal M., Kühl M., van Loosdrecht M. C. M. and Vrouwenvelder J. S. (2012). Quantitative measurement and visualization of biofilm O_2 consumption rates in membrane filtration systems. *Journal of Membrane Science,* **392–393**, 66–75. doi: 10.1016/j.memsci.2011.12.003

Qin J. J., Chen S., Maung H. O., Kekre K. A., Cornelissen E. R. and Ruiken C. J. (2010). Experimental studies and modeling on concentration polarization in forward osmosis. *Water Science and Technology,* **61**, 2897–904. doi: 10.2166/wst.2010.078

Radu A. I., Vrouwenvelder J. S., van Loosdrecht M. C. M. and Picioreanu C. (2010). Modeling the effect of biofilm formation on reverse osmosis performance: flux, feed channel pressure drop and solute passage. *Journal of Membrane Science,* **365**, 1–15. doi: 10.1016/j.memsci.2010.07.036

Radu A. I., Vrouwenvelder J. S., van Loosdrecht M. C. M. and Picioreanu C. (2012). REMOVED: modeling biofouling, scaling and combined fouling in reverse osmosis membrane devices. *Procedia Engineering,* **44**, 341–342. doi: http://dx.doi.org/10.1016/j.proeng.2012.08.409

Sagiv A., Zhu A., Christofides P. D., Cohen Y. and Semiat R. (2014). Analysis of forward osmosis desalination via two-dimensional FEM model. *Journal of Membrane Science,* **464**, 161–172. doi: 10.1016/j.memsci.2014.04.001

Schock G. and Miquel A. (1987). Mass transfer and pressure loss in spiral wound modules. *Desalination,* **64**, 339–352. doi: 10.1016/0011-9164(87)90107-X

Shannon M. A., Bohn P. W., Elimelech M., Georgiadis J. G., Mariñas B. J. and Mayes A. M. (2008). Science and technology for water purification in the coming decades. *Nature*, **452**, 301–310. doi: 10.1038/nature06599

Staal M., Prest E. I., Vrouwenvelder J. S., Rickelt L. F. and Kühl M. (2011). A simple optode based method for imaging O_2 distribution and dynamics in tap water biofilms. *Water Research*, **45**, 5027–5037. doi: http://dx.doi.org/10.1016/j.watres.2011.07.007

Su J. and Chung T.-S. (2011). Sublayer structure and reflection coefficient and their effects on concentration polarization and membrane performance in FO processes. *Journal of Membrane Science*, **376**, 214–224. doi: 10.1016/j.memsci.2011.04.031

Subramani A., Kim S. and Hoek E. (2006). Pressure, flow, and concentration profiles in open and spacer-filled membrane channels. *Journal of Membrane Science*, **277**, 7–17. doi: 10.1016/j.memsci.2005.10.021

Suh C. and Lee S. (2013). Modeling reverse draw solute flux in forward osmosis with external concentration polarization in both sides of the draw and feed solution. *Journal of Membrane Science*, **427**, 365–374. doi: 10.1016/j.memsci.2012.08.033

Tan C. H. and Ng H. Y. (2008). Modeling of external and internal concentration polarization effect on flux behaviour of forward osmosis. *Water Science and Technology: Water Supply*, **8**, 533. doi: 10.2166/ws.2008.116

Valladares Linares R., Li Z., Abu-Ghdaib M., Wei C. H., Amy G. and Vrouwenvelder J. S. (2013). Water harvesting from municipal wastewater via osmotic gradient: an evaluation of process performance. *Journal of Membrane Science*, **447**, 50–56. doi: 10.1016/j.memsci.2013.07.018

Valladares Linares R., Bucs S. S., Li Z., AbuGhdeeb M., Amy G. and Vrouwenvelder J. S. (2014a). Impact of spacer thickness on biofouling in forward osmosis. *Water Research*, **57**, 223–33. doi: 10.1016/j.watres.2014.03.046

Valladares Linares R., Li Z., Sarp S., Bucs S. S., Amy G. and Vrouwenvelder J. S. (2014b). Forward osmosis niches in seawater desalination and wastewater reuse. *Water Research*, **66**, 122–39. doi: 10.1016/j.watres.2014.08.021

Valladares Linares R., Wexler A. D., Bucs S. S., Dreszer C., Zwijnenburg A., Flemming H.-C., Kruithof J. C. and Vrouwenvelder J. S. (2015). Compaction and relaxation of biofilms. *Desalination and Water Treatment*, 1–13. doi: 10.1080/19443994.2015.1057036

Vrouwenvelder J. S., Manolarakis S. A., van der Hoek J. P., van Paassen J. A. M., van der Meer W. G. J., van Agtmaal J. M. C., Prummel H. D. M., Kruithof J. C. and van Loosdrecht M. C. M. (2008). Quantitative biofouling diagnosis in full scale nanofiltration and reverse osmosis installations. *Water Research*, **42**, 4856–4868. doi: 10.1016/j.watres.2008.09.002

Vrouwenvelder J. S., Graf von der Schulenburg D. A. A., Kruithof J. C. C., Johns M. L. L. and van Loosdrecht M. C. M. C. M. (2009). Biofouling of spiral-wound nanofiltration and reverse osmosis membranes: a feed spacer problem. *Water Research*, **43**, 583–94. doi: 10.1016/j.watres.2008.11.019

Wang Y., Wicaksana F., Tang C. Y. and Fane A. G. (2010). Direct microscopic observation of forward osmosis membrane fouling. *Environmental Science and Technology*, **44**, 7102–7109. doi: 10.1128/JVI.01705-10

Werner C. M., Logan B. E., Saikaly P. E. and Amy G. L. (2013). Wastewater treatment, energy recovery and desalination using a forward osmosis membrane in an air-cathode microbial osmotic fuel cell. *Journal of Membrane Science,* **428**, 116–122. doi: 10.1016/j.memsci.2012.10.031

West S., Wagner M., Engelke C. and Horn H. (2015). Optical coherence tomography for the in situ three-dimensional visualization and quantification of feed spacer channel fouling in reverse osmosis membrane modules. *Journal of Membrane Science,* **498**, 345–352. doi: 10.1016/j.memsci.2015.09.047

Xie M., Lee J., Nghiem L. D. and Elimelech M. (2015). Role of pressure in organic fouling in forward osmosis and reverse osmosis. *Journal of Membrane Science,* **493**, 748–754. doi: 10.1016/j.memsci.2015.07.033

Yangali-Quintanilla V., Li Z., Valladares R., Li Q. and Amy G. (2011). Indirect desalination of Red Sea water with forward osmosis and low pressure reverse osmosis for water reuse. *Desalination,* **280**, 160–166. doi: 10.1016/j.desal.2011.06.066

Yoon H., Baek Y., Yu J. and Yoon J. (2013). Biofouling occurrence process and its control in the forward osmosis. *Desalination,* **325**, 30–36. doi: 10.1016/j.desal.2013.06.018

Zhang T. C. and Bishop P. L. (1994). Density, porosity, and pore structure of biofilms. *Water Research,* **28**, 2267–2277. doi: 10.1016/0043-1354(94)90042-6

Zou S., Wang Y. N., Wicaksana F., Aung T., Wong P. C. Y., Fane A. G. and Tang C. Y. (2013). Direct microscopic observation of forward osmosis membrane fouling by microalgae: critical flux and the role of operational conditions. *Journal of Membrane Science,* **436**, 174–185. doi: 10.1016/j.memsci.2013.02.030

Part 7
Outlook

Chapter 7.1

Emerging applications for greater sustainability*

7.1.1 INTRODUCTION

Water and energy shortages have motivated the development and deployment of a diverse suite of technologies to support our growth beyond the traditional limitations of natural resources. These technologies are diverse not only with respect to method and use, but also in the anticipated consequences of their operation. Our ability to desalinate billions of gallons of seawater daily to satisfy a global water demand that exceeds accessible freshwater resources can be deemed a technological triumph. However, we also recognize that long-term solutions to energy and water shortages cannot rely on provision of one through intense consumption of the other; the vast amount of energy used daily in desalination for potable water production must also be conserved.

Methods that conserve or generate water and energy in a non-polluting and non-resource-exhausting manner are more capable of sustaining us. Utilizing wastes and abundant, low value resources as inputs for these processes also benefits us in the long term. The "engineered osmosis" processes of forward osmosis (FO) and pressure retarded osmosis have recently received increased attention as technologies that could potentially fulfill these criteria.

In FO, a water-permeable and salt-rejecting membrane is placed between two solutions of different osmotic pressures. These solutions can be natural resources, waste streams, or high purity solutions. Osmosis drives water to permeate through the membrane from the less concentrated "feed solution" to the more concentrated "draw solution". If FO is being employed to extract water from a resource, that resource acts as the feed solution, and it is paired with a draw solution of higher osmotic pressure.

---

*This chapter is based, with permission from the copyright holder, on the previously published paper. (Hoover L. A., Phillip W. A. Tiraferri A., Yip N. Y. and Elimelech M. (2011). Forward with Osmosis: Emerging Applications for Greater Sustainability. *Environmental Science & Technology*, **45**, 9824–9830. doi:10.1021/es202576h)

Engineered draw solutions are composed of pure water and specifically selected draw solutes. After extracting water from the feed solution, the engineered draw solution is separated into two streams: a concentrated draw solution, which is recycled to the membrane to facilitate continuous FO operation, and freshwater, which is the desired product. Draw solution regeneration can be an energy-intensive step in an otherwise low-energy process (McGinnis & Elimelech, 2007).

However, FO does not require draw solution regeneration in all cases. If a draw solute is used that adds value to the extracted water, the diluted draw solution can be used as is, and new draw solutes can be introduced to the system to create additional product. This is the case in commercially available personal hydration packs, which use a sugar-and-nutrient draw solution to provide energy-boosting drinks from natural waters. Alternatively, when concentration of the feed solution is the goal, an abundant and low value draw solution, such as seawater, can be used in a once-through fashion (Cath *et al.* 2006; Cath *et al.* 2010).

Avoiding draw solution regeneration can bypass a significant energy cost in FO technology implementation. Several emerging applications do just that while performing processes that enhance sustainability at the water–energy nexus. Herein, we highlight five regeneration-free FO applications under development that could benefit society by increasing energy production or efficiency, preventing pollution, utilizing waste, or relieving agriculture associated water stress (Table 7.1.1). For each, we will discuss the concept, expected benefits, and anticipated barriers or limitations. In doing so, we hope to motivate academic research, commercial implementation, and discussion of these and other potential uses of forward osmosis.

7.1.2 OSMOTIC DILUTION FOR ENERGY CONSERVATION

A surging world population, fixed freshwater resources, and limits to water conservation and reclamation make the inclusion of seawater desalination in long-term water production strategies inevitable. Desalination provides needed water, but it also requires a great deal of energy and can adversely impact the environment. The question, then, is how to make desalination as sustainable as possible. FO can help alleviate both energy costs, discussed here, and environmental concerns, discussed in the subsequent section.

The most energy efficient technology for desalination is reverse osmosis, in which a hydraulic pressure, much greater than the osmotic pressure of the saline feed water, is applied to drive water across a salt-rejecting membrane to produce freshwater. Though great gains in the energy efficiency of reverse osmosis technology have been made over the past few decades (National Research Council 2008), seawater desalination is an inherently energy-intensive process, and future gains in energy efficiency will be limited (Stoughton & Lietzke, 1965; Elimelech & Phillip, 2011). Energy costs contribute as much as 75% of the operating costs of RO desalination plants (National Research Council, 2008; Semiat, 2008). Innovations that reduce energy consumption will strengthen the suitability of reverse osmosis for addressing long-term drinking water needs.

Table 7.1.1 Potential positive impacts of the highlighted technologies on sustainability at the water – energy nexus.

Application of Forward Osmosis	Water	Energy	Waste Utilization	Environmental Protection
Dilution of input stream to reverse osmosis (RO) desalination plants.	Increases pool of available water resources.	Reduces RO operating pressure.	Reclaims impaired water for potable use.	
Dilution of desalination plant brine.		Provides low-energy concentration for chosen feed solution.	Usefully employs impaired water.	Dilutes brine and enhances its dispersion to reduce harm to marine life.
Osmotic cleaning of fouled reverse osmosis membranes.		Reduces RO operating pressure.		Replaces use of hazardous cleaning chemicals.
Cultivation and broth dewatering for production of algae biofuels.	Reduces use of freshwater in biofuel algae cultivation.	Participates in fuel production. Reduces energy used to dewater broth.	Uses wastewater nutrients to cultivate a source of biofuels.	Supports the production of alternatives to traditional fuels without increasing land use.
Use of brackish water for irrigation of crops with fertilizer.	Uses brackish water to replace or supplement freshwater use.	Reduces energy cost of desalinating brackish water for irrigation.		Enables use of local brackish water resources.

Reverse osmosis would require significantly less energy to desalinate diluted seawater than it does for full strength seawater because dilution decreases the osmotic pressure that must be overcome with applied hydraulic pressure (Cath *et al.* 2010; Bamaga *et al.* 2011). The amount of energy used in reverse osmosis desalination depends on several additional factors, including the water recovery (i.e., how much of the water in the feed stream is extracted as freshwater) (Spiegler & El-Sayed, 2001). But even without detailed calculations, one can see that less saline feed streams require less energy by comparing typical energy demands for desalination between brackish water and seawater. Energy usage at reverse osmosis desalination plants ranges from 3 to 7 kWh · m^{-3} for seawater and 0.5 to 3 kWh · m^{-3} for brackish water, even while brackish water plants operate at much higher recoveries (National Research Council, 2008), which require more energy.

Ideally, water used to dilute the seawater feed stream upstream of the reverse osmosis modules would come from a source that has little alternative use, for example, wastewater effluent or other impaired water (Cath *et al.* 2010). Such streams cannot be directly mixed into the seawater, as this would introduce additional contaminants and membrane foulants into the feed stream, compromising process efficacy. Direct use of these sources, particularly wastewater, for production of potable water also faces the significant challenge of negative public perception (Cath *et al.* 2010).

FO can safely and efficiently enable the use of impaired water to dilute the seawater entering a reverse osmosis desalination plant (Figure 7.1.1a). A low salinity wastewater, or other source, is separated by a selective FO membrane from seawater, which acts as the draw solution and extracts freshwater from the impaired source without incurring the associated contaminants. FO membranes, like reverse osmosis membranes, completely reject pathogens and large molecules (Cath *et al.* 2010; Bamaga *et al.* 2011) and pairing FO with reverse osmosis contributes an additional degree of safety for the use of impaired water. Another benefit is that membrane fouling is less problematic for FO than for pressure-driven membrane processes (Mi & Elimelech, 2010).

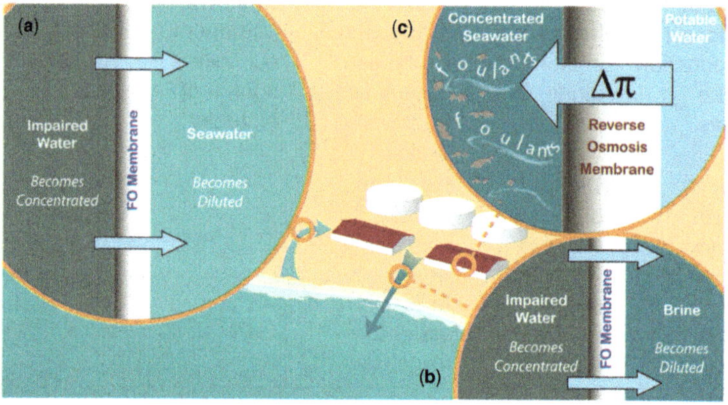

Figure 7.1.1 Desalination plant showing three locations where forward osmosis (FO) can be applied to increase sustainability (counter-clockwise): (a) upstream dilution of seawater feed stream to reduce energy consumption; (b) downstream dilution of concentrated brine before discharge to the marine environment; and (c) osmotic backwashing of reverse osmosis membranes to restore membrane performance.

Osmotic dilution of seawater feed streams uses impaired water to reduce the energy demand of reverse osmosis desalination. Though this approach offers multi-barrier protection of drinking water, the challenges associated with societal acceptance of the reclamation of impaired water for potable use may still apply. Education of the public on the benefits and quality of this system can encourage its acceptance. Successful implementation will also require co-location of a source of diluting water with the desalination plant.

7.1.3 OSMOSIS ENGINEERED FOR PROTECTION OF THE ENVIRONMENT

Osmosis can be applied elsewhere within desalination plants to reduce the negative impacts of desalination plant discharges on the environment. Greater protection of the environment is particularly needed if the discharge system cannot be constructed at a location where hydrodynamic conditions favor mixing and dispersion. At both membrane-based and thermal desalination installations, FO can be used to dilute concentrated brine before it is discharged into the environment (Figure 7.1.1b). Osmotic backwashing can replace harsh chemicals as the primary method of cleaning fouled reverse osmosis membranes (Figure 7.1.1c). To explain how these techniques can improve the sustainability of desalination, we first examine the problems to be addressed.

A primary environmental concern for desalination is the difference in salinity and temperature between the discharged brine and the surrounding natural waters receiving the discharge. Whether the desalination plant is based on thermal or membrane processes, the production of freshwater from the saline feed source necessitates the return of more concentrated feed water to the environment. The high salinity brine discharged by thermal desalination plants will also have an elevated temperature. Aquatic organisms have some ability to adjust to slight changes in salinity and temperature, but significant and sustained changes can devastate species populations at many different levels within ecosystems (Lattemann & Höpner, 2008). Furthermore, these salinity and temperature differences create density differences between receiving water and discharged brine. Density differences prevent the brine from mixing with and being broadly dispersed – and thus, diluted – by the surrounding water (Lattemann & Höpner, 2008). The problems associated with brine disposal are thus focused more strongly in the immediate vicinity of the discharge.

In addition to salinity and temperature stresses, flora and fauna at discharge sites face exposure to the diverse fleet of cleaning and pre-treatment chemicals that are used to maintain the performance of the desalination plant. Though the concentrations of these chemicals will be low, organisms may be more vulnerable to harm by them because of synergistic effects with temperature and salinity stress (Lattemann & Höpner, 2008).

Osmotic dilution of the brine could mitigate these environmental impacts. As with dilution of seawater feed streams to reverse osmosis plants, a feed solution of lower salinity can be selected to provide diluting water to the brine through an FO membrane (Figure 7.1.1b). Because this diluting water will not contribute to the potable water supply, more flexibility exists in feed solution selection. Energy savings for a partner process could be realized if desalination plant brine was diluted through FO by a feed solution that requires concentrating, such as anaerobic digester centrate (Cath *et al.* 2006; Holloway *et al.* 2007). To ensure that an environmental benefit is still created by osmotic dilution of the brine, the FO membrane used must be stable in the selected feed solution and must be able to reject all feed solution constituents of environmental concern. The source of the feed solution should also be co-located with the desalination plant, as is true for osmotic dilution of the feed stream to reverse osmosis plants.

Decreasing the salinity and temperature differences between brine and receiving water will enhance natural dispersion of the discharged brine plume, magnifying the initial dilution of salt, temperature differences, and chemicals. Many environmental threats can be mitigated by diluting these chemicals. Biocides, including chlorine and its associated disinfection byproducts, are used to pre-treat water for thermal desalination and can harm organisms in the ecosystem surrounding the discharge location (Lattemann & Höpner, 2008). Thermal desalination plants leach heavy metals, especially copper, and corrosion inhibitors (Lattemann & Höpner, 2008). Reverse osmosis desalination plants use coagulants to pre-treat feedwater, and the water used to back-flush this portion of the plant can have high turbidity. The release of this water decreases light penetration (Lattemann & Höpner, 2008), and harm to the whole food chain can result. Several different types of chemicals are also used to clean fouled reverse osmosis membranes.

Osmotic backwashing of the membranes used in reverse osmosis desalination can reduce the use and discharge of these membrane cleaning chemicals. Both biological and chemical foulants accumulate on the membrane surface during normal operation. This layer of foulants adds resistance to transport across the membrane, causing water flux and plant efficiency to decline. The harsh chemicals used to reverse this effect include metal chelating agents (such as EDTA), detergents (such as SDS), biocides, acids and bases, and scale inhibitors. Antiscalants, which are also used in thermal desalination plants, can cause eutrophication or may change the distribution of divalent metals in the discharge environment (Lattemann & Höpner, 2008). The discharge of extreme pH solutions, detergents, and oxidants used to clean membranes can disrupt aquatic ecosystems.

The concept of osmotic backwashing is relatively simple (Figure 7.1.1c). To backwash the membrane, one must reverse the dominance of the hydraulic pressure gradient, which favors permeation of water from seawater into the desalinated water side of the membrane, over the osmotic pressure gradient, which favors water flux in the opposite direction. This can be accomplished in one of two ways. Either the applied hydraulic pressure can be reduced, or the pressure can be maintained and a highly concentrated brine pulse can be introduced into the seawater side of the RO membrane (Ramon *et al.* 2010).

In addition to reducing chemical use and discharge, adoption of osmotic backwashing to remove foulants may reduce energy consumption. The harsh chemical cleaning protocols typically used require reverse osmosis train shut-down and can decrease membrane lifetime (Fritzmann *et al.* 2007), and this type of treatment can be performed only a few times each year. Between cleanings, foulants continually accumulate on the membrane surface, increasing the resistance to water permeation and requiring the applied hydraulic pressure to be increased in order to maintain a constant water flux. Operating at a higher pressure increases the energy consumption. Because osmotic backwashing involves relatively little chemical stress on the membrane or disruption to operation (Fritzmann *et al.* 2007), it can be performed frequently. This may suppress the buildup of the fouling layer, require smaller pressure increases to maintain constant flux, and ultimately reduce the consumption of energy.

Experimental investigations and modeling have demonstrated fast, effective removal of foulants with osmotic backwashing (Sagiv & Semiat, 2005; Qin *et al.* 2010; Ramon *et al.* 2010) and these findings justify further exploration. Studies involving more complex, realistic fouling conditions could elucidate the mechanisms by which osmotic backwashing removes foulants. This knowledge would aid researchers and industry in identifying the most promising uses of osmotic backwashing in addition to those aspects that require fuller development.

7.1.4 MEMBRANES AT SEA: FUEL FROM WASTE

Biofuels are one of the most promising sources of sustainable energy (Demirbas, 2009). However, the use of arable land and the clearing of carbon-rich forests for the cultivation of biofuel crops, such as corn or soy, could negatively affect the global food supply while actually increasing greenhouse gas emissions (Johansson & Azar, 2007). Algae, on the other hand, do not require fertile soil and can be cultivated using wastewater, which is nutrient rich but generally prohibited in irrigation of food crops. In other words, producing biofuels from algae reduces competition for valuable arable land (Aaronson & Dubinsky, 1982; Pienkos & Darzins, 2009) and allows us to utilize a 'waste' stream as a source for nutrients and water (Clarens *et al.* 2010; Wiley *et al.* 2011). Algae have a high photosynthetic efficiency that enables fast growth and further enhances their attractiveness as biofuel feedstocks (Miao & Wu, 2006).

The Ames Research Center of the National Aeronautics and Space Administration (NASA) has undertaken the challenge of developing a system for feasible production of biofuels from algae. The Center has proposed a process that leverages FO to optimize growth and harvesting of microalgae while simultaneously treating wastewater. The system is known as OMEGA: Offshore Membrane Enclosure for Growing Algae. It could potentially circumvent several of the limitations of terrestrial algae cultivation with an elegant, low energy, and low impact engineering solution (Aaronson & Dubinsky, 1982; Miao & Wu, 2006; Pienkos & Darzins, 2009; Wiley *et al.* 2011).

OMEGAs consist of clear plastic enclosures with water-permeable selective membranes incorporated into their surfaces (Figure 7.1.2). The enclosures are filled with nutrient-rich municipal wastewater and seeded with algae which grow and produce lipids (Trent, 2009; Trent *et al.* 2013). The bags are deployed in high salinity aquatic environments, often oceans, where the water provides both infrastructural support through flotation and heat capacity for temperature regulation (Trent *et al.* 2013), and wave action provides mixing. OMEGAs utilize seawater as an unlimited and ever-present draw solution to dewater the algae broth, an otherwise energy-intensive and expensive process (Trent, 2009; Trent *et al.* 2013). The membranes integrated in the enclosures allow the passage of sunlight and the exchange of CO_2/O_2 to support algae photosynthesis (Trent *et al.* 2013). The dense selective layer of the membranes prevents or minimizes the escape of wastewater contaminants.

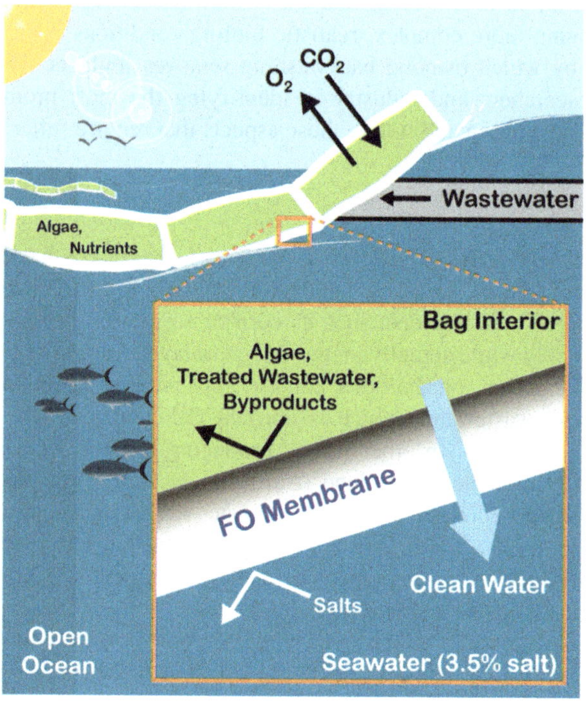

Figure 7.1.2 Schematic of the Offshore Membrane Enclosure for Growing Algae (OMEGA) system. Inset shows the permeation through and rejection by the forward osmosis (FO) membrane in contact with the seawater. The top of the enclosure, which is in contact with the atmosphere, contains specialized membranes that allow the passage of sunlight and the exchange of CO_2/O_2 to facilitate algae photosynthesis.

A significant advantage of OMEGAs is in algae biomass harvesting, a task that requires the culture broth to be dewatered. Efficient dehydration of the algal harvests is currently one of the main challenges of microalgae production, because of the low algal concentrations in the growth media (Decker, 2009). In OMEGAs, FO membranes retain algae and nutrients while facilitating removal of water from the bags through the osmotic pressure driving force provided by the higher salinity seawater (Trent *et al.* 2013). During a typical 10 to 20 day algae incubation period, more than 75% of the water is removed from the algae culture by osmosis using abundant seawater as a draw solution (Trent, 2009). This partial dewatering greatly reduces the cost of the overall dehydration process (AlgaeSsytemsLLC OMEGA).

When the algae inside an OMEGA are ready for harvesting, the slurry is pumped out of the bags and sent to a processing facility that completes the dewatering process. The bags are then available for refilling with nutrient-rich wastewater to reinitiate the growth cycle. Biofouling often plagues systems that use membranes and wastewater or natural waters. However, several aspects of the OMEGA system suggest that biofouling will be less severe in this application. The relatively low water flux through the FO membrane will lead to low rates of fouling, and the lack of hydraulic pressure will ensure that any fouling that does occur is highly reversible (Mi & Elimelech, 2010).

Despite the benefits of OMEGA systems, several challenges have been pointed out that merit further investigation. Though biofouling of the membrane's active layer inside the OMEGA may be minimal and reversible, fouling and scaling of the FO membrane support layer, which is in constant contact with seawater, may be more difficult to overcome. During the growth cycle, algae biofilms inside the OMEGA enclosure could decrease the rate of dewatering (Zou *et al.* 2011). In addition, membranes and other enclosure components need to be able to withstand pounding waves and cold temperatures (Howell, 2009; Trent, 2010; Zou *et al.* 2011).

NASA envisions the OMEGAs producing enough fuel to meet U.S. aviation needs of 21 billion gallons a year, which would require 10 million acres of ocean (Howell, 2009), an area equal in size to Switzerland and twice as large as Massachusetts. This large area would be established by locally distributing the OMEGAs, or by creating a management structure in which fishermen franchised and monitored the systems (Howell, 2009). A demonstration plant planned for Santa Cruz, California will help determine if the OMEGA process can be feasibly implemented (Trent, 2010; Wiley *et al.* 2011).

7.1.5 OSMOTIC AUGMENTATION OF WATER RESOURCES FOR AGRICULTURE

The growing world population is increasingly stressing not only water and energy resources, but also food resources. With the amount of arable land unlikely to increase (Fedoroff *et al.* 2010), irrigation will be employed more extensively and effectively to meet the rising demand for food (Plusquellec, 2002). Food production already requires large quantities of water. An estimated 70 percent of current global water withdrawals are for agricultural purposes (FAO AQUASTAT) (Figure 7.1.3a). Successfully feeding future generations will require sustainable technological improvements for irrigation and water management (Tilman, 1999; Falkenmark *et al.* 2009), including innovative ways to increase the availability of water for agriculture. Here, FO holds promise to expand the water supply for irrigation by tapping an abundant, low value water source – brackish groundwater.

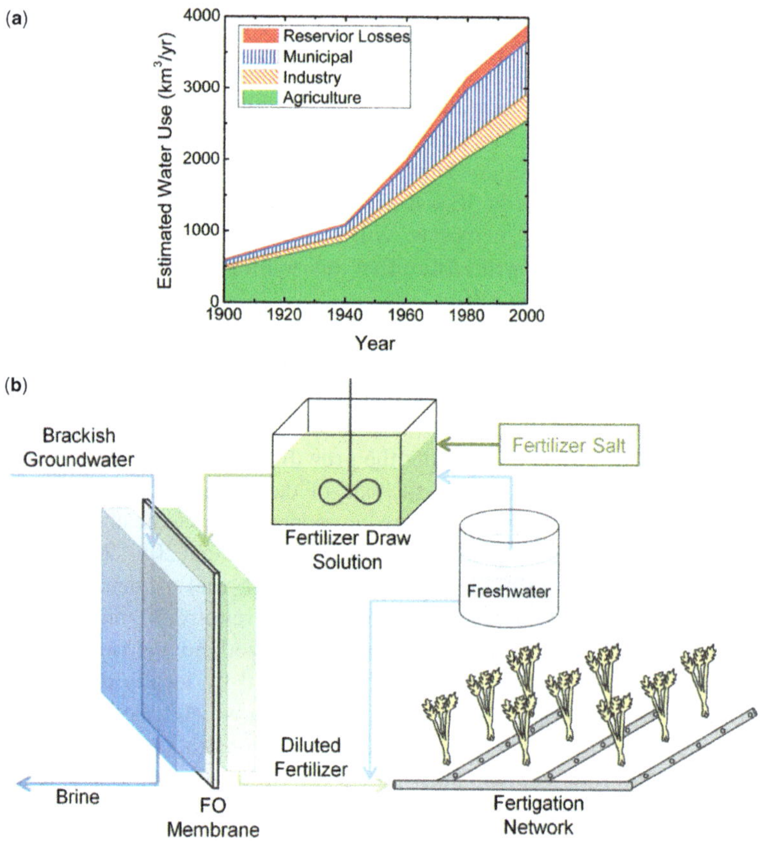

Figure 7.1.3 (a) Estimated water use over the past century. Agriculture currently accounts for 70% of global water withdrawals. (b) Schematic representation of forward osmosis (FO) application to increase the availability of water for irrigation through extraction of freshwater from brackish sources. The fertilizer is used to generate the concentrated draw solution, and brackish groundwater is used as the feed solution. The diluted fertilizer solution is subsequently applied to crops through a fertigation distribution network.

Water salinities above a crop's tolerance can stunt or prevent growth and diminish forage quality (Maas & Hoffman, 1977; Flowers, 2004). Salt tolerance imposes a ceiling on the salinity of water that can be used and restricts irrigation sources to freshwater for most crops. Due to these restrictions and the limited availability of freshwater, brackish water is increasingly being desalinated for irrigation (Yermiyahu *et al.* 2007). In some regions where freshwater is scare, brackish water abounds. For example, in the U.S. state of New Mexico, approximately 75 percent of groundwater is too saline for most uses without pretreatment (Reynolds, 1962). Long-distance diversion of freshwater to such places is costly to society. FO can overcome the desalination energy hurdle to make replacing freshwater with locally available brackish water for irrigation more feasible, allowing both energy and freshwater to be conserved or used elsewhere. This can be achieved by the innovative use of concentrated fertilizer mixtures to extract water from brackish sources (Moody & Kessler, 1976).

In this FO application, a concentrated fertilizer solution is used as the draw solution while brackish water is employed as the feed stream. Water permeates the membrane, diluting the fertilizer solution. If needed, the solution can be mixed with freshwater to achieve the desired concentration of fertilizer. The fertilizer-bearing water is applied to crops through the irrigation system in a process called fertigation (Bar-Yosef, 1999) (Figure 7.1.3b).

FO–brackish water supplied fertigation could partially relieve agricultural demand for freshwater without incurring a large energy penalty, addressing both water and energy costs in food production. But first, a few challenges must be overcome. Concentrated brackish water must be used or disposed of properly, and brackish water resources must be managed responsibly to avoid subsidence and other problems associated with overdrawing from aquifers. Of paramount importance is the selection of a suitable fertilizer to generate the draw solution. On top of fulfilling the principle objective of crop fertilization, the draw salt also needs to be water soluble, inexpensive, highly rejected by the FO membrane, and able to generate sufficient osmotic driving force for water permeation (Achilli *et al.* 2010). A recent study estimated that 1 kg of fertilizer can extract 24 to 59 liters of freshwater from a simulated brackish water of very high salinity (300 mM sodium chloride) (Phuntsho *et al.* 2011). More such studies are necessary to achieve a comprehensive evaluation of suitable fertilizer draw solutions. Additionally, the proper place of fertigation in a balanced portfolio of water management and allocation strategies used to support sustainable agriculture must be studied.

7.1.6 OUTLOOK

Unlike those currently at the focus of FO development, the FO applications described in this article avoid draw solution regeneration and thus can perform while using very little energy. These low-energy applications have the potential to enhance the energy efficiency of existing technologies, minimize pollution of aquatic and marine environments, utilize wastewater as a resource, or alleviate water shortages in agriculture. Specific limitations and research needs of each of the highlighted applications that must be overcome were indicated in the article.

At the core of the highlighted FO applications, and many more, is the FO membrane. Major progress has been made in the past few years in fabricating membranes for FO, but more remains to be done to improve the performance of these membranes. Enhancing FO membrane water flux is of paramount importance, because this will reduce the required membrane surface area and thus the capital cost of FO systems. There is also a need to develop FO membranes with high solute rejection capabilities, not only to minimize the passage of contaminants from the feed stream into the draw solution, but also to minimize the reverse permeation of draw solutes into the feed stream (Phillip *et al.* 2010).

Unique to these resource-efficient FO applications is the exposure of the membranes to feed and draw solution streams with high fouling potential. Hence, there is a pressing need to develop membranes with low propensity for irreversible fouling in order to enable continuous operation without the need for costly pretreatment of the feed and draw solutions. Furthermore, membrane module configurations that maximize mixing for the continuous removal of foulants from the membrane surface need to be developed.

The FO applications described in this article, which are currently conceptual or at the early stages of development, have great potential to increase sustainability at the water-energy nexus as mature technologies. Additional research and development as well as process demonstration at the pilot-scale must be completed to permit successful implementation and commercialization. Adapting the advances of the past few decades in membrane technology and materials science to FO can put these technological achievements within our reach.

7.1.7 SUMMARY

Many conventional practices in the production and use of water, energy, and food are unsustainable. Existing technologies and concepts can be improved with the integration of forward osmosis, a membrane-based technology that uses osmosis as its driving force. This Feature highlights five emerging applications of forward osmosis that elegantly bypass the difficult step of draw solution regeneration and make common processes more sustainable. These applications enhance the efficiency of the production and use of water, energy, and food; utilize wastes and abundant, low value resources; and better protect the environment.

7.1.8 REFERENCES

Aaronson S. and Dubinsky Z. (1982). Mass production of microalgae. In: New Trends in Research and Utilization of Solar Energy through Biological Systems. Springer, Basel, Switzerland, pp. 42–46.

Achilli A., Cath T. Y. and Childress A. E. (2010). Selection of inorganic-based draw solutions for forward osmosis applications. *Journal of Membrane Science,* **364**, 233–241.

AlgaeSsytemsLLC OMEGA. http://algaesystems.com/technology/omega

Bamaga O., Yokochi A., Zabara B. and Babaqi A. (2011). Hybrid FO/RO desalination system: preliminary assessment of osmotic energy recovery and designs of new FO membrane module configurations. *Desalination,* **268**, 163–169.

Bar-Yosef B. (1999). Advances in Fertigation. *Advances in agronomy,* **65**, 1.

Cath T. Y., Childress A. E. and Elimelech M. (2006). Forward osmosis: principles, applications, and recent developments. *Journal of Membrane Science,* **281**, 70–87.

Cath T. Y., Hancock N. T., Lundin C. D., Hoppe-Jones C. and Drewes J. E. (2010). A multi-barrier osmotic dilution process for simultaneous desalination and purification of impaired water. *Journal of Membrane Science,* **362**, 417–426.

Clarens A. F., Resurreccion E. P., White M. A. and Colosi L. M. (2010). Environmental life cycle comparison of algae to other bioenergy feedstocks. *Environmental Science & Technology,* **44**, 1813–1819.

Decker J. (2009). Blooming Biofuel: How Algae Could Provide the Solution. Renewable Energy World Magazine, June 22, London, UK.

Demirbas A. (2009). Political, economic and environmental impacts of biofuels: a review. *Applied Energy,* **86**, S108–S117.

Elimelech M. and Phillip W. A. (2011). The future of seawater desalination: energy, technology, and the environment. *Science,* **333**, 712–717.

Falkenmark M., Rockström J. and Karlberg L. (2009). Present and future water requirements for feeding humanity. *Food Security,* **1**, 59–69.

FAO AQUASTAT database. http://www.fao.org/nr/aquastat

Fedoroff N. V., Battisti D. S., Beachy R. N., Cooper P. J. M., Fischhoff D. A., Hodges C. N., Knauf V. C., Lobell D., Mazur B. J., Molden D., Reynolds M. P., Ronald P. C., Rosegrant M. W., Sanchez P. A., Vonshak A. and Zhu J.-K. (2010). Radically rethinking agriculture for the 21st century. *Science,* **327**, 833–834.

Flowers T. (2004). Improving crop salt tolerance. *Journal of Experimental Botany,* **55**, 307–319.

Fritzmann C., Löwenberg J., Wintgens T. and Melin T. (2007). State-of-the-art of reverse osmosis desalination. *Desalination,* **216**, 1–76.

Holloway R. W., Childress A. E., Dennett K. E. and Cath T. Y. (2007). Forward osmosis for concentration of anaerobic digester centrate. *Water Research,* **41**, 4005–4014.

Howell K. (2009). NASA Bags Algae, Wastewater in Bid for Aviation Fuel. New York Times.

Johansson D. J. and Azar C. (2007). A scenario based analysis of land competition between food and bioenergy production in the US. *Climatic Change,* **82**, 267–291.

Lattemann S. and Höpner T. (2008). Environmental impact and impact assessment of seawater desalination. *Desalination,* **220**, 1–15.

Maas E. and Hoffman G. (1977). Crop salt tolerance\-current assessment. *Journal of the Irrigation and Drainage Division,* **103**, 115–134.

McGinnis R. L. and Elimelech M. (2007). Energy requirements of ammonia–carbon dioxide forward osmosis desalination. *Desalination,* **207**, 370–382.

Mi B. and Elimelech M. (2010). Organic fouling of forward osmosis membranes: fouling reversibility and cleaning without chemical reagents. *Journal of Membrane Science,* **348**, 337–345.

Miao X. and Wu Q. (2006). Biodiesel production from heterotrophic microalgal oil. *Bioresource Technology,* **97**, 841–846.

Moody C. and Kessler J. (1976). Forward osmosis extractors. *Desalination* **18**, 283–295.

National Research Council. (2008). Desalination: a National Perspective. Committee on Advancing Desalination Technology. National Academies Press, Washington, DC.

Phillip W. A., Yong J. S. and Elimelech M. (2010). Reverse draw solute permeation in forward osmosis: modeling and experiments. *Environmental Science & Technology,* **44**, 5170–5176.

Phuntsho S., Shon H. K., Hong S., Lee S. and Vigneswaran S. (2011). A novel low energy fertilizer driven forward osmosis desalination for direct fertigation: evaluating the performance of fertilizer draw solutions. *Journal of Membrane Science*, **375**, 172–181.

Pienkos P. T. and Darzins A. (2009). The promise and challenges of microalgal-derived biofuels. *Biofuels, Bioproducts and Biorefining,* **3**, 431–440.

Plusquellec H. (2002). Is the daunting challenge of irrigation achievable? *Irrigation and Drainage,* **51**, 185–198.

Qin J.-J., Oo M. H., Kekre K. A. and Liberman B. (2010). Development of novel backwash cleaning technique for reverse osmosis in reclamation of secondary effluent. *Journal of Membrane Science,* **346**, 8–14.

Ramon G., Agnon Y. and Dosoretz C. (2010). Dynamics of an osmotic backwash cycle. *Journal of Membrane Science,* **364**, 157–166.

Reynolds S. (1962). Twenty-Fifth Biennial Report of the State Engineer of New Mexico for the 49th and 50th Fiscal Years July 1 1960, to June 30 (1962). The Vaillant Co., Albuquerque, NM.

Sagiv A. and Semiat R. (2005). Backwash of RO spiral wound membranes. *Desalination,* **179**, 1–9.

Semiat R. (2008). Energy issues in desalination processes. *Environmental Science & Technology,* **42**, 8193–8201.

Spiegler K. and El-Sayed Y. (2001). The energetics of desalination processes. *Desalination,* **134**, 109–128.

Stoughton R. and Lietzke M. (1965). Calculation of Some Thermodynamic Properties of Sea Salt Solutions at Elevated Temperatures from Data on NaCI Solutions. *Journal of Chemical and Engineering Data,* **10**, 254–260.

Tilman D. (1999). Global environmental impacts of agricultural expansion: the need for sustainable and efficient practices. *Proceedings of the National Academy of Sciences,* **96**, 5995–6000.

Trent J. D. (2009). Win Sea Algae. In: International Workshop on Offshore Algae Cultivation, April 20–22, Maribo, Denmark.

Trent J. D. (2010). OMEGA (Offshore Membrane for Enclosing Algae). NASA-NAVY: A Strategic Planning Discussion.

Trent J. D., Gormly S. J., Embaye T. N., Delzeit L. D., Flynn M. T., Liggett T. A., Buckwalter P. W. and Baertsch R. (2013). Algae bioreactor using submerged enclosures with semi-permeable membranes. Google Patents.

Wiley P. E., Campbell J. E. and McKuin B. (2011). Production of biodiesel and biogas from algae: a review of process train options. *Water Environment Research,* **83**, 326–338.

Yermiyahu U., Tal A., Ben-Gal A., Bar-Tal A., Tarchitzky J. and Lahav O. (2007). Rethinking Desalinated Water Quality and Agriculture. *Science,* **318**, 920–921.

Zou S., Gu Y., Xiao D. and Tang C. Y. (2011). The role of physical and chemical parameters on forward osmosis membrane fouling during algae separation. *Journal of Membrane, Science* **366**, 356–362.

Chapter 7.2

Life cycle cost assessment*

7.2.1 INTRODUCTION

Along with the growing demand for fresh water there is an increasing need to resort to non-conventional water sources. Seawater desalination and wastewater recovery present a promising solution to the increasing pressure on water resources. However, the high costs of desalinating/treating water can impact decision making on implementation of conventional technologies. The use of energy still remains the main component of the costs of these systems (Younos, 2005).

The energy consumption for desalination using conventional seawater reverse osmosis (SWRO) systems lies between 2.5 and 4 $kWh \cdot m^{-3}$ depending on many parameters (i.e. intake type, pretreatment, seawater salinity, etc.) (Fritzmann *et al.* 2007). Typical costs of water desalination by SWRO is in the range of 0.5–1 $USD \cdot m^{-3}$, which has been achieved by advances in energy recovery devices and membranes with improved performance; however, a decrease in costs due to technological developments is not foreseen as equipment and energy costs will increase (Fritzmann *et al.* 2007; Ghaffour *et al.* 2013). At the same time, brine discharge regulations are getting more stringent, raising the costs for new projects (Lattemann & Höpner, 2008).

Water production costs from wastewater recovery and reuse typically lie in the range between 0.40–1.26 $USD \cdot m^{-3}$ (Guo *et al.* 2014), depending on which level the treatment is initiated (i.e. primary or secondary wastewater), and the treatment level required for its reuse (i.e. direct/indirect potable or non-potable reuse, industrial water, irrigation).

*This chapter is based, with permission from the copyright holder, on the previously published paper. (Valladares Linares R., Li Z., Yangali-Quintanilla V., Ghaffour N., Amy G., Leiknes T. and Vrouwenvelder J. S. (2016). Life cycle cost of a hybrid forward osmosis – low pressure reverse osmosis system for seawater desalination and wastewater recovery. *Water Research*, **88**, 225–234. doi:10.1016/j.watres.2015.10.017)

Forward osmosis (FO) is a membrane process that can reduce the cost of desalination by extracting water from impaired sources, integrating both processes into a hybrid system. FO utilizes the osmotic dilution concept which relies on the salinity difference between two solutions to drive water permeation through a membrane capable of rejecting solutes. In osmotic dilution, a dilute stream becomes concentrated and a concentrated stream is diluted as permeation occurs across a semipermeable FO membrane (Cath *et al.* 2010; Hancock *et al.* 2011). A hybrid systems uses wastewater on one side of the FO membrane and seawater on the other side of the membrane, thus recovering water from the wastewater stream. By eliminating a draw solution and energy intensive water recovery from the draw solution, osmotic dilution becomes a low energy FO process (Shaffer *et al.* 2015). This FO process achieves two objectives: (i) volume-reduction treatment of wastewater, and (ii) reduction of osmotic pressure of seawater prior to RO desalination. Benefits of reducing the volume of wastewater are reduced energy consumption for treatment, lower volume transported, lower chemical use, and the possibility of harvesting energy (e.g. biogas) and nutrients (e.g. phosphates) from the concentrated wastewater more efficiently. The big opportunity relies in the use of a low-value wastewater effluent, i.e. primary effluent, which at the same time is high in organics for further concentration. In contrast, secondary effluent is a higher-value water with lower organics for biogas production.

Osmotic dilution can also be adapted in a conventional seawater desalination facility as a forward osmosis – low pressure reverse osmosis unit (FO-LPRO) (Valladares Linares *et al.* 2013a), offering the potential for energy and cost savings in a SWRO facility by lowering the operating hydraulic pressure, enabling the use of brackish water RO membranes (BWRO) instead of SWRO membranes, and increasing the water recovery ratio of the whole system (higher flux). Environmental impacts may be diminished by reducing electricity requirements, and also by discharging brines with lower salinity and lower volumes to the aquatic ecosystem (Lattemann & Höpner, 2008). Moreover, reducing the volume of the impaired water offers additional benefits, previously described (Wei *et al.* 2014).

The driving factor for considering implementing a FO-LPRO system versus a reverse osmosis (RO) system (for desalination purposes) or versus an ultrafiltration/nanofiltration (UF/NF) – advance oxidation process (AOP) (for secondary wastewater recovery) or a membrane bioreactor-reverse osmosis-advanced oxidation process (MBR-RO-AOP) hybrid system (for primary wastewater recovery) should be the energy savings compared to the capital expenses. FO has been depicted as a near horizon low-energy desalination technology considering that the recovery rate of actual desalination/treatment processes is changed (Amy *et al.* 2013).

Energy savings associated with the integrated FO-LPRO system compared to a conventional SWRO system are mainly related to the reduction in the osmotic pressure of the partially desalinated water and the hydraulic operational pressure required by the recovery process (i.e. low pressure RO system) to produce fresh water. Lower energy consumption is needed as the dilution rate increases; however, this requires a higher capital cost for the membrane area (Cath *et al.* 2010). For a hybrid FO-LPRO seawater desalination system, the specific energy consumption (SEC) associated to the FO-LPRO process, after an energy consumption analysis based on a conservative estimate, ranged between 1.3 to 1.5 $kWh \cdot m^{-3}$ using a secondary wastewater effluent as feed and seawater as draw solution (total production capacity of 2400 $m^3 \cdot d^{-1}$) (Yangali-Quintanilla *et al.* 2011), which is lower than the energy consumption of conventional SWRO.

It is important to compare similar processes in terms of influent and effluent water quality. A previous study compared RO for both seawater desalination and tertiary wastewater treatment, which cannot produce water with the same quality (Dolnicar & Schäfer, 2006). The study reports that capital costs for a plant producing water from seawater are about twice the costs of a plant reusing secondary effluent (not considering the costs of the primary/secondary wastewater treatment facility). Similarly, the operation and maintenance (O&M) costs for producing RO water from seawater are 2 times higher than the cost of reusing secondary sewage. The total life cycle costs for producing RO water from secondary effluent and seawater are 0.28 and 0.62 $USD \cdot m^{-3}$, respectively (Côté *et al.* 2005). The final cost of water can differ by a factor of 2 due to inaccuracies (i.e. not considering the cost of treating raw wastewater effluent) in the calculation method. Several studies have shown that an MBR-RO-AOP system is a multi-barrier approach that could be/has been implemented in water reuse projects (Comerton *et al.* 2005; Pisarenko *et al.* 2012; Gerrity *et al.* 2013).

A report prepared by CH2MHILL in collaboration with Colorado School of Mines (Gomez *et al.* 2011) did a cost modeling of a FO-RO system for seawater desalination and wastewater recovery in Texas. The results indicated that the use of a FO-RO system is not cost competitive when compared to tertiary treatment of wastewater using a RO membrane unit followed by AOP (ultraviolet (UV) light) for disinfection. However, a critical aspect in the cost of FO-RO hybrid systems was identified as the FO membrane. If FO membrane modules can be commercially produced at a reasonable price (i.e. comparable to production costs of RO modules with the same packing density), it is anticipated that use of FO-RO may be viable in the future (Gomez *et al.* 2011).

Hancock *et al.* (2012) did a comparative life cycle assessment to investigate the comparative environmental impacts of established and novel processes for seawater desalination and water reclamation, including a hybrid FO-RO and a conventional SWRO system. Improvements in the FO technology (i.e. higher water permeation of FO membrane and FO packing density) could reduce the environmental impact of the hybrid FO-RO system by 25% compared to SWRO scenario. A recent study by Coday *et al.* (2015) determined that the cost of a FO hybrid system to recover pit water from well drilling operations and hydraulic fracturing in the oil and gas industry could reduce the cost of water management by 60% compared to conventional deep well disposal.

A life cycle cost analysis covers the cost of an asset, or its part throughout its cycle life, while fulfilling the performance requirements. It includes construction and O&M (Mezher *et al.* 2011). On the other hand, the desalinated water production cost depends on a number of factors affecting both capital (CAPEX) and operational (OPEX) costs. Some technologies have high CAPEX (e.g. land, engineering, unit purchase, transportation, installation, etc., until commissioning) whereas other technologies are higher in OPEX (e.g. labor, maintenance and spare parts replacement, energy, and chemicals). The unit cost used in the life cycle cost assessment is $USD \cdot m^{-3}$ of water produced. The differences in water cost estimation, in the literature, can be attributed to factors such as differences in (i) fuel or energy cost, (ii) material and construction costs, (iii) feed water properties (e.g. salinity and turbidity), (iv) land cost, (v) subsidies and (iv) methods of cost calculation (Mezher *et al.* 2011; Zekri *et al.* 2013). This study used a module-scale approach (Deshmukh *et al.* 2015) including literature information with approximations based on global trends, real data from desalination/wastewater treatment markets, industrial reports and commercially available products.

The objective of this manuscript is to present an accurate and practical life cycle cost comparison between conventional water treatment technologies (i.e. SWRO, MBR-RO-AOP) and a proposed FO-LPRO hybrid membrane system for seawater desalination and wastewater recovery. A detailed analysis is presented on CAPEX and OPEX of each technology, as well as a sensitivity analysis for the FO-LPRO system on FO membrane flux and FO module cost. Additionally, biogas production from concentrated wastewater is explored as a benefit from the use of such FO system coupled with an anaerobic membrane bioreactor (AnMBR). The discussion focuses as well on the benefits of hybrid systems, the integration of water management sectors (water production and wastewater treatment), and successful cases for wastewater reuse in relation to public perception.

7.2.2 METHODOLOGY

7.2.2.1 Life cycle cost methodology

The comparison between a hybrid FO-LPRO system and other water treatment technologies is based on the assumption that both systems will produce high quality product water and they can be used for the same purpose. In this case, the comparison includes two approaches: (i) seawater desalination and (ii) wastewater treatment and recovery.

This work presents a detailed economic analysis on CAPEX and OPEX of 4 different scenarios, detailed in Section 7.2.2.2. The total cost of each technology is compared based on a production capacity of 100,000 $m^3 \cdot d^{-1}$ of potable water, as well as the total water cost per cubic meter. For the wastewater treatment processes, the corresponding population equivalent (P.E.) (referring to the relation of wastewater volume generated by one person in a day) for a 100,000 $m^3 \cdot d^{-1}$ capacity plant is approximately 530,000 P.E. (1 P.E. = 0.188 $m^3 \cdot d^{-1}$) (Henze, 2002).

A sensitivity analysis was made taking into consideration parameters such as the FO membrane pure water flux ($L \cdot m^{-2} \cdot h^{-1}$), FO module cost, and the break-even point between the extra CAPEX of the proposed FO-LPRO technology related to the reduction in OPEX when compared to conventional scenarios.

All costs are shown in present value (*PV*), calculated as:

$$PV(USD) = C\frac{1-(1+i)^{-n}}{i}$$

where *C* is the cost in USD, *i* is the interest rate in percentage and *n* is the lifetime of the project in years (Yangali-Quintanilla *et al.* 2011). PV calculates the value of an amount of money in present time compared to the same amount received at a future date, considering that the money value increases with time based on an interest rate.

A calculation spreadsheet was generated taking into consideration the methodology mentioned and the calculations described in Section 7.2.2.3 of this manuscript. No commercially available software was used due to the fact that no tool has been designed to simulate an FO system or a hybrid FO-LPRO system.

7.2.2.2 Technologies analyzed

The technologies compared in this manuscript are based on current best practice technologies used for seawater desalination and wastewater reclamation.

a) Hybrid forward osmosis – low pressure reverse osmosis (FO-LPRO) process for seawater desalination and wastewater recovery. The feed water for the FO process is considered to be a primary municipal wastewater effluent with an osmotic potential of approximately 0.50 bar, calculated using the van't Hoff equation or Pitzer model, both based on the osmotic pressure generated by small organic molecules and ions present in a wastewater sample (van't Hoff, 1887; Pitzer, 1973). The draw solution is considered to be seawater with total dissolved solids (TDS) of 40,625 $mg \cdot L^{-1}$ and an osmotic potential of 27.6 bars. The diluted seawater (50% dilution) is then fed to the LPRO unit with a TDS of 20,313 $mg \cdot L^{-1}$ and an osmotic potential of 13.8 bars (Valladares Linares *et al.* 2014a).
b) Seawater reverse osmosis (SWRO) desalination process. The feed water is considered to be seawater with a TDS of 40,625 $mg \cdot L^{-1}$ (value reported for the Red Sea water) and an osmotic potential of 27.6 bars. A 50% total recovery was considered for the SWRO system (Shaffer *et al.* 2012).
c) Membrane bioreactor – reverse osmosis – advance oxidation (MBR-RO-AOP) process for wastewater treatment and reuse. The advance oxidation process (AOP) was composed of a UV irradiation system with the addition of hydrogen peroxide (H_2O_2). AOP has been selected because of its ability to remove organic pollutants from wastewater streams not treatable by conventional techniques. More information on AOP can be found elsewhere (Oller *et al.* 2011). Reclaimed water has an average TDS of 210 $mg \cdot L^{-1}$.

7.2.2.3 OPEX and CAPEX calculations

The calculations in this work are based on the following assumptions: (i) all calculations are present values with a 20 year lifetime for the project and a 6% interest rate (Yangali-Quintanilla *et al.* 2011), (ii) electrical energy costs were set at 0.08 USD $\cdot$ kWh^{-1} (Yangali-Quintanilla *et al.* 2011), (iii) the calculated electricity consumption for the FO-LPRO system was set to 2.50 kWh $\cdot$ m^{-3} , based on previous calculations on electrical consumption for water abstraction, pretreatment, pumping and product transfer (Yangali-Quintanilla *et al.* 2011; Shaffer *et al.* 2012), though a conservative estimate was taken driven by the use of a primary wastewater effluent as feed solution, (iv) the recovery of the FO process was set to 50% in relation to initial volume of the draw solution (seawater) and the final dilution to feed the LPRO unit, (v) the AOP was assumed to be a combination of UV irradiation dosages between 300 to 650 mJ $\cdot$ cm^{-2} and 6 g of H_2O_2 per m^3 (Scheideler *et al.* 2011), (vi) the calculated electricity consumption for the conventional SWRO system was 3.5 kWh $\cdot$ m^{-3} (Fritzmann *et al.* 2007; Ghaffour *et al.* 2013), (vii) the cost of each FO membrane module was assumed to be USD 1500, containing a membrane area of 27 m^2, based on estimations for a real module to be developed by Porifera Inc. (Porifera Inc. 2014) (no pressure vessels are considered for this type of FO modules since they have their own modular housing), (viii) the flux for the FO membrane was considered to be 10 L $\cdot$ m^{-2} $\cdot$ h^{-1} (based on Porifera PFO-150SUB), and the flux for the SWRO membrane was considered to be 15 L $\cdot$ m^{-2} $\cdot$ h^{-1} (based on RO spiral wound modules SW 30–8040 from Dow Filmtec membranes), (ix) RO and LPRO pressure vessels contain 7 spiral wound elements, (x) SWRO and FO membrane replacement rate per year was considered as 10% with a membrane lifetime of 5 years (Avlonitis *et al.* 2003), (xi) the average relative engineering, procurement and construction (EPC) cost per m^3 $\cdot$ d^{-1} of the desalination process for the FO-LPRO system was set to USD 1409 (Fritzmann *et al.* 2007; Gomez *et al.* 2011), and (xii) the average relative EPC cost per m^3 $\cdot$ d^{-1} of the desalination process for the SWRO system was set to USD 1207 (Fritzmann *et al.* 2007; Gasson *et al.* 2010).

Table 7.2.1 presents the detailed EPC cost for a SWRO, FO stand-alone unit, LPRO and hybrid FO-LPRO desalination plant. The total cost of the proposed hybrid system is higher than that of other treatment systems chosen due to the additional need for FO membrane elements, additional costs on pumps and materials. FO stand-alone unit or LPRO cannot individually be used for seawater desalination; nevertheless, FO-LPRO can be compared to SWRO desalination. In addition, it reduces the wastewater volume which impacts treatment costs.

The OPEX for SWRO and FO-LPRO units are presented in Figure 7.2.1. Calculations are based on actual percentages for running desalination plants. Energy consumption represents on average 38% of the total cost of the plant through the lifetime of the project (Gasson *et al.* 2010).

A summary of calculated total costs and operational expenditures of the MBR-RO-AOP process for wastewater treatment and reuse is given in Table 7.2.2.

Table 7.1.1 Engineering, procurement and construction (EPC) cost for a SWRO, FO stand-alone unit, LPRO and hybrid FO-LPRO desalination plant with a total production capacity of 100,000 $m^3 \cdot d^{-1}$.

	SWRO		FO		LPRO		FO-LPRO	
Average relative EPC cost per $m^3 \cdot d^{-1}$ (USD)	1207		787		1000		1461	
Plant capacity ($m^3 \cdot d^{-1}$)	100,000		100,000		100,000		100,000	
	% of Total Cost	**Cost (USD)**	**% of Total Cost**	**Cost (USD)**	**% of Total Cost**	**Cost (USD)**	**% of Total Cost**	**Cost (USD)**
Equipment and materials	25.0%	$30,175,000	22.5%	$17,750,000	25.0%	$25,000,000	29.3%	$42,750,000
Membranes	5.5%	$6,638,500	29.4%	$23,148,148	5.5%	$5,500,000	19.6%	$28,648,148
Pressure vessels	1.5%	$1,810,500	–	–	1.5%	$1,500,000	1.0%	$1,500,000
Pumps	7.3%	$8,811,100	6.5%	$5,183,000	7.3%	$7,300,000	8.5%	$12,483,000
Energy recovery	2.0%	$2,414,000	–	–	2.0%	$2,000,000	1.4%	$2,000,000
Piping and high-grade alloy metals	12.5%	$15,087,500	–	–	12.5%	$12,500,000	8.6%	$12,500,000
Others*	46.0%	$55,763,400	41.5%	$32,660,000	46.0%	$46,200,000	31.6%	$46,200,000
Equipment + materials + membrane	30.5%		51.9%		30.5%		48.9%	
Construction**	69.5%		48.1%		69.5%		51.1%	
Total cost (USD)		**120,700,000**		**78,741,148**		**100,000,000**		**146,081,148**
Element membrane area (m^2)	28		27		34		*n.a.*	
Capacity of each module ($m^3 \cdot d^{-1}$)	10.08		6.48		12.24		*n.a.*	
Water flux ($L \cdot m^{-2} \cdot h^{-1}$)	15		10		15		*n.a.*	
Total number of elements	9921		15,432		8170		*n.a.*	
Cost per element (USD)	675		1500		675		*n.a.*	

n.a. not applicable.

*Includes pretreatment, civil engineering, design, legal services, installation, and intakes/outfalls.

**Considers pressure vessels, pumps, energy recovery, piping and high-grade alloy metals.

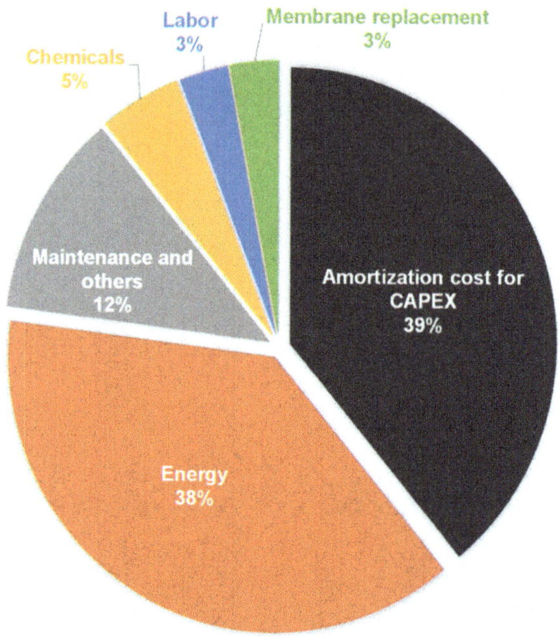

Figure 7.2.1 Operational expenses (OPEX) for a SWRO and FO-LPRO unit in percentage from the total annual cost of a 100,000 $m^3 \cdot d^{-1}$ plant (Gasson *et al.* 2010). 39% of the cost represents the amortization of CAPEX.

Table 7.2.2 Specific energy consumption (SEC) and engineering, procurement and construction (EPC) cost per m^3 of advanced wastewater treatment technologies for wastewater reclamation (MBR-RO-AOP).

Technology	SEC ($kWh \cdot m^{-3}$)	EPC cost ($USD \cdot m^{-3}$)	References
MBR	0.114	686	(Goldstein & Smith, 2000; EPA 2007; Helble & Möbius, 2008)
RO	0.910	550	(Goldstein & Smith, 2000; Palacios *et al.* 2008)
AOP	0.420	296	(Goldstein & Smith, 2000;Lehr & Wiley, 2004; Scheideler *et al.* 2011; Chatzisymeon *et al.* 2013)
Total	1.444	1532	

The 4th scenario presented in the results and discussion section, referred to as SWRO + MBR-RO-AOP, consists of two projects built and operated simultaneously for seawater desalination and wastewater treatment and recovery, each one with a capacity of treating 50,000 $m^3 \cdot d^{-1}$, to sum up a total of 100,000 $m^3 \cdot d^{-1}$. By analyzing this scenario, a more accurate comparison can be made with the proposed hybrid FO-LPRO system for desalination and wastewater recovery.

7.2.3 RESULTS AND DISCUSSION

Different scenarios were analyzed for current and proposed wastewater recovery and reuse and seawater desalination technologies. The comparison between OPEX and CAPEX related to the total cost of a treatment plant and the total cost per m^3 of water produced is presented, along with a sensitivity analysis on the FO membrane water flux, cost of the FO module, and the total savings throughout the lifetime.

The economic cost of a project has a great impact on the final decision of which technology/scenario should be adopted. However, there are other selection criteria that need to be taken into consideration for an integrated water management project in a coastal region. Details on the production of biogas from concentrated wastewater effluent when post-treated are discussed, as well as successful wastewater recovery and reuse projects, final water quality, public perception on direct potable reuse and co-location of seawater desalination and wastewater treatment facilities.

7.2.3.1 Life cycle cost analysis and sensitivity evaluation

7.2.3.1.1 OPEX and CAPEX: total cost comparison

Figure 7.2.2 presents the total costs of the treatment plants for each scenario analyzed, calculated as the sum of the CAPEX and total OPEX throughout the lifetime of the project.

Compared to SWRO, the FO-LPRO hybrid membrane system has a 21% higher CAPEX and a 56% lower OPEX due to savings in energy consumption and fouling control. FO-LPRO system was considered to have a 30% reduction in the chemical cleaning cost for both FO and RO membrane modules compared to the SWRO scenario. This reduction is mostly based on the use of hydraulic cleaning to mitigate FO membrane fouling (Mi & Elimelech, 2010; Valladares Linares *et al.* 2013b). Compared to the MBR-RO-AOP, the FO-LPRO systems had a 7% lower CAPEX and 9% higher OPEX. When a simultaneous desalination and wastewater treatment and recovery project was considered (SWRO + MBR-RO-AOP), the CAPEX increased by 13% compared to the proposed hybrid FO-LPRO, and the OPEX is 21% higher.

For the total cost of the project (CAPEX + OPEX), the most economically feasible scenario is the FO-LPRO plant, with a total cost of 266 million USD for producing 100,000 $m^3 \cdot d^{-1}$. The wastewater treatment and recovery scenario has a similar total cost. The total cost of SWRO scenario is 16% more expensive than that of the proposed FO-LRPO scenario. The combined SWRO + MBR-RO-AOP scenario had the highest cost with a total of 310 million USD, an increase of 17% compared to the FO-LPRO scenario.

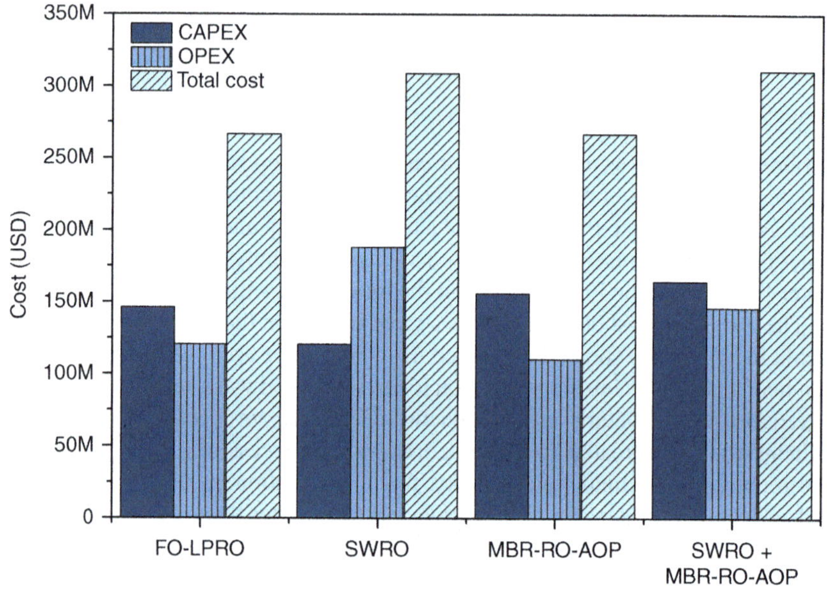

Figure 7.2.2 Capital expenses (CAPEX) and operational expenses (OPEX) for the 4 water treatment scenarios analyzed for seawater desalination and wastewater recovery and reuse. Both expenses are shown in present value and summed up throughout the 20-year lifetime of the project to give the total cost. Plant water production capacity: 100,000 $m^3 \cdot d^{-1}$.

After obtaining the total costs of the treatment scenarios, the total cost per produced m^3 of water was calculated based on the capacity of the respective plant(s). Figure 7.2.3 shows the total cost of water production per m^3. The FO-LPRO scenario has the lowest cost at USD 0.636 per m^3. The MBR-RO-AOP scenario has a very similar cost of USD 0.637 per m^3. The benchmark for SWRO desalination is USD 0.737 per m^3. The average cost of water for a simultaneous desalination and wastewater treatment and recovery project is USD 0.742 per m^3, the highest cost for all water treatment scenarios analyzed.

From Figure 7.2.3, the proposed hybrid FO-LPRO system has a comparable cost to wastewater treatment and recovery scenario (MBR-RO-AOP). The MBR-RO-AOP has the disadvantage of public perception on wastewater recovery for any direct application (i.e. direct potable reuse, irrigation, industrial use, etc.), which may be solved by the approach of a double-barrier hybrid FO-LPRO where freshwater is ultimately produced from both wastewater and seawater.

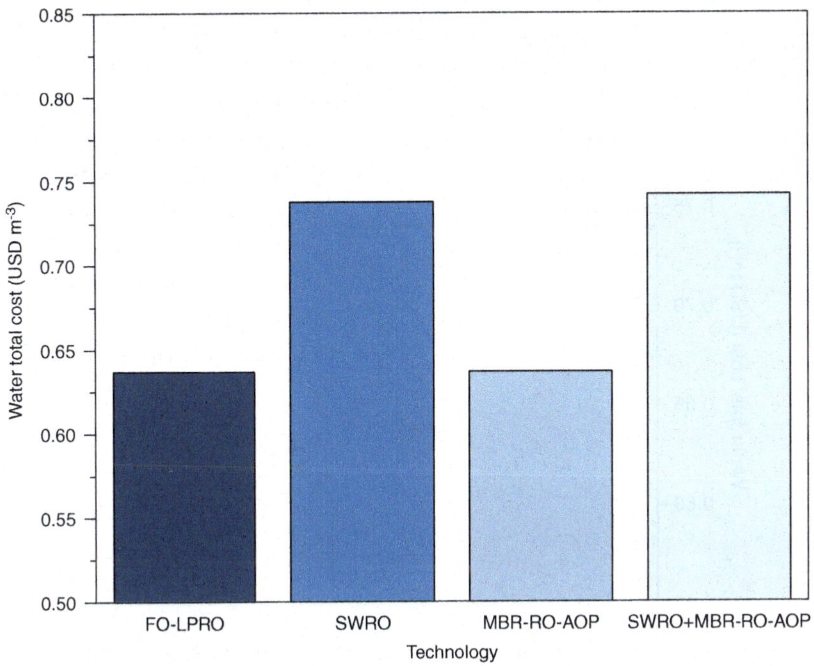

Figure 7.2.3 Total cost of water production per m^3 for each water treatment technology assessed based on a plant water production capacity of 100,000 $m^3 \cdot d^{-1}$.

Results from the calculations in this work show a potential economic advantage in the use of a hybrid FO-LPRO system over conventional desalination technologies such as SWRO. Most of the elements in the analysis are based on currently commercially available technology, particularly concerning FO membranes and modules, underlining the significance to the results presented.

7.2.3.1.2 Sensitivity analysis based on FO membrane flux and module cost

In order to determine the importance of the FO membrane water flux and the FO module cost in relation to the total water costs for the proposed hybrid FO-LPRO system, a sensitivity analysis was made based on 3 different FO module costs (for the same module type) over 4 different expected water fluxes. The results of the analysis are shown in Figure 7.2.4. Considering a membrane area of 27 m^2 in each module and varying the FO membrane water flux, it appears that even at low fluxes (5 $L \cdot m^{-2} \cdot h^{-1}$), the total water cost is still below the average cost for conventional SWRO desalination for the 3 different FO module costs considered. Moreover, water fluxes in several bench-scale studies have shown to be higher than

$5\,L \cdot m^{-2} \cdot h^{-1}$ when using seawater and municipal secondary wastewater effluent (Valladares Linares *et al.* 2012).

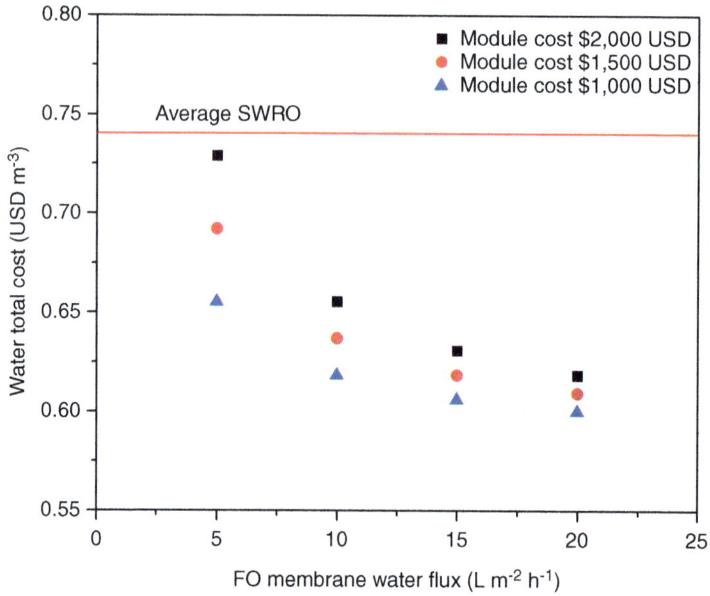

Figure 7.2.4 Sensitivity analysis on the FO module cost and FO membrane water flux related to the total cost of produced water per m^3 for a FO-LPRO hybrid system for seawater desalination and wastewater treatment and recovery. Plant water production capacity: $100{,}000\ m^3 \cdot d^{-1}$.

The FO membrane water flux shows an asymptotic behavior when related to the total water costs (Figure 7.2.4). As the water flux increases, the reduction in the total water cost is lower, even when the flux doubles. For an average cost of USD 1,500 for the FO module, and considering a flux of $5\,L \cdot m^{-2} \cdot h^{-1}$, the cost of each m^3 of water produced is USD 0.692. When the flux doubles to $10\,L \cdot m^{-2} \cdot h^{-1}$, the water total cost decreases 9% (USD 0.637). As the flux increases, the effect on the cost is reduced: when the flux doubles from 10 to $20\,L \cdot m^{-2} \cdot h^{-1}$, the total water cost decreases only by 4%.

The module cost is seen to highly influence the total water cost when the FO membrane flux is lower than $10\,L \cdot m^{-2} \cdot h^{-1}$. For the three different module costs analyzed (USD 2000, USD 1500 and USD 1,000), the difference for the total water costs for a flux of $5\,L \cdot m^{-2} \cdot h^{-1}$ is 11% and 6% compared to the lowest module price (USD 1000), respectively. If the flux considered increases to $10\,L \cdot m^{-2} \cdot h^{-1}$, the difference in cost compared to the lowest module cost decreases to 6% and 3%, respectively. When the flux is set to $20\,L \cdot m^{-2} \cdot h^{-1}$, the difference reduces to 3% and 1%. This shows that, for a higher FO membrane water flux, the cost of the module starts to be less significant for the final water cost of the whole FO-LPRO system. If the FO membrane flux is low (≈5 L·m^{-2}·h^{-1}), the cost of the FO module has a higher impact on the total water cost.

Previous estimations set the cost of a FO module as 100 USD·m^{-2} (Yangali-Quintanilla *et al.* 2011), which is approximately double the price considered for the analysis presented in Section 7.2.3.1. With recent advancements in the field of FO membranes and modules, the cost nowadays is lower and it is expected to reduce further in the future. One of the main factors to reduce the cost is mass production of FO modules, since nowadays most of the FO modules commercially available are produced in small-scale fabrication lines that include a significant amount of manual labor. This will have a considerable impact on the economic feasibility of FO-based water treatment technology.

Figure 7.2.5 shows the relation between the total cost of a FO-LPRO plant and the cost of the FO membrane module. The module cost represents an important parameter in the calculation of the CAPEX. When a 10 L·m^{-2}·h^{-1} water flux was considered for the FO membrane, there was a considerable increase in the total cost of the plant when the module cost doubled. For example, when the FO module cost was USD 1500, the total cost of the FO-LPRO plant was around USD 146 million; if the module cost doubles to USD 3000 without changing any other parameter in the calculation, the total cost of the FO-LPRO plant rises to USD 170 million, an increase of 16%.

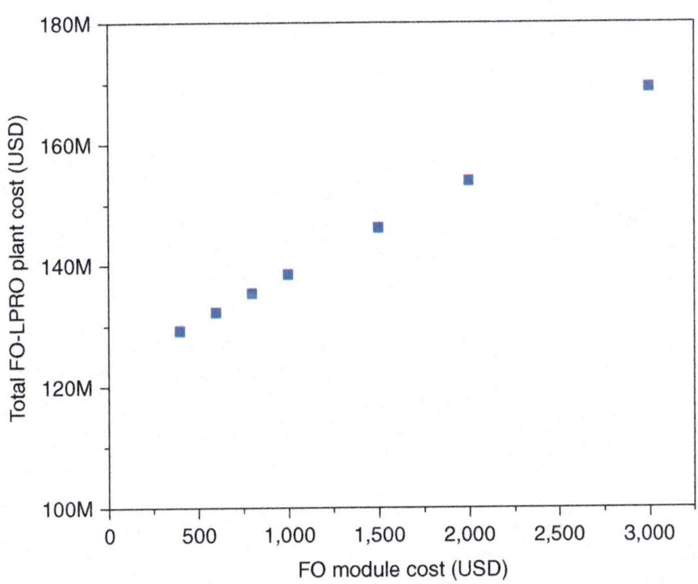

Figure 7.2.5 Sensitivity analysis on the FO module cost related to the total cost of a FO-LPRO hybrid system with a water production capacity of 100,000 m^3·d^{-1}. A water flux of 10 L·m^{-2}·h^{-1} was considered for the FO membrane.

Relating the module cost with the savings in OPEX is critical to understand the importance of the module cost in the overall economic feasibility of the project. Figure 7.2.6 shows the total savings for the FO-LPRO project compared to the SWRO desalination project (considered as the flat line at USD 0). The starting period shows negative values, which represents the higher CAPEX required for the FO-LPRO system compared to a conventional SWRO system (Figure 7.2.2). As the OPEX is being calculated for every year, and considering that these expenses are lower for the FO-LPRO system than for the SWRO system (Figure 7.2.2), the total savings will increase, eventually breaking even approximately at year 6, 8 and 10 (for a 20 year lifetime project) for a module cost of USD 1000, USD 1500 and USD 2000, respectively.

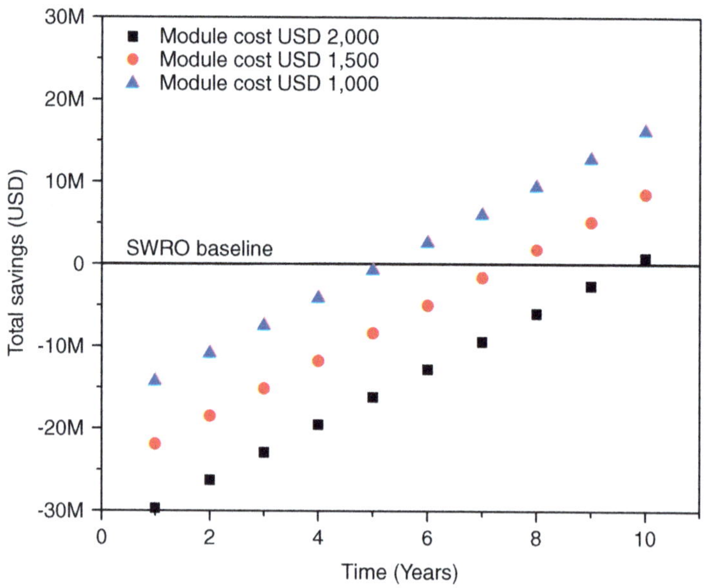

Figure 7.2.6 Total savings throughout the first 10 years of the project, calculated as the difference in total cost (CAPEX + OPEX) between a SWRO desalination plant and a FO-LPRO hybrid plant. Note that in the 3 scenarios (varying the FO module cost), the total savings are positive after 10 years. A linear increase continues over the 20-year lifetime of the project.

Scaling up the production of FO modules is expected to reduce costs, increase the financial benefits in FO-LPRO projects, and reduce the time frame to capitalize on the OPEX savings (overcoming the higher CAPEX required). A scenario can be seen where the total water costs may be further reduced thereby making this technology a viable option to treat and recover wastewater when compared to a MBR-RO-AOP system.

7.2.3.2 Biogas production

The hybrid FO-LPRO system proposed in this analysis considers a concentration ratio of 2:1 (i.e. 50% volume reduction) for the wastewater effluent. Considering that the initial volume is 100,000 $m^3 \cdot d^{-1}$, the volume to be treated after the FO process would be 50,000 $m^3 \cdot d^{-1}$. Therefore, a calculation on the amount of biogas that could be produced from a 50,000 $m^3 \cdot d^{-1}$ capacity AnMBR was made (≈ 266,000 P.E.). This is an attempt to quantify one of the added benefits of using a hybrid system that reduces the volume of wastewater to be treated, increasing the concentration of carbon (mixed liquor suspended solids) in the feed, and recovering the energy within the wastewater. Recently Wei *et al.* (2014) suggested that the addition of an FO concentration step before the AnMBR was a promising way for net energy recovery from typical municipal wastewater in temperate areas. Additional information on the operation and performance of an AnMBR can be found elsewhere (He *et al.* 2005; Hu & Stuckey, 2006; Huang *et al.* 2011). Table 7.2.3 describes the parameters taken into consideration for the calculation of the total energy production ($kWh \cdot y^{-1}$) from an ideal AnMBR.

Table 7.2.3 Yearly biogas production calculation (Sax *et al.* 1980; Jardine *et al.* 2004).

Capacity ($m^3 \cdot d^{-1}$)	50,000
COD in concentrated wastewater ($kg \cdot m^{-3}$)	0.72
Biogas production ($m^3 \cdot kg^{-1}$ COD)*	0.2625
Methane calorific value ($kWh \cdot m^{-3}$)	5.5
Calorific efficiency	0.8
Energy cost ($USD \cdot kWh^{-1}$)	0.08
Total COD ($kg \cdot y^{-1}$)	13,140,000
Total methane ($m^3 \cdot y^{-1}$)	3,449,250
Energy production ($kWh \cdot y^{-1}$)	15,176,700
Value of energy production ($USD \cdot y^{-1}$)	1,214,136

*Average value for an operating temperature between 10–30 °C.

The total content of chemical oxygen demand (COD) per year available after processing an average concentrated municipal wastewater effluent is 3 tons, which have the potential to produce approximately 3.5 million cubic meters of methane per year. Considering a methane calorific value of 5.5 $kWh \cdot m^{-3}$ and energy cost of 0.08 $USD \cdot kWh^{-1}$, the total value of the energy production from the anaerobic bioreactor is around USD 1.2 million per year. This energy can be utilized to reduce the OPEX of the hybrid FO-LPRO system and reduce the final cost of water.

The MBR-RO-AOP system could potentially be used as well to produce biogas if the sludge produced is process in anaerobic reactor. Table 7.2.4 shows the calculation of the total energy production ($kWh \cdot y^{-1}$) for a conventional

100,000 $m^3 \cdot d^{-1}$ capacity MBR, considering an average sludge yield of 0.5 Kg of volatile solids (VS) per Kg of COD and a conversion rate of 0.4 from VS to biogas (de Mes *et al.* 2003).

Table 7.2.4 Yearly biogas production calculation for MBR-RO-AOP system using an anaerobic process to treat sludge (Sax *et al.* 1980; de Mes *et al.* 2003; Jardine *et al.* 2004).

Capacity ($m^3 \cdot d^{-1}$)	100,000
COD in wastewater ($kg \cdot m^{-3}$)	0.36
COD in wastewater ($kg \cdot d^{-1}$)	36,000
Average sludge yield in MBR ($kg\ VS \cdot kg^{-1}$ COD)	0.5
Volatile solids (VS) in wastewater ($kg \cdot d^{-1}$)	18,000
Conversion rate from VS to biogas	0.4
VS converted ($kg \cdot d^{-1}$)	7200
Biogas production ($m^3 \cdot kg^{-1}$ VS converted)	1.1
Total methane ($m^3 \cdot y^{-1}$)	2,890,800
Energy production ($kWh \cdot y^{-1}$)*	12,719,520
Value of energy production ($USD \cdot y^{-1}$)*	1,017,562

*Considering the same methane calorific value, calorific efficiency and energy cost in Table 7.2.3.

The MBR-RO-AOP system has the potential to produce nearly 2.9 million cubic meters of methane per year, 17% less than the amount calculated for the FO-LPRO system coupled with an AnMBR (Table 7.2.3). The FO-LPRO system has the potential to extract more of the energetic value of the wastewater compared to the energy produced from treating the sludge in the MBR-RO-AOP system.

Moreover, Wei *et al.* (2014) have shown that, based on the integration of a heat pump and FO concentration technology into an AnMBR for municipal wastewater treatment, one can achieve a net energy recovery for the system. Table 7.2.5 compares the net energy recovery ($kWh \cdot m^{-3}$) from municipal wastewater (concentration factor of 1) to the FO-concentrated municipal wastewater (concentration factor of 5 and 10). At 3 different temperatures (10 °C, 20 °C and 30 °C), an increased concentration of 5 times in volume can achieve positive net energy recoveries. A smaller volume of concentrated wastewater can produce energy and it needs less energy to transport it.

7.2.3.3 Water quality versus public perception

In terms of water quality, it has been shown that high quality water can be produced, rejecting most of the pollutants and nutrients found in the wastewater and the seawater due to the double barrier system FO-LPRO (Cartinella *et al.* 2006; Cath *et al.* 2010; Valladares Linares *et al.* 2011; Alturki *et al.* 2012; Valladares Linares *et al.* 2013a; Valladares Linares *et al.* 2014b). One of the most successful projects for water reuse is NEWater in Singapore, where the quality of water produced by the Bedok Water Reclamation Plant was found to be better than the water supplied by Public Utility Board (PUB) of Singapore, and also met the water quality standards of the Environmental Protection Agency of the United States and the World Health Organization (Tortajada, 2006).

Table 7.2.5 Energy balance projection of mesophilic AnMBR integrating heat pump and FO technology (adapted from (Wei *et al.* 2014)).

Feed Temperature (°C)	Heat Energy Consumption ($kWh \cdot m^{-3}$)	FO Concentrating Factor	Concentrated COD ($mg \cdot L^{-1}$)	Equivalent Methane Energy Production ($kWh \cdot m^{-3}$)	Net Energy Recovery ($kWh \cdot m^{-3}$)
10	7.29	1	500	1.57	−5.72
		5	2500	7.84	0.55
		10	5000	15.68	8.38
20	4.38	1	500	1.57	−2.81
		5	2500	7.84	3.46
		10	5000	15.68	11.30
30	1.46	1	500	1.57	0.11
		5	2500	7.84	6.38
		10	5000	15.68	14.22

Although wastewater treatment is available to achieve recycled water qualities often superior to current potable water standards, public perception of water recycling activities is negative (Duong & Saphores, 2015). Some of the problems linked to this negative perception are the lack of infrastructure to supply recycled water, a highly subsidized and very cheap potable water resource, and the lack of community awareness about the limitations of freshwater resources, in particular in urban areas (Dolnicar & Schäfer, 2006), as well as the very high quality water produced by the system.

Even if direct potable reuse is not considered as the main objective of a reuse project, there are other uses for which the recycled water can be intended for (Levine & Asano, 2004). For example, artificial groundwater recharge, indirect potable reuse and industrial process water production are now part of successful wastewater recovery and reuse projects (Wintgens *et al.* 2005). These first attempts to integrate fresh water production and wastewater treatment and recovery will pave the way towards the creation of direct potable reuse projects in water-stressed areas around the world.

7.2.3.4 Wastewater recovery and reuse: successful projects

Although public perception has been one of the major obstacles towards wastewater recovery and reuse, there are successful projects that lead the change in the water management sector, showing the possibilities and benefits of these water recycling systems. Some examples in the practice of wastewater recovery are:

i) Groundwater replenishment scheme (Orange County, USA) – used for indirect potable reuse. The system comprises a microfiltration (MF) unit that treats a secondary municipal wastewater effluent, followed by an RO unit and an AOP consisting of UV and H_2O_2 for disinfection (Dawes *et al.* 1999). Total cost of water produced is 1.26 USD·m^{-3} (Guo *et al.* 2014). Capacity: 265,000 m^3·d^{-1}.

ii) Direct potable reuse (Windhoek, Namibia) – a treatment scheme of pre-ozonation, coagulation, dual media filtration, main ozonation, biological activated carbon adsorption and a two- stage granular activated carbon adsorption as well as UF prior to chlorine disinfection. Total cost of the water reclamation scheme is 0.76 USD·m^{-3} (including CAPEX and OPEX) (Wintgens *et al.* 2005; du Pisani 2006). Capacity: 21,000 m^3·d^{-1}.

iii) NEWater Project (Singapore) – There are currently four plants producing NEWater at Seletar, Bedok, Kranji and Ulu Pandan treatment plants. The system utilizes MF, RO and UV to treat municipal secondary wastewater effluent. The water produced is used for industrial and commercial purposes, and a small percentage is now being blended in fresh water reservoirs for further indirect potable reuse (Seah *et al.* 2003; Tortajada, 2006). Capacity: 316,000 m^3·d^{-1}.

iv) Western Corridor Recycled Water Project (Brisbane, Australia) – the scheme collects effluent from three advanced water treatment plants (Bundamba, Luggage Point and Gibson Island), which incorporate MF, RO and AOP with residual disinfection. The recycled water is used to supply water to power stations, industrial applications and the remaining directed to the main drinking water supply storage (used only as indirect potable reuse when combined water storage is below 40%) (Apostolidis *et al.* 2011). Capacity: 182,000 m^3·d^{-1}.

v) Torreele reuse plant for indirect potable reuse (Wulpen, Belgium) – the system was constructed as part of Wulpen wastewater treatment plant. It consists of UF and RO filtration followed by UV disinfection, producing infiltration water that is recharged in the dune aquifer. In 2005 the average cost of water was 0.60 USD·m^{-3} (Van Houtte & Verbauwhede, 2008). Capacity: 6850 m^3·d^{-1}.

vi) Sulaibiya plant for non-potable uses (Kuwait City, Kuwait) – built in 2004 to substitute Ardiya conventional wastewater treatment plant. Sulaibiya plant uses UF and RO membranes to produce high quality water used in agriculture, industry, with plans for aquifer recharge. It is the biggest wastewater reclamation plant in the region (Darwish & Al Awadhi, 2009; Alhumoud *et al.* 2010). Capacity: 375,000 $m^3 \cdot d^{-1}$.

As the integration of drinking water and wastewater treatment systems continues, the proposed FO-LPRO system represents a feasible alternative both in terms of price and improved public acceptance due to the double barrier approach towards direct potable reuse scenarios in coastal regions.

7.2.3.5 Co-location

A FO-LPRO hybrid system for seawater desalination and wastewater recovery requires co-location of the desalination plant and the wastewater treatment plant, to reduce the cost of transporting one of the two water sources to be treated. This can potentially have an impact on future projects to be built in the same location, either for municipalities or new industrial areas accompanying urbanization in coastal regions. At the same time, a reduction of the treatment plant footprint is feasible with a co-location scenario, possibly reducing the cost of the entire project. Li *et al.* (2014) have presented a good approach towards co-location of hybrid FO-LPRO systems.

7.2.4 SUMMARY

In recent years, forward osmosis (FO) hybrid membrane systems have been investigated as an alternative to conventional high-pressure membrane processes (i.e. reverse osmosis (RO)) for seawater desalination and wastewater treatment and recovery. Nevertheless, their economic advantage in comparison to conventional processes for seawater desalination and municipal wastewater treatment has not been clearly addressed. This work presents a detailed economic analysis on capital and operational expenses (CAPEX and OPEX) for: (i) a hybrid forward osmosis – low-pressure reverse osmosis (FO-LPRO) process, (ii) a conventional seawater reverse osmosis (SWRO) desalination process, and (iii) a membrane bioreactor – reverse osmosis – advanced oxidation process (MBR-RO-AOP) for wastewater treatment and reuse. The most important variables affecting economic feasibility are obtained through a sensitivity analysis of a hybrid FO-LPRO system. The main parameters taken into account for the life cycle costs are the water quality characteristics (similar feed water and similar water produced), production capacity of 100,000 $m^3 \cdot d^{-1}$ of potable water, energy consumption, materials, maintenance, operation, RO and FO module costs, and chemicals. Compared to SWRO, the FO-LPRO systems have a 21% higher CAPEX and a 56% lower OPEX due to savings in energy consumption and fouling control. In terms of the total water cost per cubic meter of water produced, the hybrid FO-LPRO desalination system has a 16% cost reduction compared to the benchmark for desalination, mainly SWRO. Compared to the MBR-RO-AOP, the FO-LPRO systems have a

7% lower CAPEX and 9% higher OPEX, resulting in no significant cost reduction per m^3 produced by FO-LPRO. Hybrid FO-LPRO membrane systems are shown to have an economic advantage compared to current available technology for desalination, and comparable costs with a wastewater treatment and recovery system. Based on development on FO membrane modules, packing density, and water permeability, the total water cost could be further reduced.

7.2.5 REFERENCES

Alhumoud J. M., Al-Humaidi H., Al-Ghusain I. N. and Alhumoud A. M. (2010). Cost/benefit evaluation of Sulaibiya Wastewater treatment plant in Kuwait. *International Business & Economics Research Journal (IBER)*, **9**(2), 23–32.

Alturki A., McDonald J., Khan S. J., Hai F. I., Price W. E. and Nghiem L. D. (2012). Performance of a novel osmotic membrane bioreactor (OMBR) system: flux stability and removal of trace organics. *Bioresource Technology*, **113**, 201–206.

Amy G., Ghaffour N., Ng K. C. and Li Z. (2013). Advances in Low-Energy Desalination. Water Reuse and Desalination. WateReuse Association, USA.

Apostolidis N., Hertle C. and Young R. (2011). Water recycling in Australia. *Water*, **3**, 869–881.

Avlonitis S. A., Kouroumbas K. and Vlachakis N. (2003). Energy consumption and membrane replacement cost for seawater RO desalination plants. *Desalination*, **157**, 151–158.

Cartinella J. L., Cath T. Y., Flynn M. T., Miller G. C., Hunter K. W. and Childress A. E. (2006). Removal of natural steroid hormones from wastewater using membrane contactor processes. *Environmental Science & Technology*, **40**, 7381–7386.

Cath T. Y., Hancock N. T., Lundin C. D., Hoppe-Jones C. and Drewes J. E. (2010). A multi-barrier osmotic dilution process for simultaneous desalination and purification of impaired water. *Journal of Membrane Science*, **362**, 417–426.

Chatzisymeon E., Foteinis S., Mantzavinos D. and Tsoutsos T. (2013). Life cycle assessment of advanced oxidation processes for olive mill wastewater treatment. *Journal of Cleaner Production*, **54**, 229–234.

Coday B. D., Miller-Robbie L., Beaudry E. G., Munakata-Marr J. and Cath T. Y. (2015). Life cycle and economic assessments of engineered osmosis and osmotic dilution for desalination of Haynesville shale pit water. *Desalination*, **369**, 188–200.

Comerton A. M., Andrews R. C. and Bagley D. M. (2005). Evaluation of an MBR–RO system to produce high quality reuse water: microbial control, DBP formation and nitrate. *Water Research*, **39**, 3982–3990.

Côté P., Siverns S. and Monti S. (2005). Comparison of membrane-based solutions for water reclamation and desalination. *Desalination*, **182**, 251–257.

Darwish M. and Al Awadhi F. M. (2009). The need for integrated water management in Kuwait. *Desalination and Water Treatment*, **11**, 204–214.

Dawes T. M., Mills W., McIntyre D., Anderson B., Kennedy J., Snow T. and Leslie G. (1999). Meeting thc Demand for Potable Water in Orange County in the 21st Century: the Role of Membrane Processes. In: AWWA Membrane Conference Proceedings, March, Chicago, IL.

de Mes T. Z. D., Stams A. J. M., Reith J. H. and Zeeman G. (2003). Methane production by anaerobic digestion of wastewater and solid wastes. In: Bio-Methane & Bio-Hydrogen,

J. H. Reith, R. H. Wijffels and H. Barten, Dutch Biological Hydrogen Foundation, Petten, The Netherlands, p. 165.

Deshmukh A., Yip N. Y., Lin S. and Elimelech M. (2015). Desalination by forward osmosis: identifying performance limiting parameters through module-scale modeling. *Journal of Membrane Science*, **491**, 159–167.

Dolnicar S. and Schäfer A. (2006). Public Perception of Desalinated versus Recycled Water in Australia. In: AWWA Desalination Symposium, 7–9 May, Honolulu, Hawaii.

du Pisani P. L. (2006). Direct reclamation of potable water at Windhoek's Goreangab reclamation plant. *Desalination*, **188**, 79–88.

Duong K. and Saphores J.-D. M. (2015). Obstacles to wastewater reuse: an overview. *Wiley Interdisciplinary Reviews: Water*, **2**, 199–214.

EPA. (2007). Wastewater Technology Fact Sheets: Membrane Bioreactors. United States Environmental Protection Agency, Washington, DC.

Fritzmann C., Löwenberg J., Wintgens T. and Melin T. (2007). State-of-the-art of reverse osmosis desalination. *Desalination*, **216**, 1–76.

Gasson C., Gonzalez-Manchon C. and Alvarado-Revilla F. (2010). Desalination Markets 2010 – Global Forecast and Analysis. Global Water Intelligence, Oxford, UK.

Gerrity D., Pecson B., Trussell R. S. and Trussell R. R. (2013). Potable reuse treatment trains throughout the world. *Journal of Water Supply: Research & Technology, AQUA*, **62**, 321–338.

Ghaffour N., Missimer T. M. and Amy G. L. (2013). Technical review and evaluation of the economics of water desalination: current and future challenges for better water supply sustainability. *Desalination*, **309**, 197–207.

Goldstein R. and Smith W. (2000). Water and Sustainability: U.S. Electricity Consumption for Water Supply & Treatment – The Next Half Century. EPRI, Palo Alto, CA.

Gomez J. D., Huehmer R. P. and Cath T. (2011). Assessment of Osmotic Mechanisms Pairing Desalination Concentrate and Wastewater Treatment. Texas Water Development Board, Austin, Texas.

Guo T., Englehardt J. and Wu T. (2014). Review of cost versus scale: water and wastewater treatment and reuse processes. *Water Science & Technology*, **69**, 223–234.

Hancock N., Cath T., Xu P., Heil D. and Black N. (2011). Novel Dual-Barrier Hybrid Osmotic Dilution–RO System: Pilot-Scale Demonstration of Technical and Ecological Merit. Proceedings of the American Water Works Association, Membrane Technology Conference, 31 March 2011, Long Beach, CA.

Hancock N. T., Black N. D. and Cath T. Y. (2012). A comparative life cycle assessment of hybrid osmotic dilution desalination and established seawater desalination and wastewater reclamation processes. *Water Research*, **46**, 1145–1154.

He Y., Xu P., Li C. and Zhang B. (2005). High-concentration food wastewater treatment by an anaerobic membrane bioreactor. *Water Research*, **39**, 4110–4118.

Helble A. and Möbius C. (2008). Comparing aerobic and anaerobic wastewater treatment processes for papermill effluent considering new developments. *Vortrag ZELLCHEMING-Hauptversammlung*, **25**, 1–23.

Henze M. (2002). Wastewater Treatment: Biological and Chemical Processes. Springer Science & Business Media, New York.

Hu A. Y. and Stuckey D. C. (2006). Treatment of dilute wastewaters using a novel submerged anaerobic membrane bioreactor. *Journal of Environmental Engineering*, **132**, 190–198.

Huang Z., Ong S. L. and Ng H. Y. (2011). Submerged anaerobic membrane bioreactor for low-strength wastewater treatment: effect of HRT and SRT on treatment performance and membrane fouling. *Water Research*, **45**, 705–713.

Jardine C. N., Boardman B., Osman A., Vowles J. and Palmer J. (2004). Methane UK. Research Report 30, Environmental Change Institute, Citeseer, Oxford.

Lattemann S. and Höpner T. (2008). Environmental impact and impact assessment of seawater desalination. *Desalination*, **220**, 1–15.

Lehr J. H. and Wiley J. (2004). Wiley's Remediation Technologies Handbook: Major Contaminant Chemicals and Chemical Groups. John Wiley & Sons, New Jersey, USA.

Levine A. D. and Asano T. (2004). Peer reviewed: recovering sustainable water from wastewater. *Environmental Science & Technology*, **38**, 201A–208A.

Li Z., Valladares Linares R., Abu-Ghdaib M., Zhan T., Yangali-Quintanilla V. and Amy G. (2014). Osmotically driven membrane process for the management of urban runoff in coastal regions. *Water Research*, **48**, 200–209.

Mezher T., Fath H., Abbas Z. and Khaled A. (2011). Techno-economic assessment and environmental impacts of desalination technologies. *Desalination*, **266**, 263–273.

Mi B. and Elimelech M. (2010). Organic fouling of forward osmosis membranes: fouling reversibility and cleaning without chemical reagents. *Journal of Membrane Science*, **348**, 337–345.

Oller I., Malato S. and Sánchez-Pérez J. A. (2011). Combination of advanced oxidation processes and biological treatments for wastewater decontamination – a review. *Science of the Total Environment*, **409**, 4141–4166.

Palacios M., Fernandez-Vera J., Fernández-Pinazo F., Echevarria F. and Hernandez-Moreno J. (2008). Desalination versus Reclaimed Water Reuse Under Sustainability Criteria: the Energy Point of View. In: UNESCO UCI International Conference on Water Scarcity, 1–5 December, Global Changes, and Groundwater Management, Irvine California.

Pisarenko A. N., Stanford B. D., Yan D., Gerrity D. and Snyder S. A. (2012). Effects of ozone and ozone/peroxide on trace organic contaminants and NDMA in drinking water and water reuse applications. *Water Research*, **46**, 316–326.

Pitzer K. S. (1973). Thermodynamics of electrolytes. I. Theoretical basis and general equations. *The Journal of Physical Chemistry*, **77**, 268–277.

Porifera Inc. (2014). http://porifera.com/products/pfo-elements/

Sax R. I., Holtz M. and Pette K. (1980). Production of Biogas from Wastewaters of Food Processing Industries. In Conf. on Industrial Energy Conservation Technology and Exhibition, 13–16 April, Houston, TX.

Scheideler J., Lekkerkerker-Teunissen K., Knol T., Ried A., Verberk J. and Van Dijk H. (2011). Combination of O_3/H_2O_2 and UV for multiple barrier micropollutant treatment and bromate formation control – an economic attractive option. *Water Practice & Technology*, **6**, 1–2.

Seah H., Poon J., Leslie G. and Law I. (2003). Singapore's NeWater demonstration project: another milestone in indirect potable reuse. *Water(Australia)*, **30**, 74–77.

Shaffer D. L., Yip N. Y., Gilron J. and Elimelech M. (2012). Seawater desalination for agriculture by integrated forward and reverse osmosis: improved product water quality for potentially less energy. *Journal of Membrane Science*, **415–416**, 1–8.

Shaffer D. L., Werber J. R., Jaramillo H., Lin S. and Elimelech M. (2015). Forward osmosis: where are we now? *Desalination*, **356**, 271–284.

Tortajada C. (2006). Water management in Singapore. *International Journal of Water Resources Development*, **22**, 227–240.

Valladares Linares R., Yangali-Quintanilla V., Li Z. and Amy G. (2011). Rejection of micropollutants by clean and fouled forward osmosis membrane. *Water Research*, **45**, 6737–6744.

Valladares Linares R., Yangali-Quintanilla V., Li Z. and Amy G. (2012). NOM and TEP fouling of a forward osmosis (FO) membrane: foulant identification and cleaning. *Journal of Membrane Science*, **421–422**, 217–224.

Valladares Linares R., Li Z., Abu-Ghdaib M., Wei C.-H., Amy G. and Vrouwenvelder J. S. (2013a). Water harvesting from municipal wastewater via osmotic gradient: an evaluation of process performance. *Journal of Membrane Science*, **447**, 50–56.

Valladares Linares R., Li Z., Yangali-Quintanilla V., Li Q. and Amy G. (2013b). Cleaning protocol for a FO membrane fouled in wastewater reuse. *Desalination and Water Treatment*, **51**, 4821–4824.

Valladares Linares R., Li Z., Sarp S., Bucs S. S., Amy G. and Vrouwenvelder J. S. (2014a). Forward osmosis niches in seawater desalination and wastewater reuse. *Water Research*, **66**, 122–139.

Valladares Linares R., Li Z. Y., Sarp S., Park Y. G., Amy G. and Vrouwenvelder J. S. (2014b). Higher boron rejection with a new TFC forward osmosis membrane. *Desalination and Water Treatment*, **55**, 2734–2740.

Van Houtte E. and Verbauwhede J. (2008). Operational experience with indirect potable reuse at the Flemish Coast. *Desalination*, **218**, 198–207.

van't Hoff J. H. (1887). The function of osmotic pressure in the analogy between solutions and gases. *Proceedings of the Physical Society of London*, **9**, 307.

Wei C.-H., Harb M., Amy G., Hong P.-Y. and Leiknes T. (2014). Sustainable organic loading rate and energy recovery potential of mesophilic anaerobic membrane bioreactor for municipal wastewater treatment. *Bioresource Technology*, **166**, 326–334.

Wintgens T., Melin T., Schäfer A., Khan S., Muston M., Bixio D. and Thoeye C. (2005). The role of membrane processes in municipal wastewater reclamation and reuse. *Desalination*, **178**, 1–11.

Yangali-Quintanilla V., Li Z., Valladares R., Li Q. and Amy G. (2011). Indirect desalination of Red Sea water with forward osmosis and low pressure reverse osmosis for water reuse. *Desalination*, **280**, 160–166.

Younos T. (2005). The economics of desalination. *Journal of Contemporary Water Research & Education*, **132**, 39–45.

Zekri S., Ahmed M., Chaieb R. and Ghaffour N. (2013). Managed aquifer recharge using quaternary-treated wastewater: an economic perspective. *International Journal of Water Resources Development*, **30**, 246–261.

Chapter 7.3

Niches in seawater desalination and wastewater reuse*

7.3.1 INTRODUCTION

7.3.1.1 Increasing need for fresh water along coasts

Due to the development of coastal regions in many countries, two-fifths of cities with populations of 1 million to 10 million people are located near coastlines (Tibbetts, 2002). Moreover, 14 of the largest 17 cities in the world are situated along coasts (Figure 7.3.1) (Creel, 2003). More than 3 billion people will live under water-scarce or water-stressed conditions by year 2025 (The Barila Group *et al.* 2009). Since more than 97% of the water in the world is seawater, desalination technologies have the potential to solve the fresh water crisis, particularly in coastal areas. Present conventional desalination technologies such as high pressure reverse osmosis (RO) involve expensive and energy intensive processes. A similar situation is faced when a wastewater effluent is treated with advanced technologies to produce high quality water. Therefore, there is an urgent need for a better technology that can recover water from impaired sources, providing an economic supply of fresh water to the increasing global human population.

7.3.1.2 Current membrane systems in the water industry: reverse osmosis

Desalination is a general term for the process of removing salt from water to produce fresh water. Fresh water is defined as containing less than 1000 $mg \cdot L^{-1}$ of salts or total dissolved solids (TDS) (Miller & Evans, 2006). The most common membrane desalination process in the world is reverse osmosis (RO) (Fritzmann *et al.* 2007).

---

*This chapter is based, with permission from the copyright holder, on the previously published paper. (Valladares Linares R., Li Z., Sarp A., Bucs Sz. S., Amy G. and Vrouwenvelder J. S. (2014). Forward osmosis niches in seawater desalination and wastewater reuse. *Water Research*, **66**, 122–139. doi:10.1016/j.watres.2014.08.021)

It is based on a property of certain polymers called semi-permeability, in which the polymer membrane has high permeability to water, but not for dissolved solids (i.e. salts). By applying a pressure difference across the membrane higher than the osmotic pressure of the feed water (i.e., seawater), fresh water is forced to permeate through the membrane. Common operating pressures for seawater desalination systems range between 55 to 68 bar (Liberman, 2003). For water reuse, where the feed water is a treated wastewater effluent or a brackish groundwater, lower pressures are used to produce fresh water, but the pretreatment steps are usually similar to those used in seawater desalination installations.

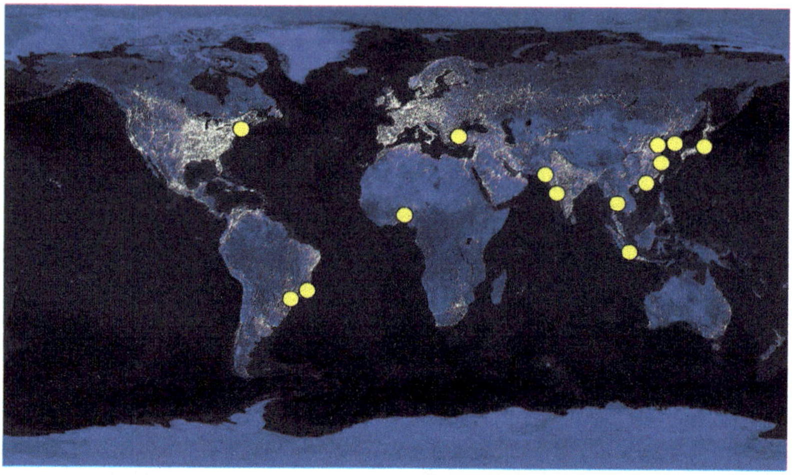

Figure 7.3.1 Fourteen of the world's largest cities are located along the coast (indicated in yellow marker), which translates into an opportunity to economically integrate drinking water and wastewater management, especially in water-stressed areas (Creel, 2003) (Image adapted from: (NASA, 2000)).

The three major drawbacks of using RO filtration systems to produce fresh water are: (i) energy intensive operation due to high hydraulic pressure used to drive the process (≈60 bar), (ii) extensive pretreatment needed to maintain long-term operation of membrane modules, and (iii) membrane fouling, resulting in decreasing membrane permeability and in feed channel pressure drop increase. To compensate for the loss in water production, systems require an increase in hydraulic pressure applied and more frequent chemical cleanings, all of which translate into higher operational costs.

7.3.1.3 Forward osmosis hybrid systems: an opportunity

Forward osmosis (FO) is an emerging membrane technology with a range of possible water treatment applications. Almost no external hydraulic pressure is required to run the process. Potential FO applications are a function of the respective feed and draw solutions selected, and the water quality objectives. In FO, water is extracted from a lower osmotic pressure feed solution (FS) into a higher osmotic pressure draw solution (DS); the process is driven by the osmotic pressure difference between the two aqueous solutions on the opposite sides of the semi-permeable membrane and results in concentration of the FS and dilution of the DS. Osmotic pressure of an aqueous solution can be calculated by using the Van't Hoff relation (van't Hoff, 1887):

$$\pi = R\,T\,\Sigma i\,M \tag{7.3.1}$$

where i is the dimensionless van't Hoff factor for the specific ion, M is the molarity of the specific ion, R is the gas constant (0.08206 $L \cdot atm \cdot mol^{-1} \cdot K^{-1}$), and T is the temperature in Kelvin. The water flux through a semi-permeable membrane by the osmotic pressure difference is given as (McCutcheon *et al.* 2005):

$$J_w = A(\pi_{D,b} - \pi_{F,b}) \tag{7.3.2}$$

where, J_w is the water flux through the semi-permeable membrane, A is the pure water permeability coefficient of the semi-permeable membrane, $\pi_{D,b}$ and $\pi_{F,b}$ are the bulk osmotic pressures of the draw and feed solutions, respectively.

For FO membranes, water flux decline due to fouling is lower when compared to RO systems, especially when wastewater with high fouling propensity is used as feed solution, because the FO process itself does not induce suspended solids and other organic contaminants into the membrane (Holloway *et al.* 2007), reducing as well the need for an extensive pretreatment. Experiments by Lee *et al.* (2010) comparing fouling behavior under identical hydrodynamic operating conditions (i.e., initial permeate flux and cross-flow velocity) and feed water chemistries (i.e., pH, ionic strength, and calcium concentration) of FO and RO systems showed that the thickness and compactness of the fouling layers of FO and RO are significantly different. Permeate flux during organic fouling in FO was after hydraulic cleaning almost completely recovered, while no noticeable change was observed for the RO system. In another study, data showed that FO fouling is governed by the coupled influence of chemical and hydrodynamic interactions. Calcium binding, permeation drag, and hydrodynamic shear force are the major factors governing the development of a fouling layer on the FO membrane surface (Mi & Elimelech, 2008).

Consequently, research has identified the potential for hybrid forward osmosis/low pressure reverse osmosis (FO/LPRO) systems for several applications, including seawater desalination (Choi *et al.* 2009), to reduce the cost and fouling propensity of producing fresh water from impaired-quality water sources (Chang *et al.* 2002; Achilli *et al.* 2009; Hoover *et al.* 2011; Boo *et al.* 2013). Recently, studies have shown the potential of FO/LPRO systems to produce low cost high quality fresh water using low pressure desalination, while simultaneously recovering impaired water from a recycled feed water (Cath *et al.* 2009; Yangali-Quintanilla *et al.* 2011). In a recent review, Hoover *et al.* (2011) described three ways in which FO can be applied in current reverse osmosis desalination plants to increase sustainability: (i) upstream dilution of seawater feed stream to reduce energy consumption, (ii) downstream dilution of concentrated brine before discharge to the marine environment, and (iii) osmotic backwashing of reverse osmosis membranes to restore membrane performance.

There are mainly two clusters of applications concerning FO in the water production and water treatment industry (Zhao *et al.* 2012b) (Figure 7.3.2): (i) desalination, and (ii) water reuse. Specific applications and present status for both application types are described. Additionally, an evaluation of the energy and economics of forward osmosis, major challenges, and the future direction of research in the field are presented.

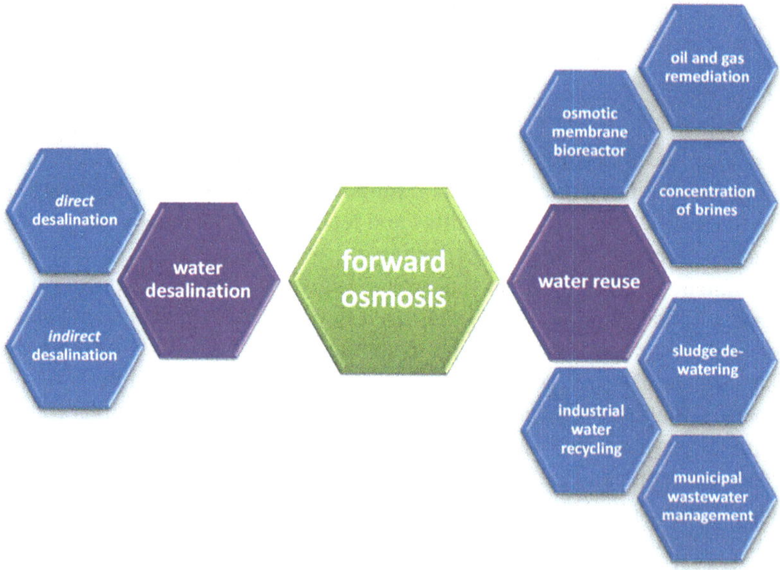

Figure 7.3.2 Forward osmosis applications in the water treatment industry clustered in two main areas: water desalination (left) and water reuse (right).

7.3.2 DESALINATION APPLICATIONS

There are two FO desalination approaches: (i) *direct* FO desalination, and (ii) *indirect* FO desalination. Figure 7.3.3 describes the general layout of the processes.

7.3.2.1 *Direct* desalination

The *direct* FO desalination concept is similar to other conventional membrane based desalination processes (e.g. reverse osmosis, nanofiltration) in which fresh water is directly extracted from a saline water (seawater or brackish water). *Direct* FO desalination uses saline water as the feed solution (FS) and an osmotic reagent such as a non-volatile salt like NaCl, or a volatile salt such as ammonia-carbon dioxide solution, among others (Chekli *et al.* 2012; Ge *et al.* 2013), as the draw solution (DS). In this process, an additional step, a draw solution recovery process, is needed to separate the DS from the solution in the diluted DS to recover fresh water (McCutcheon *et al.* 2005; Li *et al.* 2012a; Li *et al.* 2013a).

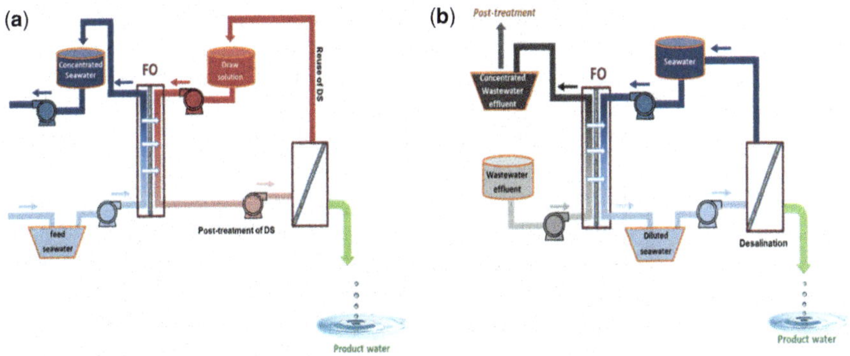

Figure 7.3.3 Layout of the two forward osmosis processes for water desalination: (a) *direct* and (b) *indirect*. Adapted from (Li *et al.* 2013b).

Various work efforts have been done from patents (McGinnis, 2009), lab tests (Choi *et al.* 2009), process modeling (Shaffer *et al.* 2012) and pilot scale demonstrations (Hancock *et al.* 2011). Table 7.3.1 presents the most recent studies/ patents on *direct* FO desalination.

One of the most studied processes for *direct* desalination involves the use of an ammonia-carbon dioxide solution (a mixture of ammonium bicarbonate and ammonium hydroxide to reach a desired ratio between ammonia and carbon dioxide) as a draw solution, and a thermal process to recover fresh water and regenerate the osmotic agent (McCutcheon *et al.* 2005; Cath *et al.* 2006; Gray *et al.* 2006; McCutcheon *et al.* 2006; Chanukya *et al.* 2013). McCutcheon and Elimelech (2006) demonstrated that a water flux of 36 $L \cdot m^{-2} \cdot h^{-1}$ could be obtained when using a 6 M ammonia-carbon dioxide draw solution and a 0.05 M NaCl feed solution. Another study showed that the total equivalent work requirements of this process are significantly lower than the work required in conventional desalination technologies (i.e. RO), translated to energy savings ranging from 72% to 85% (McGinnis & Elimelech, 2007). Chanukya *et al.* (2013) studied the concentration polarization effect using ammonium bicarbonate as draw solution in *direct* desalination experiments, concluding that forward osmosis is a feasible technique for the recovery of water from saline water.

Table 7.3.1 Overview of *direct* FO desalination investigations reported in the last two decades.

Year	Feed Solution	Draw Solution	Post-treatment	Status	References
1992	Saline water	Sugar cane	Reverse osmosis	Patent	(Yaeli, 1992)
2002	Seawater	KNO_3, SO_2, and NH_4NO_3	Precipitation (cooling) and separation through thermal waste heat	Patent	(McGinnis, 2002)
2005–2011	NaCl (0.05–2 M)	Ammonia-carbon dioxide solution (ammonium bicarbonate and ammonium hydroxide)	Thermal decomposition	Bench, pilot, and patent	(McCutcheon *et al.* 2005, 2006; McGinnis & Elimelech, 2007; McGinnis, 2009, 2011)
2006	N.D.	Magnetoferritin particles	Magnetic field	Patent	(Oriard & Haggerty, 2006)
2010	Contaminated water	Crosslinked superabsorbent polymer	Microfiltration	Patent	(Jones *et al.* 2010)
2011–2013	NaCl (0.034 M)	Ionic polymer hydrogels	Dewatering hydrogels via external pressure	Bench	(Li *et al.* 2011; Li *et al.* 2013a)
2011	Synthetic seawater	Hydrophilic nanoparticles	Ultrafiltration	Bench	(Ling & Chung, 2011)
2012	Brackish water	Divalent salts (i.e. Na_2SO_4 or $MgSO_4$)	Nanofiltration	Bench	(Zhao *et al.* 2012a)

2012	Seawater and brackish water	Cloud point solutes (i.e. polyethylene glycols)	Cloud point extraction (thermal process)	Patent	(Iyer, 2012)
2012	Seawater, brackish water and contaminated water	Retrograde soluble solutes (i.e. polyoxy random copolymer)	Coalescer (thermal process) and nanofiltration	Pilot and patent	(Carmignani *et al.* 2012) Trevi Systems Inc.
2013	NaCl (0.034 M)	Thermally responsive hydrogels with a semi-interpenetrating network	Dewatering hydrogels via thermal process	Bench	(Cai *et al.* 2013)
2013	NaCl (0.086 M)	$CuSO_4$	Metathesis precipitation	Bench	(Alnaizy *et al.* 2013)
2013	Saline water and synthetic seawater	Thermoresponsive magnetic nanoparticles	Magnetic field	Bench	(Zhao *et al.* 2013)
2013	Synthetic brackish water	NaCl 0.2–1 M	Solar-powered electrodialysis	Bench	(Zhang *et al.* 2013)

Direct FO desalination has also been conducted using different novel draw solutions in order to achieve an easier and more sustainable draw solution regeneration process. Generally, a favorable DS should be highly soluble, high osmotic pressure, low cost, non-toxic, easily separable and reusable, and eco-friendly. The reviews of DS are available in the literature and has stated the various aspects of the DS in process performance and selection criteria of suitable DS (Chekli *et al.* 2012; Zhao *et al.* 2012b). Li *et al.* (2011) used special hydrogels with carbon filler particles incorporated as DS to enhance the water flux through the FO membrane, achieving fluxes of approximately 1.2 $L \cdot m^{-2} \cdot h^{-1}$ for the highest DS concentration. Another study on hydrogels used as DS for the FO process showed lower fluxes ($\approx$ 0.2 $L \cdot m^{-2} \cdot h^{-1}$), thermally regenerating the absorption capacity of the hydrogels (Cai *et al.* 2013). Hydrophilic nanoparticles have been shown to produce a water flux slightly higher than 6 $L \cdot m^{-2} \cdot h^{-1}$ using synthetic seawater as a feed solution; the DS regeneration process (ultrafiltration) achieved a recovery of 92.7% (Ling & Chung, 2011). Recently, divalent salts such as Na_2SO_4 showed to achieve fluxes between 8 to 10 $L \cdot m^{-2} \cdot h^{-1}$ when desalinating brackish water; the DS recovery process (nanofiltration) had a rejection of 98% (Zhao *et al.* 2012a). Most of the draw solutes investigated for *direct* FO desalination are still not commercially feasible due to their cost, maximum water fluxes that they can produce, or the relatively low efficiency of the regeneration step (further discussed in Section 7.3.6).

Carmignani *et al.* (2012) developed a patent used by Trevi Systems Inc. to commercialize a technology that can desalinate seawater using 87.5% less electrical energy than current RO membrane systems, using a retrograde soluble solute (based on a diol, a compound which contains two hydroxyl groups) as osmotic agent and a two-stage recovery process: in the first stage a coalesce (used to separate emulsions into their components) is used to heat, agglomerate and cool down the draw solution to produce a cooled single-phase water rich stream, which is purified in the second stage by a nanofiltration system, producing pure water. The concentrated single-phase draw solute stream is recirculated back to the coalescer. The limited availability of this type of draw solute in the market might increase the operational costs, particularly due to the loss of draw solute in the recovery process.

Modern Water has deployed FO as a platform technology to produce desalinated water via *direct* FO desalination. It is the only company in the world with a commercial FO plant for direct seawater desalination (located in Al Najdah, Oman). Construction of the plant was completed in September 2012 and it is currently under operation. The process has shown to deliver significant operating and capital expenditure savings, reduced chemical consumption, robust fouling resistant membranes and a lower carbon footprint than competing technologies such as conventional high pressure RO membrane systems (Modern Water, 2013). These benefits are mainly associated to the reduction in RO membrane fouling due to the use of FO as a pretreatment step.

In direct FO desalination, the feed seawater may require a pre-treatment which is similar to RO desalination process. So far, there has been very little research using natural seawater in direct FO desalination. Therefore, the fouling propensity is not sufficiently studied. Li *et al.* (2012a) reported that a fouling matrix containing natural organic matter and polymerized silica was formed on the membrane in direct FO desalination using natural seawater as the FS. The result from this study may help to develop a suitable pre-treatment protocol in direct FO desalination.

7.3.2.2 *Indirect* desalination

Indirect desalination uses a high salinity water as the draw solution and an impaired-quality water source such as wastewater effluent (Valladares Linares *et al.* 2013a) or urban stormwater runoff (Li *et al.* 2014) is used as feed solution. Seawater and brackish water are potential DS for indirect desalination. In addition to the free-of-charge draw solution (seawater/brackish water), the attractiveness of this process is to extract clean water from the feed using free osmotic energy, leading to partially desalinated water (diluted DS) which can be further desalinated by a subsequent low-pressure reverse osmosis (LPRO) step as part of an FO-LPRO hybrid process (Cath *et al.* 2010; Yangali-Quintanilla *et al.* 2011), and thus reduce the cost of the entire desalination process (Table 7.3.2).

Table 7.3.2 Relation between osmotic pressure and specific energy consumption for desalination in seawater reverse osmosis (SWRO) systems. Adapted from (Li *et al.* 2014).

	Natural Seawater	Osmotically Diluted Seawater
TDS (ppm)[a]	35,627	11,876
Osmotic pressure (bar)[b]	26.14	8.71
Specific energy consumption for SWRO (kWh · m^{-3})[c]		
Number of elements: 1	12.23	5.49
Number of elements: 7	7.74	2.80

[a]The calculations were made using a recovery rate of 90% and rejection rate of 100% in the osmotic detention pond, and a dilution factor of 3 for seawater in each batch.
[b]The osmotic pressure was evaluated by Lenntech Osmotic Pressure Calculator (www.lenntech.com/calculators/osmotic/osmotic-pressure.htm).
[c]The calculations were made by DOW Reverse Osmosis System Analysis (ROSA) using membrane SW 30–4040 from DOW Filmtec with a water production of 4.50 $m^3 \cdot day^{-1}$, 1 stage and 1 vessel operation.

In addition, the promising role of the FO process to integrate wastewater treatment and seawater desalination has been shown, aiming for a sustainable solution of the water-energy nexus for coastal cities (Valladares Linares *et al.*

2013a; Li *et al.* 2014). These studies, particularly the use of primary effluent as feed for the FO process, have provided insight into the concept of using FO to avoid costly treatment of wastewater effluents in conventional treatment processes. The concentrated primary wastewater effluent can provide an opportunity for subsequent, more cost-effective wastewater treatment, including the recovery of bio-energy (bio-gas) when an anaerobic process is used to treat the concentrated FS (McCarty *et al.* 2011).

Indirect desalination experiments showed the ability of FO membranes to reject nutrients from wastewater, particularly chemical oxygen demand (COD) and phosphate. Rejection of nitrogen compounds was moderate (Cath *et al.* 2009; Valladares Linares *et al.* 2013a). Additionally, Valladares Linares *et al.* (2013a) studied partial desalination of seawater with the use of a submerged membrane module which makes it possible to adapt the process to a primary clarifier tank. This work also showed that FO membranes have high rejection of heavy metals present in the wastewater (>98%). An analysis of fractional organic carbon composition in the fouling layer on the FO membrane showed it is mainly formed by biopolymers and protein-like substances. Similar result has been observed in a study of osmotic membrane bioreactor for municipal wastewater treatment (Zhang *et al.* 2012). However, the pre-treatment of wastewater has not been reported in the study of FO process. The reason is probably that FO is mostly considered to be a pre-treatment step to concentrate wastewater and concentrated wastewater has added value, which can be used to recover biogas or other valuable compounds (Valladares Linares *et al.* 2013a; Xie *et al.* 2013, 2014).

When the FO system was coupled with a low pressure RO system it was found that the hybrid process worked as a double barrier against selected micropollutants, including pharmaceutically active compounds, hormones and other organic micropollutants (Cath *et al.* 2009). In practice conditions, using municipal secondary effluent as feed solution and seawater as draw solution, FO membranes were able to reject most of the micropollutants; rejections were moderate for hydrophilic neutral compounds (44–95%), moderate for hydrophobic neutral contaminants (48–92%), and high for the hydrophilic ionic micropollutants (96–99%). FO coupled with low pressure RO was effective in rejecting low molecular weight hydrophilic neutral micropollutants, with rejections that went beyond 89.1%. For the rest of the compounds, rejections were greater than 99% (Valladares Linares *et al.* 2011).

A cleaning protocol was studied and the cleaning efficiency of several methods to remove NOM-fouling on the active layer (AL) and transparent exopolymeric particles (TEP) from the support layer (SL) of a FO membrane immersed in a municipal secondary wastewater effluent was reported, using seawater as draw solution (Figure 7.3.4). Osmotic backwash did not help to recover flux of the FO membrane. Nevertheless, NOM-fouling showed high hydraulic reversibility (89.5% flux recovery) when air scouring for 15 minutes within the concentrated wastewater effluent was used as a cleaning technique. Chemical cleaning of the AL with Alconox (and industrial cleaning agent based on phosphate compounds) and sodium ethylenediaminetetraacetic acid (EDTA) proved to slightly increase the fouling reversibility (93.6%). Chemical cleaning of the SL removed the reversible fouling up to 94.5%. The irreversible fouling in this set of experiments was 5.5%, which may be attributed to biopolymers that cannot be removed from the membrane surface, along with the trace TEP left on the membrane surface (Valladares Linares *et al.* 2013b). Cath *et al.* (2009) suggested that the nature of the FO membrane, in conjunction with the minimal compaction of the fouling layer, is responsible for the negligible irreversible fouling observed after chemical membrane cleaning.

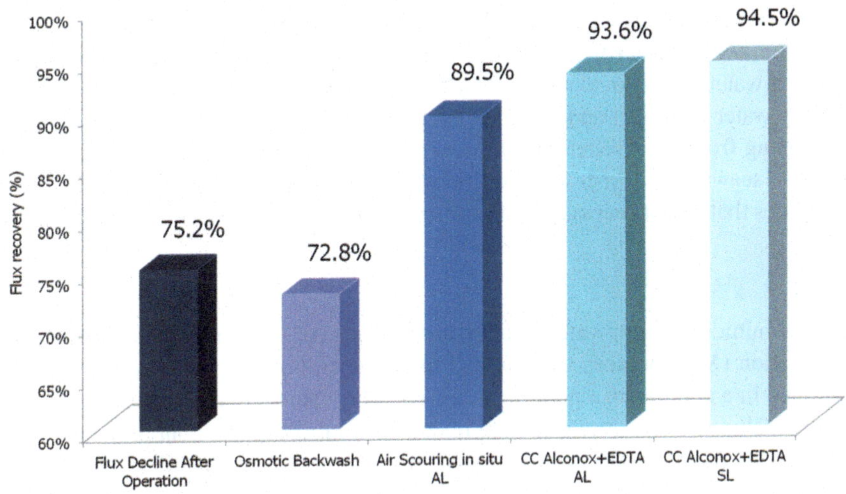

Figure 7.3.4 Flux recovery compared to initial flux achieved after consecutive cleaning steps to the active layer (AL) and support layer (SL) of an FO membrane used for *indirect* desalination using a secondary wastewater effluent as feed solution and seawater as draw solution.

The influence of feed spacer thickness (28, 31 and 46 mil) on biofouling development and membrane performance in a FO system was assessed (Valladares Linares *et al.* 2014). For all spacer thicknesses, operated at the same feed flow and the same run time, the same amount of biomass was found, while the flux reduction decreased with thicker spacers. These observations are in good agreement with biofouling studies for RO systems, considering the key differences between FO and RO. These findings contradict previous FO cross-flow studies on particulate/colloidal fouling, where higher cross-flow velocities improved system performance. At the end of the study it could be determined that the biomass amount alone does not determine the flux decline.

In either case, *direct* or *indirect* FO *desalination*, the baseline for energy comparison is seawater reverse osmosis (SWRO) with a specific energy of 2.5–4 $kWh \cdot m^{-3}$ (Fritzmann *et al.* 2007). Recent work (Yangali-Quintanilla *et al.* 2011) has suggested that the FO-LPRO hybrid can approach a specific energy threshold of 2 kWh · m^3 and possible pass this threshold with a new higher flux FO membrane of about 10 L · m^{-2} · h^{-1} (further discussed in Section 7.3.4).

7.3.3 IMPAIRED-QUALITY WATER TREATMENT AND REUSE APPLICATIONS

Most FO approaches for impaired-quality water treatment and reuse are similar to the *direct* seawater desalination method described in Section 7.3.2.1, where impaired-quality water is used as feed, while a draw solution is used to reduce the volume of feed; the DS is further treated by the other post-treatment process for the recovery of clean water and recirculation of DS. The applications of FO process for impaired-quality water treatment and reuse are highlighted in Table 7.3.3. Examples of water harvesting from wastewater have been described in the *indirect* desalination section because seawater/brackish water was used as draw solution. Section 7.3.3 describes processes that do not necessarily involve seawater or brackish water as draw solutions.

7.3.3.1 Water harvesting from municipal wastewater

In the municipal wastewater treatment processes, FO integrated with membrane distillation (MD) process was applied for sewer mining. A stable water flux was attained in a continuous operation at the recovery rate up to 80% (Xie *et al.* 2013). The FO showed a moderate to high rejection of most organic contaminants while MD rejected the residue contaminants to achieve a near complete rejection in this hybrid process. To recover clean water from secondary wastewater effluent, a photovoltaic powered forward osmosis – electrodialysis (FO-ED) process was tested as well. The process showed a good efficiency for removal of total organic carbon from feed wastewater and production of fresh water (Zhang *et al.* 2013). By using the natural energy (osmotic pressure and solar energy), this hybrid system is favorable to supply the potable water in isolated areas, remote areas and islands.

Membrane bioreactor (MBR), which involves both biological–activated sludge process and membrane filtration, has become one of the most commonly applied technologies in the treatment of many wastewater types. The integration of FO and MBR could reduce the energy consumption in the conventional MBR. The study of novel FO-MBR or osmotic MBR (OsMBR) has been initiated in the last five years (Cornelissen *et al.* 2008; Achilli *et al.* 2009; Qin *et al.* 2010; Lay *et al.* 2011; Yangali-Quintanilla *et al.* 2011; Li *et al.* 2012b; Yap *et al.* 2012; Zhang *et al.* 2012). The OsMBR process can not only reduce the energy cost in the pressure driven membrane process (e.g. microfiltration or ultrafiltration) as well as fouling control by air scouring in the conventional MBR, but also provides a more sustainable flux and reliable removal of contaminants. This is guaranteed by the lower fouling propensity without applied hydraulic pressure and separation capacity of FO membranes in OsMBR.

Table 7.3.3 Overview of most recent approaches of forward osmosis used for reclamation and treatment of impaired-quality water.

Feed Water	Draw Solution	Process Configuration	FO Membranematerial	Membrane Module and Testing Status	Process Purpose	References
Raw sewage	NaCl solution	FO-MD	CTA	Flat-sheet, lab scale	Direct sewer mining	(Xie *et al.* 2013)
Anaerobically digested sludge centrate	$MgCl_2$ solution	FO-MD	CTA	Flat-sheet, lab scale	Water and phosphorus recovery	(Xie *et al.* 2014)
Secondary wastewater effluent	NaCl solution	FO-ED	CTA	Flat-sheet, parallel plate-and-frame module, lab scale	Potable water production, utilization of natural energy for water treatment and reuse	(Zhang *et al.* 2013)
Activated sludge solution	$MgSO_4$, $NaSO_4$, $ZnSO_4$, NaCl solutions	OsMBR	TFC NF, TFC RO and CTA FO membranes	Flat-sheet, lab scale	Water reclamation from activated sludge	(Cornelissen *et al.* 2008)

(*Continued*)

Table 7.3.3 Overview of most recent approaches of forward osmosis used for reclamation and treatment of impaired-quality water (*Continued*).

Feed Water	Draw Solution	Process Configuration	FO Membranematerial	Membrane Module and Testing Status	Process Purpose	References
Domestic sewage	NaCl/ $MgSO_4$ solution	OsMBR	CTA	Flat-sheet, pilot scale	Pilot system composing the anoxic/ aerobic and FO processes for domestic sewage treatment	(Qin *et al.* 2010)
Synthetic wastewater with seed sludge from WWTPs	NaCl solution	Submerged OMBR	CTA	Flat-sheet, lab scale	Long-term OMBR testing for water reclamation from wastewater	(Lay *et al.* 2011)
Activated sludge	NaCl solution	Submerged OsMBR	TFC	Hollow fiber, lab scale, submerged module	Reclamation of water from activated sludge	(Zhang *et al.* 2012)
Synthetic wastewater	NaCl solution	OsMBR-RO	CTA	Flat-sheet, lab scale, immersed plate-and-frame module	Water reclamation from wastewater, potable water production	(Achilli *et al.* 2009)

Synthetic wastewater with seed sludge from WWTPs	Natural seawater	FO-SBR-LPRO	CTA		Flat-sheet, lab scale, submerged plat-and-frame module	Water reclamation from wastewater and potable water production	(Yangali-Quintanilla *et al.* 2011)
Activated sludge	Synthetic RO brine	Single FO	CTA		Flat-sheet, lab scale	Sludge thickening, digestion and dewatering	(Zhu *et al.* 2012)
High nutrient sludge	EDTA sodium salt	FO-NF	CTA		Flat-sheet, lab scale	Dewatering of sludge, fresh water recovery	(Hau *et al.* 2014)
Synthetic dye wastewater	Poly(acrylic acid) sodium salt solution	FO-MD	CA		Hollow fiber, lab scale	Dye wastewater treatment and fresh water recovery	(Ge *et al.* 2012)
Synthetic industrial wastewater containing heavy metals	NaCl solution	FO-RO	CTA		Flat-sheet, lab scale	Heavy metal removal from industrial wastewater	(Rassoul *et al.* 2012)
Synthetic oil-water emulsion	NaCl solution	Single FO	TFC		Flat-sheet, lab scale	Water reclamation and treatment of oily wastewater	(Duong & Chung, 2014)

(*Continued*)

Table 7.3.3 Overview of most recent approaches of forward osmosis used for reclamation and treatment of impaired-quality water (*Continued*).

Feed Water	Draw Solution	Process Configuration	FO Membranematerial	Membrane Module and Testing Status	Process Purpose	References
Drilling wastewater	NaCl solution	FO/RO closed loop system	CTA	Flat-sheet, spiral-wound module, pilot to industrial scale	oilfield wastewater reclamation	(HTI, 2011)
Produced water from oil/ natural gas processing	NH_4HCO_3 solution	FO-Thermal distillation	CTA	Flat-sheet, Mechanically Enhanced Circular Raceway FO module, pilot scale	Treatment of produced water from oil and/ or natural gas operations	(Nelson & Ghosh, 2011)
Drilling wastewater from shale gas filed	NaCl solution	Single FO	CTA	Flat-sheet, plate-and-frame module, lab scale	treatment of drilling mud and fracturing wastewater from oil and gas operations	(Hickenbottom *et al.* 2013)

Produced water from shale gas filed	NH_3/CO_2 solution	FO-Thermal distillation-RO	TFC		Flat-sheet, spiral-wound module, pilot scale	Water reclamation of frac flowback and produced waters from natural gas extraction operations	(McGinnis *et al.* 2013)
Concentrated RO brine	NaCl solution	FO-RO	CTA		Flat sheet, lab scale	Post-treatment of RO brine for ZLD	(Martinetti *et al.* 2009)
Saline water or other impaired quality water	Fertilizer solution	FO-Irrigation	CTA		Flat-sheet, lab scale	Water reclamation and reuse for irrigation	(Phuntsho *et al.* 2011)
Synthetic wastewater	NaCl solution	OsMFC	CTA		Flat-sheet, lab scale	Wastewater Treatment, Water Extraction and Bioelectricity Generation	(Zhang *et al.* 2011; Werner *et al.* 2013)
Sewage	Seawater	Floatable FO enclosure	CTA		Flat-sheet, transparent plastic bag with inserts of FO membrane, pilot scale	Wastewater treatment, bio-energy production	(NASA, 2012)

FO has been tested for dewatering of nutrient-rich anaerobic digester centrate (Holloway *et al.* 2007). A RO process could be used to recover the fresh water from the clean and diluted DS (i.e. NaCl solution) while organic compounds were rejected in the FO step. The FO process was also used to dewater activated sludge (Zhu *et al.* 2012; Hau *et al.* 2014). EDTA sodium salt was tested as a possible DS for dewatering of high nutrient sludge. The nutrients in the sludge were successfully removed by the FO process. The macromolecular DS (i.e. EDTA based salt) can be post-treated by a NF process for fresh water recovery. Alternatively, RO brine was used as the DS in Zhu's research. A good thickening efficiency of sludge was reached. In addition to reduction of sludge volume for lower cost of further transportation and handling, the concentrated RO brine was also osmotically diluted and could be safely disposed to limit the negative impact to the environment which is a problem in the conventional RO process.

7.3.3.2 Industrial wastewater reclamation and reuse

Another major part of impaired-quality water treatment and reclamation is industrial wastewater. A company in the US has successfully piloted a system, based on forward osmosis, that recycles wastewater generated by textile and carpet mills using dyeing processes (Catalyx, 2009a, b). Another FO application is to remove heavy metals from industrial wastewater or other impaired-quality water (Rassoul *et al.* 2012; Li *et al.* 2014). The effects of hydrodynamic conditions, organic fouling, temperature, feed and draw solution properties have been studied. FO membrane shows a nearly complete rejection to the metals tested (e.g. Pb, Zn, Cu, Cd), indicating its potential to be a convenient and economic method for industrial wastewater treatment.

Nowadays, most efforts of FO applications for industrial wastewater treatment have been dedicated to the treatment and reclamation of produced water from oil and gas (O&G) industry. Bench scale testing indicates the capacity of FO process for the separation of emulsified oil-water (Duong & Chung, 2014). Fresh water can be recovered from high concentration (up to 200,000 ppm) synthetic oily solution by a customized TFC FO membrane, at a reasonable flux of about $11.8\ L \cdot m^{-2} \cdot h^{-1}$. Some tests from bench to commercial scale have been conducted directly to the real produced water from O&G processing. FO combined with RO in a closed loop has been used to reclaim drilling wastewater from gas exploration operations (HTI, 2011), and it is capable of recycling 242,000 gallons drilling wastewater per day, consequently reducing the need for additional fresh water and truck traffic. Similar studies and applications have been conducted by individual research teams and companies using different membrane materials, membrane modules, draw solutions and process configurations (Nelson & Ghosh, 2011; Abousnina, 2012; Hickenbottom *et al.* 2013). Major benefits were achieved by the use of FO: the volume of the waste stream and the need for a fresh water source were greatly reduced in all cases, for feed solutions having a dissolved contaminant concentration greater than 10,000 ppm, a well-designed FO process can economically outperform an RO process, and a high rejection of organic and inorganic contaminants was obtained. However, due to the complexity of both organic and inorganic compounds in the produced water, a relatively complicated pre-treatment process may be required (McGinnis *et al.* 2013).

7.3.3.3 Other applications for impaired-quality water treatment

In contrast to the osmotic dilution of desalination brine for safe dispose mentioned in section 3.1, FO can be used to minimize the volume and increase the concentration of brine by directly extracting water from the brine. Easier post-treatment (such as crystallization) of the FO concentrated brine can help to attain a zero liquid discharge (ZLD) or near-ZLD system. Bench scale testing shows that FO can reach water recovery up to 90% from RO brine in brackish water desalination, and the total water recovery of brackish water desalination can be enhanced to 98% by a hybrid RO-FO process (Martinetti *et al.* 2009). Other studies have also shown the potential of FO to concentrate and recover water from brines (Tang & Ng, 2008; Saito *et al.* 2012).

Furthermore, fertilizer driven FO is a novel application of FO in water reclamation and reuse. A blend of fertilizers was used as the DS (i.e., fertigation). The preliminary tests show that 1 Kg of commercial fertilizer can extract 11–29 L of fresh water from seawater. The technology shows its potential for the irrigation in coastal areas with water scarcity (Phuntsho *et al.* 2011). In addition to seawater, other impaired-quality water could also be utilized as the feed water to osmotically dilute the fertilizer DS, while FO membranes provide a reliable barrier to reject the feed water contaminants.

Since FO process relies on natural osmotic pressure as the driving force, it has a promising potential for the emergency water supply where electric power or other energy sources are not available. A single FO process with a blend of sugar and salts as DS extracts water from impaired quality water, and diluted DS could be used as an energy drink. The practical applications of this concept have been explored in disaster relief operations, military drills and space mission (HTI, 2010a, b; Flynn *et al.* 2012).

Integrating FO into microbial fuel cells (MFC) has gained interest in recent years. A new concept of osmotic microbial fuel cell (OsMFC) has been introduced by several researchers has shown its promising future even though the testing has been conducted under a small bench scale (Zhang *et al.* 2011; Werner *et al.* 2013). Compared to indirect FO desalination, in which wastewater is concentrated and seawater is diluted for future post-treatment, OsMFC can achieve water extraction, biological wastewater treatment, and bio-energy production, simultaneously. OsMFC also provides a high power density than conventional MFC.

Different from OsMFC, NASA has been creating another novel route involving FO, namely Offshore Membrane Enclosure for Growing Algae (OMEGA), for clean bio-energy production (NASA, 2012). In this process, the freshwater algae and sewage is contained by the transparent plastic bags with inserts of FO membrane. The algae absorb carbon dioxide from atmosphere and nutrients from sewage to produce biofuel and oxygen. The FO membrane allows the exchange of carbon dioxide and oxygen between the atmosphere and enclosures, and fresh water from sewage into the surrounding seawater, while nutrients remain in the enclosed bags.

7.3.4 ENERGY AND ECONOMICS OF FO SYSTEMS

The driving factor for considering implementing a FO system versus a RO system (for desalination purposes) or versus a ultrafiltration/nanofiltration (UF/NF) – advance oxidation process (AOP) (for secondary wastewater recovery) or a MBR – RO – AOP hybrid process (for primary wastewater recovery) should be the energy savings related to the capital expenses. FO has been depicted as a near horizon low-energy desalination technology considering that the recovery rate of actual desalination processes is changed (see Figure 7.3.5, (Amy *et al.* 2013)).

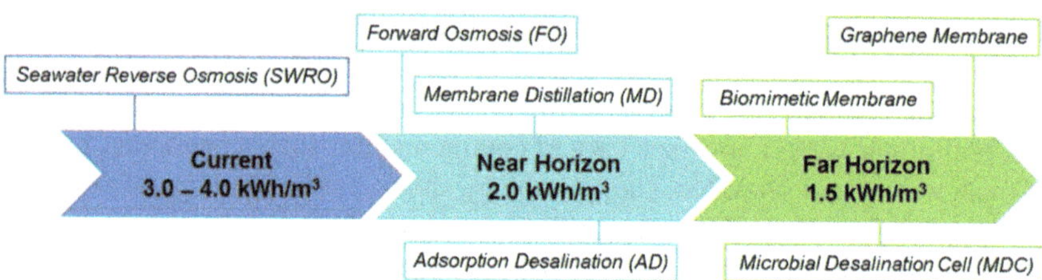

Figure 7.3.5 Technology roadmap for low-energy desalination at a different recovery rate than conventional processes such as reverse osmosis. Note that the specific energy consumption for thermal technologies such as adsorption desalination and membrane distillation considers the availability of a waste heat stream. Adapted from Amy *et al.* (2013).

The energy consumption for desalinating water with RO membranes lies between 2.5 and 4 $kWh \cdot m^{-3}$ (Fritzmann *et al.* 2007), as a result of the development of new efficient membranes and the use of energy recovery devices over the last decade.

According to Shaffer *et al.* (2012) (Table 7.3.4), the specific energy consumption (SEC) of a *direct* desalination FO system coupled with a RO process is lower when compared to a two-pass RO process for a total system recovery higher than 25% (total production capacity of 100,000 $m^3 \cdot d^{-1}$).

For an *indirect* desalination system, the SEC associated to the FO-RO process, after an energy consumption analysis based on a conservative estimate, ranged between 1.3 to 1.5 $kWh \cdot m^{-3}$ using a secondary wastewater effluent as feed and seawater as draw solution (total production capacity of 2400 $m^3 \cdot d^{-1}$) (Yangali-Quintanilla *et al.* 2011).

Table 7.3.4 Numerically modelled specific energy consumption (SEC) for an integrated *direct* desalination FO-RO and a RO system. Adapted from Shaffer *et al.* (2012).

	Integrated FO/RO	Two-pass RO
Total System Recovery (%)	**SEC (kWh·m^{-3})**	**SEC (kWh·m^{-3})**
10	3.03	3.39
25	2.40	2.88
50	2.93	3.79
70	4.49	6.43

Energy savings associated with the integrated FO-RO system compared to a two-pass RO system are mainly related to the reduction in the osmotic pressure of the partially desalinated water and the hydraulic operational pressure required by the recovery process (i.e. low pressure RO system) to produce fresh water, compared to a traditional RO system (Table 7.3.5). It is clear that lower energy consumption is possible as the dilution rate increases. This, however, represents a higher capital cost for the membrane area required.

Table 7.3.5 Osmotic pressure and operational system pressure calculations for different seawater dilution scenarios in a RO system.

Dilution (%)	TDS (mg·L^{-1})	Osmotic Pressure* (bar)	Operational System Pressure* (bar)
0	40,625	27.6	68.9
25	30,471	20.7	56.9
40	24,378	16.5	45.5
50	20,313	13.8	46.5
60	16,250	11.0	37.2

*Osmotic pressure was calculated by Lenntech Osmotic Pressure Calculator, considering 1 membrane with a production of 0.2 m^3·h^{-1}. The membrane used for seawater is SW 30–4040 from Dow/Filmtec; for brackish water the BW 30–4040 membrane for Dow/Filmtec is considered.

With current commercial technology, hybrid FO systems for both desalination and water recovery applications have proven to have higher capital cost compared to conventional technologies (Yangali-Quintanilla *et al.* 2011; Aydiner *et al.* 2013). Nevertheless, due to the demonstrated lower operational costs of hybrid FO systems, the payback time for the construction and the unit cost for each m^3 of fresh water produced with the FO system is shorter than conventional desalination/water recovery technologies (i.e. ultrafiltration/RO systems). Minimum water flux considerations are made for these evaluations, where the average FO water flux should be 10.5 L·m^{-2}·h^{-1} when competing against water reuse technologies

(UF-LPRO), and a water flux of 5.5 $L \cdot m^{-2} \cdot h^{-1}$ is needed to recover and desalinate water at a lower cost compared to RO systems (Yangali-Quintanilla *et al.* 2011).

Recently, a company in the US, Trevi Systems Inc., developed a patent (Carmignani *et al.* 2012) which can desalinate seawater using 87.5% less electrical energy than current RO systems. Furthermore, the low pressure system can use waste heat or low grade thermal heat at 75°C to drive the FO desalination process. The operating cost for each $m^3 \cdot d^{-1}$ of fresh water using natural gas as heat source can be 30% lower than cost of conventional RO technologies; savings increase to more than 50% if a co-generation heat source is used. For this case, both the capital and operational expenses are lower for the FO hybrid system.

Besides the costs associated with the system, there are key benefits of using FO hybrid systems compared to RO: (i) chemical storage and feed systems may be reduced for capital, operational and maintenance cost, (ii) water quality is improved, (iii) reduced process piping costs, (iv) more flexible treatment units, and (v) higher overall sustainability of the desalination process.

7.3.5 PRESSURE RETARDED OSMOSIS: SPECIAL FO APPLICATION FOR ENERGY RECOVERY IN WATER INDUSTRY

Pressure retarded osmosis (PRO) is a special type of FO application, where the primary and end product is energy instead of water. Potential energy of a natural or engineered salinity gradient system can be utilized by a PRO process in the form of electricity or hydraulic pressure. Even though the idea of using salinity gradient for energy production was suggested in the early 50's (Pattle, 1954; Norman, 1974), some of the first known applications of PRO were suggested and tested by Sydney Loeb in the 70's (Loeb, 1975, 1976; Loeb *et al.* 1976; Mehta & Loeb, 1979). The main difference between a modern PRO and an FO process is the applied pressure on the high salinity draw solution. The pressure of the high salinity draw solution can be kept relatively constant during the PRO process, even though the volumetric flow rate is to be increased. Therefore, the draw side of the PRO process can be assumed to be isobaric, in most cases. The near-isobaric behavior of the draw side of the PRO process can be explained by the harvested Gibbs free energy of mixing and volumetric expansion. Thus, PRO can be used to increase the internal energy of the draw solution with respect to the ratio of the permeated water flux (Thorsen & Holt, 2009; Achilli & Childress, 2010; La Mantia *et al.* 2011; Ramon *et al.* 2011; Yip & Elimelech, 2012).

7.3.5.1 Generating power with PRO

In addition to the osmotic pressure, the applied pressure also impacts the water flux in the PRO process. The water flux equation should be modified in order to include the effect of applied pressure on the water flux:

$$J_w = A\left[(\pi_{D,b} + P_{F,b}) - (\pi_{F,b} + P_{D,b})\right] = A(\Delta\pi - \Delta P) \qquad (7.3.3)$$

where, $P_{D,b}$ and $P_{F,b}$ are the bulk hydraulic pressures of the draw and feed solutions, respectively. "$\Delta\pi$" and "ΔP" are the effective osmotic and hydraulic pressures, respectively.

PRO can be used to generate electricity or recover energy (hydraulic pressure) by utilizing the Gibbs free energy of mixing with respect to the salinity difference of two aqueous solutions (Sandler, 1999). In a continuous PRO system, a constant hydraulic pressure is applied on a high salinity aqueous solution and permeation of water from low salinity aqueous solution continues while the osmotic pressure difference of two solutions is higher than the applied hydraulic pressure. It was theorized that, when seawater and river water were used for draw and feed solutions, respectively, the highest extractable work in a constant-pressure PRO process is 0.75 $kWh \cdot m^{-3}$ (Yip & Elimelech, 2012). Power density (W), which can be extracted from a PRO system per membrane area (m^2), can be defined as a function of water flux and hydraulic pressure difference across the membrane (Achilli & Childress, 2010):

$$W = J_w \Delta P = A(\Delta\pi - \Delta P)\Delta P \tag{7.3.4}$$

If differentiating W is plotted (Figure 7.3.6) with respect to ΔP, it can be seen that W reaches a maximum when:

$$\Delta P = \frac{\Delta\pi}{2} \tag{7.3.5}$$

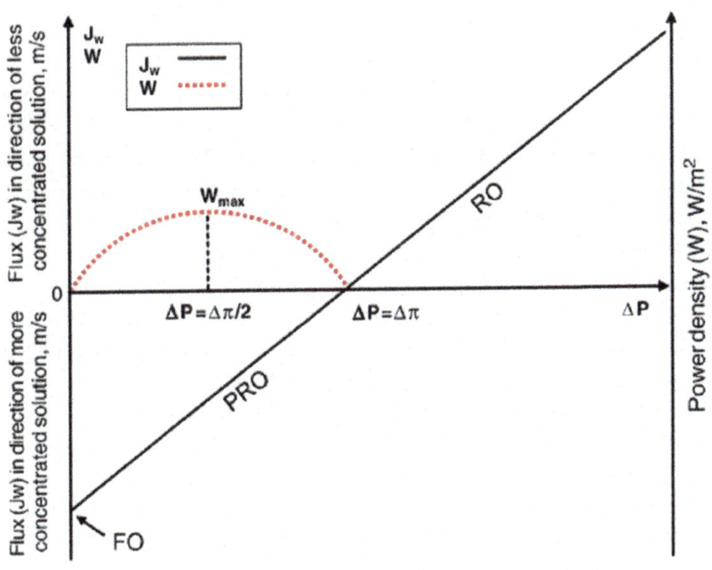

Figure 7.3.6 Theoretical water flux and power density behavior with respect to hydraulic pressure difference across the membrane. Adapted from Lee *et al.* (1981).

substituting this value for ΔP in equation, yields:

$$W_{max} = A\frac{(\Delta\pi)^2}{4} \tag{7.3.6}$$

In the case of energy production from PRO systems; a hydraulic turbine can be used to generate electricity by utilizing the pressure and volumetric flux of the aqueous solution. The power generated by a hydraulic turbine can be calculated by:

$$\text{Power} = \eta \cdot \rho \cdot Q \cdot g \cdot P \tag{7.3.7}$$

where, η is the efficiency of the turbine, ρ is the water density, Q is the flow, g is the acceleration due to gravity, and P is the pressure of the aqueous solution. Modern Pelton turbines can reach up to 92% efficiency; however the average efficiency is generally below 90% even though high pressures are applied. In PRO systems, where the applied pressures are in the range from 10 to 30 bars, this efficiency value further drops to between 60 to 80%. Therefore, when seawater is selected as the draw solution, the net produced power per m^3 of draw solution could be around 0.37 kWh (Yip & Elimelech, 2012).

7.3.5.2 Large-scale applications of PRO

Until recently, the Norwegian electric company, Statkraft, had been the pioneer of large scale PRO applications by constructing and operating the first prototype PRO facility in the world (Statkraft, 2014). Statkraft adopted a system where seawater was used as the draw solution in their system (see Figure 7.3.7) (Skilhagen *et al.* 2008). Spiral wound membrane modules, which were supplied by Hydranautics, were used with a projected power density of 5 $W \cdot m^{-2}$ at around a 12.5 bar hydraulic pressure difference across the membrane. However, low extractable work values of the selected system configuration and parameters (seawater as the draw solution), inefficiencies in the hydraulic power generation systems and high capital investment requirements, forced the company to cease their osmotic power operations in December 2013.

The Japanese Mega-ton Water System project, however, adopted a PRO system which uses seawater reverse osmosis (SWRO) brine as the draw solution (see Figure 7.3.8) (Kurihara & Hanakawa, 2013), in which the highest theoretical extractable work was calculated as 2.75 kWh per m^3 of draw solution used as opposed to 0.75 kWh per m^3 in seawater systems. The Megaton project selected hollow fiber PRO membrane modules, which were supplied by the Toyobo company, with a predicted power density of more than 12 $W \cdot m^{-2}$ at around a 30 bar hydraulic pressure difference across the membrane. The achievement of 13.3 $W \cdot m^{-2}$ for a 10 inch diameter module was announced during the "Mega-ton Water System" International Symposium (November 21–22, 2013) (Mega-ton Water System, 2014). The Mega-ton project, as Statkraft did, used hydraulic turbines to convert the salinity gradient potential of SWRO brine to electricity in its 460 $m^3 \cdot day^{-1}$ brine flow rate capacity pilot plant, in Fukuoka, Japan (Kurihara & Hanakawa, 2013).

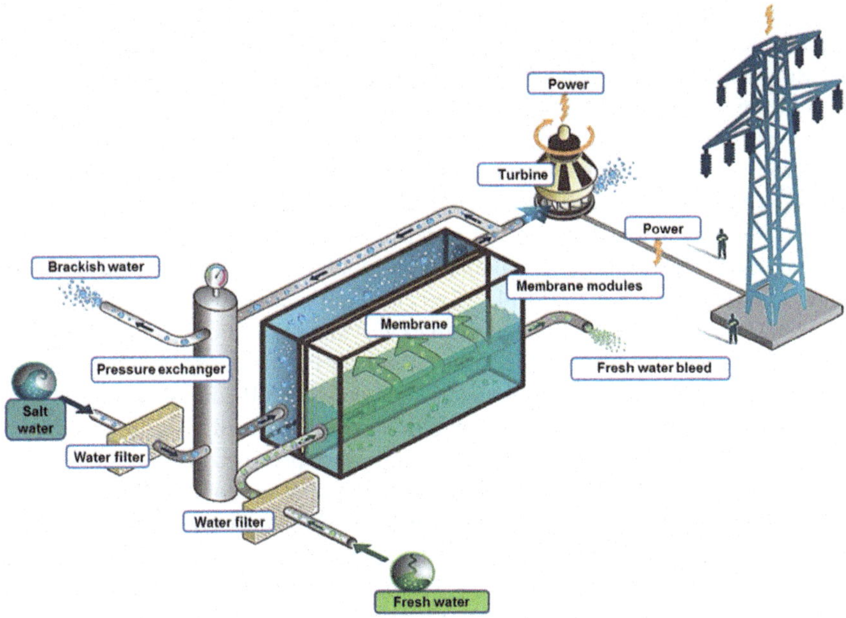

Figure 7.3.7 Pressure retarded osmosis (PRO) system adopted by Statkraft (2014).

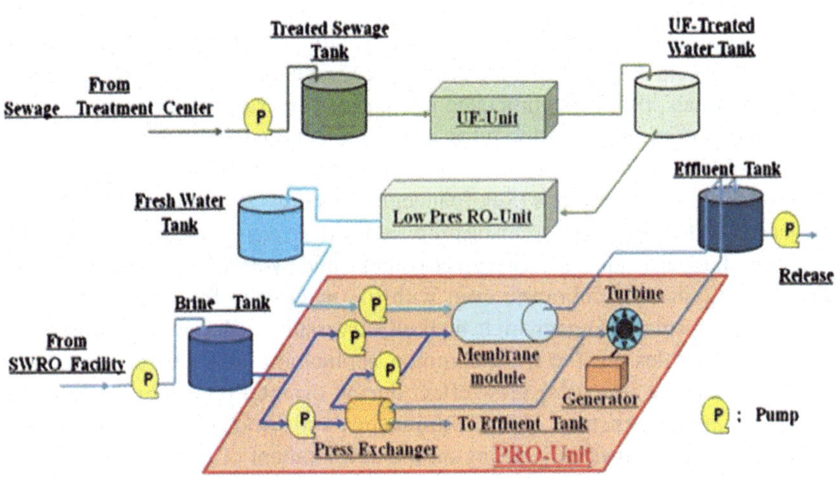

Figure 7.3.8 Pressure retarded osmosis (PRO) adopted by the Mega-ton Project (Mega-ton Water System, 2014).

So far, there have not been many PRO applications where the salinity gradient is harvested and converted to the hydraulic pressure in order to be used in different pressure dependent systems. H_2O Innovation, headquartered in Quebec, Canada, is working on a project, namely OSMOP, with Canadian utility company Hydro-Quebec to construct a mobile PRO pilot plant along the Saint Lawrence River for *in situ* proof of concept for osmotic power in the province of Quebec. Hydro-Quebec has identified 12 GW of osmotic potential from scientific research conducted along the Hudson Bay, James Bay and St Lawrence estuary, which would add 25% to its annual power capacity and the company will look into operating its own osmotic plant by 2020 (Helfer *et al.* 2014). A recently initiated Korean national research project, Global MVP (Membrane Distillation, Valuable Source Recovery, PRO), is one of the first multimillion dollar projects, where PRO is used to utilize hydraulic pressure without converting it to electricity. Global MVP adopts an SWRO-PRO hybrid system, where diluted draw solution from PRO is used to pre-pressurize the seawater prior to the SWRO process. Seawater pre-pressurizing is established by deploying isobaric pressure exchangers, which can reach up to 97% efficiency in terms of energy recovery. Woongjin Chemical (CSM membranes) is the membrane producer partner in the project and supplies spiral wound PRO membrane modules with a predicted power density of $7.5\ W \cdot m^{-2}$. Energy Recovery Inc. (USA) is another key partner in the project in order to optimize and supply high efficiency isobaric pressure exchangers which are customized for the specified purpose.

7.3.5.3 PRO membranes

There have been several companies and research groups working intensively on the PRO membrane development in order to increase the power density of these membranes at relatively high pressures (Arena *et al.* 2011; Bui *et al.* 2011; Tiraferri *et al.* 2011; Alsvik & Hägg, 2013). During the recent evolution of high performance PRO membranes, thin-film composite membranes emerged as a better choice for relatively higher mechanical durability and salt rejection. Furthermore, macrovoid-free support layers were accepted as a solution to membrane deformation during high pressure PRO operations (Sukitpaneenit & Chung, 2012; Chou *et al.* 2013).

Flat sheet PRO membranes can be produced with a very thin support layer, which enables higher permeate fluxes at a given osmotic and hydraulic pressure difference. However, flat sheet membranes, even with the macrovoid free structure, can be subject to deformation under high hydraulic pressures. Moreover, spiral wound membrane modules are the most common application of flat sheet membranes, which can cause a serious shadow effect during osmotic processes, with respect to its mechanical structure. HTI, Hydranautics and Woongjin Chemical (CSM) are three global companies working on the development of the efficient and high power density PRO membranes. Hollow fiber PRO membranes, on the other hand, have distinct advantages over flat sheet membranes as being distinguished to have a self-supported mechanical structure, high packing density, and outstanding flow patterns. However, Toyobo is the sole globally recognized membrane producer, which has the proven capability, background, and experience to develop high efficiency hollow fiber PRO membranes.

7.3.6 MAJOR CHALLENGES FOR COMMERCIALIZATION

FO is not yet a mature technology, but it is expected to help reduce the cost of producing fresh water from impaired-quality sources in the next decade. According to the 25th Desal Data IDA Worldwide Desalting Plant Inventory, the installed SWRO capacity of desalination plants around the world as of June 30, 2012 was 74.8 million $m^3 \cdot d^{-1}$. The cumulative contracted capacity, which includes plants that are contracted or under construction, reached 80.5 million $m^3 \cdot d^{-1}$. It is expected that the installed capacity will reach 120 million $m^3 \cdot d^{-1}$ by 2020 (Bennett, 2013). Clearly there is a need for fresh water; therefore, new generation FO systems will play a significant role in the desalination and water treatment industries. The main advantages, disadvantages and challenges of FO technology compared to present RO standards are depicted in Table 7.3.6.

Table 7.3.6 Comparison of FO and RO membrane processes. Adapted from (Chung *et al.* 2012).

Process	Advantages	Disadvantages	Challenges
FO	Less energy intensive for water transport across the membrane; reversible fouling; lower operational costs	Permeate water requires a second separation step; higher capital costs; ICP* in FO membranes	Low water flux membranes; lack of cost-effective draw solutes; limited studies on fouling and biofouling; reduction of ICP
RO	Permeate water is high quality product; lower capital costs	High energy consumption; irreversible fouling; higher operational costs	How to improve energy recovery efficiency; How to mitigate membrane fouling

*ICP – internal concentration polarization.

The major challenges of FO to be a commercially viable technology are: (i) developing a high flux membrane, capable of maintaining an elevated salt rejection and a reduced internal concentration polarization (ICP) effect, (ii) for PRO applications, the membrane has to additionally give adequate mechanical support to withstand hydraulic pressure, (iii) in some applications, the availability of appropriate draw solutions, which can be recirculated via an efficient recovery process, (iv) better understanding of fouling and biofouling occurrence in osmotically driven processes, (v) assuring the high quality of the water produced, (vi) hybridization with other technologies that can increase the benefits of FO use (i.e. water recovery, energy production, etc.). Chung *et al.* (2012) suggested the use of RO concentrate instead of seawater as the draw solute for water reuse to lower the operational cost and bring FO closer to commercialization. Choi *et al.* (2009) found that the integration of FO and RO may result in a high flux of fresh water and a stable performance under some operating conditions; however, the internal concentration polarization in FO is an important factor affecting the overall performance, since it reduces the effective osmotic pressure that drives the filtration process. These are just examples of listed technologies that integrate FO with other processes (Table 7.3.3).

From the listed challenges above, research has largely focused on producing a better FO membrane (Wang *et al.* 2007; Chou *et al.* 2010; Wang *et al.* 2010a; Wang *et al.* 2010b; Zhang *et al.* 2010; Setiawan *et al.* 2011; Fu *et al.* 2013; You *et al.* 2013; Zhong *et al.* 2013; Romero-Vargas Castrillón *et al.* 2014). Some excellent reviews in the aspect of FO and PRO membranes are available in the literature covering the milestones of FO/PRO membrane development, membrane material, fabrication and methods to improve membrane property (Zhao *et al.* 2012c). The first commercially available FO membrane was developed by HTI. It is a cellulose acetate membrane embedded about a polyester mesh. Nowadays, the secondary generation TFC FO membranes are also available in the market, such as HTI, Oasys, and Porifera. The water flux of commercial FO membrane has been improved significantly from about 10 up to 30–40 $L \cdot m^{-2} \cdot h^{-1}$. It should be noted particularly that hollow fiber and biomimetic FO membranes have shown considerable potential in the future applications of FO process (Wang *et al.* 2010b; Yip *et al.* 2010; Zhao *et al.* 2012c; Sun *et al.* 2013; Zhong *et al.* 2013). In addition to the competitive flux comparing to flat-sheet membrane, hollow fiber membrane can offer much higher packing density in membrane module. Biomimetic membrane incorporates aquaporin, a natural biological water channel and molecular gate, into lipid film or polymeric membrane. The exceptional selectivity and permeability that aquaporin demonstrate towards water molecules in natural process makes it possible to design a high performance biomimetic membrane for desalination (Tang *et al.* 2013). In FO application, biomimetic aquaporin membrane showed a water flux of 16.4 $L \cdot m^{-2} \cdot h^{-1}$ and selectivity (water to NaCl) as high as 7000 using 0.3 M sucrose as the DS and 200 ppm NaCl as the FS (Wang *et al.* 2012). However, the long-term stability of biomimetic membrane should be addressed in future studies including biodegradation of aquaporin, effect of chemical cleaning, effect of solution pH and temperature, etc.

Biofouling is, on the contrary, one of the least understood and investigated topics in forward osmosis systems (Valladares Linares *et al.* 2014). Valladares Linares *et al.* (2012) studied the fouling layer on the FO membrane operated in an *indirect* desalination system. The natural organic matter (NOM) fouling layer was found to be composed of biopolymers and protein-like substances, which adversely affect the flux of water through the FO membrane. On the support layer of the membrane, in contact with seawater, transparent exopolymeric particles (TEP) formed clusters clearly identifiable with an optical microscope and a TEP-specific dye.

In the case of PRO process, availability of commercial PRO membrane limits the application of PRO. Although power density of PRO membrane has reached higher than 20 $W \cdot m^{-2}$ (Zhang & Chung, 2013; Straub *et al.* 2014) in lab-scale test, there are no high performing and cheap commercially available PRO membrane in the market today. In addition, the studies of fouling and scaling propensity in PRO process are still limited (She *et al.* 2013; Thelin *et al.* 2013; Yip & Elimelech, 2013; Zhang *et al.* 2014). Recently, a number of studies has been conducted to understand the difference between theoretical and practical power density in PRO process in terms of membrane structure and process configuration (Almansoori & Saif, 2014; Altaee *et al.* 2014; Helfer *et al.* 2014). Both commercial membrane performance and fouling are the crucial reasons that have limited the power density of PRO pilot plants, and led Statkraft to discontinuing development of osmotic power. The osmotic power generated by PRO process is not competitive in the power market until technology gap of membrane performance and fouling control can be filled. However, higher salinity gradient in RO-PRO system is likely to make RO-PRO a more promising process of renewable energy generation than river-to-sea PRO system (Achilli *et al.* 2014; Prante *et al.* 2014). In addition, PRO-based energy in industrial water processes is one area that could see commercial activity within the near future.

7.3.7 SUMMARY

This review focuses on the present status of forward osmosis (FO) niches in two main areas: seawater desalination and wastewater reuse. Specific applications for desalination and impaired-quality water treatment and reuse are described, as well as the benefits, advantages, challenges, costs and knowledge gaps on FO hybrid systems are discussed. FO can play a role as a bridge to integrate upstream and downstream water treatment processes, to reduce the energy consumption of the entire desalination or water recovery and reuse processes, thus achieving a sustainable solution for the water-energy nexus. FO hybrid membrane systems showed to have advantages over traditional membrane process like high pressure reverse osmosis and nanofiltration for desalination and wastewater treatment: (i) chemical storage and feed water systems may be reduced for capital, operational and maintenance cost, (ii) water quality is improved, (iii) reduced process piping costs, (iv) more flexible treatment units, and (v) higher overall sustainability of the desalination and wastewater treatment process. Nevertheless, major challenges make FO systems not yet a commercially viable technology, the most critical being the development of a high flux membrane, capable of maintaining an elevated salt rejection and a reduced internal concentration polarization effect, and the availability of appropriate draw solutions (cost effective and non-toxic), which can be recirculated via an efficient recovery process. This review article highlights the features of hybrid FO systems and specifically provides the state-of-the-art applications in the water industry in a novel classification and based on the latest developments towards scaling up these systems.

7.3.8 REFERENCES

Abousnina R. M. (2012). Oily Wastewater Treatment: Removal of Dissolved Organic Components by Forward Osmosis. University of Wollongong, Wollongong, Australia.

Achilli A. and Childress A. E. (2010). Pressure retarded osmosis: from the vision of Sidney Loeb to the first prototype installation—Review. *Desalination*, **261**, 205–211.

Achilli A., Cath T. Y., Marchand E. A. and Childress A. E. (2009). The forward osmosis membrane bioreactor: a low fouling alternative to MBR processes. *Desalination*, **239**, 10–21.

Achilli A., Prante J. L., Hancock N. T., Maxwell E. B. and Childress A. E. (2014). Experimental results from RO-PRO: a next generation system for low-energy desalination. *Environmental Science & Technology*, **48**, 6437–6443.

Almansoori A. and Saif Y. (2014). Structural optimization of osmosis processes for water and power production in desalination applications. *Desalination*, **344**, 12–27.

Alnaizy R., Aidan A. and Qasim M. (2013). Copper sulfate as draw solute in forward osmosis desalination. *Journal of Environmental Chemical Engineering*, **1**, 424–430.

Alsvik I. L. and Hägg M.-B. (2013). Pressure retarded osmosis and forward osmosis membranes: materials and methods. *Polymers*, **5**, 303–327.

Altaee A., Zaragoza G. and Sharif A. (2014). Pressure retarded osmosis for power generation and seawater desalination: performance analysis. *Desalination*, **344**, 108–115.

Amy G., Ghaffour N., Ng K. C. and Li Z. (2013). Advances in Low-Energy Desalination. Water Reuse and Desalination. WateReuse Association, USA.

Arena J. T., McCloskey B., Freeman B. D. and McCutcheon J. R. (2011). Surface modification of thin film composite membrane support layers with polydopamine: enabling use of reverse osmosis membranes in pressure retarded osmosis. *Journal of Membrane Science*, **375**, 55–62.

Aydiner C., Sen U., Topcu S., Sesli D., Ekinci D., Altınay A. D., Ozbey B., Koseoglu-Imer D. Y. and Keskinler B. (2013). Techno-economic investigation of water recovery and whey powder production from whey using UF/RO and FO/RO integrated membrane systems. *Desalination and Water Treatment*, **52**(1-3), 123–133.

Bennett A. (2013). 50th anniversary: desalination: 50 years of progress. *Filtration + Separation*, **50**, 32–39.

Boo C., Elimelech M. and Hong S. (2013). Fouling control in a forward osmosis process integrating seawater desalination and wastewater reclamation. *Journal of Membrane Science*, **444**, 148–156.

Bui N.-N., Lind M. L., Hoek E. and McCutcheon J. R. (2011). Electrospun nanofiber supported thin film composite membranes for engineered osmosis. *Journal of Membrane Science*, **385**, 10–19.

Cai Y., Shen W., Loo S. L., Krantz W. B., Wang R., Fane A. G. and Hu X. (2013). Towards temperature driven forward osmosis desalination using Semi-IPN hydrogels as reversible draw agents. *Water Research*, **47**, 3773–3781.

Carmignani G., Sitkiewitz S. and Webley J. W. (2012). Recovery of retrograde soluble solute for forward osmosis water treatment. US Patent 20120267308 A1. Trevi Systems Inc. (Petaluma, CA, US), United States.

Catalyx (2009a). Forward osmosis for recycling dye wastewater. *Filtration & Separation*, **46**, 14.

Catalyx (2009b). High BOD and COD carpetdyeing wastewater recycled using forward osmosis. *Membrane Technology*, **4**, 8.

Cath T. Y., Childress A. E. and Elimelech M. (2006). Forward osmosis: principles, applications, and recent developments. *Journal of Membrane Science*, **281**, 70–87.

Cath T. Y., Drewes J. E. and Lundin C. D. (2009). A Novel Hybrid Forward Osmosis Process for Drinking Water Augmentation Using Impaired Water and Saline Water Sources. In: Proceedings of the 24th Annual WateReuse Symposium, September 13–16, Seattle, Washington.

Cath T. Y., Hancock N. T., Lundin C. D., Hoppe-Jones C. and Drewes J. E. (2010). A multi-barrier osmotic dilution process for simultaneous desalination and purification of impaired water. *Journal of Membrane Science*, **362**, 417–426.

Chang I.-S., Le Clech P., Jefferson B. and Judd S. (2002). Membrane fouling in membrane bioreactors for wastewater treatment. *Journal of Environmental Engineering*, **128**(11), 1018–1029.

Chanukya B. S., Patil S. and Rastogi N. K. (2013). Influence of concentration polarization on flux behavior in forward osmosis during desalination using ammonium bicarbonate. *Desalination*, **312**, 39–44.

Chekli L., Phuntsho S., Shon H. K., Vigneswaran S., Kandasamy J. and Chanan A. (2012). A review of draw solutes in forward osmosis process and their use in modern applications. *Desalination and Water Treatment*, **43**, 167–184.

Choi Y.-J., Choi J.-S., Oh H.-J., Lee S., Yang D. R. and Kim J. H. (2009). Toward a combined system of forward osmosis and reverse osmosis for seawater desalination. *Desalination*, **247**, 239–246.

Chou S., Shi L., Wang R., Tang C. Y., Qiu C. and Fane A. G. (2010). Characteristics and potential applications of a novel forward osmosis hollow fiber membrane. *Desalination*, **261**, 365–372.

Chou S., Wang R. and Fane A. G. (2013). Robust and High performance hollow fiber membranes for energy harvesting from salinity gradients by pressure retarded osmosis. *Journal of Membrane Science*, **448**, 44–54.

Chung T.-S., Li X., Ong R. C., Ge Q., Wang H. and Han G. (2012). Emerging forward osmosis (FO) technologies and challenges ahead for clean water and clean energy applications. *Current Opinion in Chemical Engineering*, **1**, 246–257.

Cornelissen E. R., Harmsen D., de Korte K. F., Ruiken C. J., Qin J.-J., Oo H. and Wessels L. P. (2008). Membrane fouling and process performance of forward osmosis membranes on activated sludge. *Journal of Membrane Science*, **319**, 158–168.

Creel L. (2003). Ripple Effects: Population and Coastal Regions. Population Reference Bureau, Washington D.C.

Duong P. H. H. and Chung T.-S. (2014). Application of thin film composite membranes with forward osmosis technology for the separation of emulsified oil–water. *Journal of Membrane Science*, **452**, 117–126.

Flynn M. T., Soler M., Shull S., Broyan J., Chambliss J., Howe A. S., Gormly S., Hammoudeh M., Shaw H. and Howard K. (2012). Forward Osmosis Cargo Transfer Bag. In: 42nd International Conference on Environmental Systems, 15–19 July, San Diego, CA.

Fritzmann C., Löwenberg J., Wintgens T. and Melin T. (2007). State-of-the-art of reverse osmosis desalination. *Desalination*, **216**, 1–76.

Fu F.-J., Zhang S., Sun S.-P., Wang K.-Y. and Chung T.-S. (2013). POSS-containing delamination-free dual-layer hollow fiber membranes for forward osmosis and osmotic power generation. *Journal of Membrane Science*, **443**, 144–155.

Ge Q., Wang P., Wan C. and Chung T.-S. (2012). Polyelectrolyte-promoted forward osmosis–membrane distillation (FO–MD) hybrid process for dye wastewater treatment. *Environmental Science & Technology*, **46**, 6236–6243.

Ge Q., Ling M. and Chung T.-S. (2013). Draw solutions for forward osmosis processes: developments, challenges, and prospects for the future. *Journal of Membrane Science*, **442**, 225–237.

Gray G. T., McCutcheon J. R. and Elimelech M. (2006). Internal concentration polarization in forward osmosis: role of membrane orientation. *Desalination*, **197**, 1–8.

Hancock N., Cath T., Xu P., Heil D. and Black N. (2011). Novel Dual-Barrier Hybrid Osmotic Dilution–RO System: Pilot-Scale Demonstration of Technical and Ecological Merit. Proceedings of the American Water Works Association, Membrane Technology Conference, 31 March 2011, Long Beach, CA.

Hau N. T., Chen S.-S., Nguyen N. C., Huang K. Z., Ngo H. H. and Guo W. (2014). Exploration of EDTA sodium salt as novel draw solution in forward osmosis process for dewatering of high nutrient sludge. *Journal of Membrane Science*, **455**, 305–311.

Helfer F., Lemckert C. and Anissimov Y. G. (2014). Osmotic power with pressure retarded osmosis: theory, performance and trends – a review. *Journal of Membrane Science*, **453**, 337–358.

Hickenbottom K. L., Hancock N. T., Hutchings N. R., Appleton E. W., Beaudry E. G., Xu P. and Cath T. Y. (2013). Forward osmosis treatment of drilling mud and fracturing wastewater from oil and gas operations. *Desalination*, **312**, 60–66.

Holloway R. W., Childress A. E., Dennett K. E. and Cath T. Y. (2007). Forward osmosis for concentration of anaerobic digester centrate. *Water Research*, **41**, 4005–4014.

Hoover L. A., Phillip W. A., Tiraferri A., Yip N. Y. and Elimelech M. (2011). Forward with osmosis: emerging applications for greater sustainability. *Environmental Science & Technology*, **45**, 9824–9830.

HTI (2010a). Humanitarian Water. http://www.htiwater.com/divisions/humanitarian/about.html

HTI (2010b). Military Water Filters & Military Desalination Filters. http://www.htiwater.com/divisions/military_regulatory/about.html

HTI (2011). Oil Wastewater Treatment & Gas Wastewater Treatment: Lead Story. http://www.htiwater.com/divisions/oil-gas/lead_story.html

Iyer S. (2012). Systems and methods for forward osmosis fluid purification using cloud point extraction. US Patent 20120267297 A1.

Jones R. A. L., Ryan A. J., Storey H., Butler M. and Crook C. J. (2010). Apparatus and method for purifying water by forward osmosis. US Patent 20100213129 A1.

Kurihara M. and Hanakawa M. (2013). Mega-ton Water System: Japanese national research and development project on seawater desalination and wastewater reclamation. *Desalination*, **308**, 131–137.

La Mantia F., Pasta M., Deshazer H. D., Logan B. E. and Cui Y. (2011). Batteries for efficient energy extraction from a water salinity difference. *Nano Letters*, **11**, 1810–1813.

Lay W. C. L., Zhang Q., Zhang J., McDougald D., Tang C., Wang R., Liu Y. and Fane A. G. (2011). Study of integration of forward osmosis and biological process: membrane performance under elevated salt environment. *Desalination*, **283**, 123–130.

Lee K. L., Baker R. W. and Lonsdale H. K. (1981). Membranes for power generation by pressure-retarded osmosis. *Journal of Membrane Science*, **8**, 141–171.

Lee S., Boo C., Elimelech M. and Hong S. (2010). Comparison of fouling behavior in forward osmosis (FO) and reverse osmosis (RO). *Journal of Membrane Science*, **365**, 34–39.

Li D., Zhang X., Yao J., Simon G. P. and Wang H. (2011). Stimuli-responsive polymer hydrogels as a new class of draw agent for forward osmosis desalination. *Chemical Communications*, **47**, 1710–1712.

Li Z.-Y., Yangali-Quintanilla V., Valladares-Linares R., Li Q., Zhan T. and Amy G. (2012a). Flux patterns and membrane fouling propensity during desalination of seawater by forward osmosis. *Water Research*, **46**, 195–204.

Li Z. Y., Valladares-Linares R., Yangali-Quintanilla V. and Amy G. (2012b). A Sequential Batch Reactor – Forward Osmosis System for Water Reuse. In: AWWA and AMTA Membrane Technology Conference and Exposition, Phoenix.

Li D., Zhang X., Simon G. P. and Wang H. (2013a). Forward osmosis desalination using polymer hydrogels as a draw agent: influence of draw agent, feed solution and membrane on process performance. *Water Research*, **47**, 209–215.

Li Z., Valladares Linares R., Muhannad A. and Amy G. (2013b). Comparative Assessment of Forward Osmosis (FO) Niches in Desalination. In: International Desalination Association (IDA) World Congress, Tianjin, China.

Li Z., Valladares Linares R., Abu-Ghdaib M., Zhan T., Yangali-Quintanilla V. and Amy G. (2014). Osmotically driven membrane process for the management of urban runoff in coastal regions. *Water Research*, **48**, 200–209.

Liberman B. (2003). The importance of energy recovery devices in reverse osmosis desalination. The Future of Desalination in Texas Vol. 2, Texas Water Development Board. Biennial Report on Seawater Desalination, December 2004.

Ling M. M. and Chung T.-S. (2011). Desalination process using super hydrophilic nanoparticles via forward osmosis integrated with ultrafiltration regeneration. *Desalination*, **278**, 194–202.

Loeb S. (1975). Fresh Water. US Patent 3,906,250.

Loeb S. (1976). Production of energy from concentrated brines by pressure-retarded osmosis: I. Preliminary technical and economic correlations. *Journal of Membrane Science*, **1**, 49–63.

Loeb S., Van Hessen F. and Shahaf D. (1976). Production of energy from concentrated brines by pressure-retarded osmosis: II. Experimental results and projected energy costs. *Journal of Membrane Science*, **1**, 249–269.

Martinetti C. R., Childress A. E. and Cath T. Y. (2009). High recovery of concentrated RO brines using forward osmosis and membrane distillation. *Journal of Membrane Science*, **331**, 31–39.

McCarty P. L., Bae J. and Kim J. (2011). Domestic wastewater treatment as a net energy producer–can this be achieved? *Environmental Science & Technology*, **45**, 7100–7106.

McCutcheon J. R. and Elimelech M. (2006). Influence of concentrative and dilutive internal concentration polarization on flux behavior in forward osmosis. *Journal of Membrane Science*, **284**, 237–247.

McCutcheon J. R., McGinnis R. L. and Elimelech M. (2005). A novel ammonia—carbon dioxide forward (direct) osmosis desalination process. *Desalination*, **174**, 1–11.

McCutcheon J. R., McGinnis R. L. and Elimelech M. (2006). Desalination by ammonia–carbon dioxide forward osmosis: influence of draw and feed solution concentrations on process performance. *Journal of Membrane Science*, **278**, 114–123.

McGinnis R. L. (2002). Osmotic desalinization process. US Patent 6391205 B1.

McGinnis R. L. (2009). Osmotic desalination process. US Patent 7560029 B2.

McGinnis R. L. (2011). Osmotic desalination process. US Patent 8002989 B2.

McGinnis R. L. and Elimelech M. (2007). Energy requirements of ammonia–carbon dioxide forward osmosis desalination. *Desalination*, **207**, 370–382.

McGinnis R. L., Hancock N. T., Nowosielski-Slepowron M. S. and McGurgan G. D. (2013). Pilot demonstration of the NH3/CO2 forward osmosis desalination process on high salinity brines. *Desalination*, **312**, 67–74.

Mega-ton Water System. (2014). http://www.megatonwater.com/en/index.html

Mehta G. D. and Loeb S. (1979). Internal polarization in the porous substructure of a semipermeable membrane under pressure-retarded osmosis. *Journal of Membrane Science*, **4**, 261–265.

Mi B. and Elimelech M. (2008). Chemical and physical aspects of organic fouling of forward osmosis membranes. *Journal of Membrane Science*, **320**, 292–302.

Miller J. E. and Evans L. R. (2006). Forward Osmosis: A New Approach to Water Purification and Desalination.

Modern Water. (2013). Membrane Processes – Forward Osmosis: Desalination. www.modernwater.com

NASA (2000). Astronomy Picture of the Day. http://apod.nasa.gov/apod/ap001127.html

NASA (2012). NASA Showcases Innovative Method To Grow Algae-Based Biofuels. http://www.nasa.gov/centers/ames/news/features/2012/omega_algae_feature.html

Nelson C. E. and Ghosh A. K. (2011). Oil & Natural Gas Technology – Membrane Technology for Produced Water in Lea County. Lea County Government and New Mexico Institute of Mining and Technology, Socorro, NM.

Norman R. S. (1974). Water salination: a source of energy. *Science*, **186**, 350–352.

Oriard T. and Haggerty P. D. (2006). Forward osmosis utilizing a controllable osmotic agent. US Patent 0278153.

Pattle R. (1954). Production of electric power by mixing fresh and salt water in the hydroelectric pile. *Nature*, **174**(4431), 660–660.

Phuntsho S., Shon H. K., Hong S., Lee S. and Vigneswaran S. (2011). A novel low energy fertilizer driven forward osmosis desalination for direct fertigation: evaluating the performance of fertilizer draw solutions. *Journal of Membrane Science*, **375**, 172–181.

Prante J. L., Ruskowitz J. A., Childress A. E. and Achilli A. (2014). RO-PRO desalination: an integrated low-energy approach to seawater desalination. *Applied Energy*, **120**, 104–114.

Qin J.-J., Kekre K. A., Oo M. H., Tao G., Lay C. L., Lew C. H., Cornelissen E. R. and Ruiken C. J. (2010). Preliminary study of osmotic membrane bioreactor: effects of draw solution on water flux and air scouring on fouling. *Water Science and Technology*, **62**, 8.

Ramon G. Z., Feinberg B. J. and Hoek E. M. (2011). Membrane-based production of salinity-gradient power. *Energy & Environmental Science*, **4**, 4423–4434.

Rassoul G., Al–Alawy A. F. and Khudair W. N. (2012). Reduction of concentrating poisonous metallic radicals from industrial wastewater by forward and reverse osmosis. *Journal of Engineering*, **7**(18), 784–798.

Romero-Vargas Castrillón S., Lu X., Shaffer D. L. and Elimelech M. (2014). Amine enrichment and poly(ethylene glycol) (PEG) surface modification of thin-film composite forward osmosis membranes for organic fouling control. *Journal of Membrane Science*, **450**, 331–339.

Saito K., Irie M., Zaitsu S., Sakai H., Hayashi H. and Tanioka A. (2012). Power generation with salinity gradient by pressure retarded osmosis using concentrated brine from SWRO system and treated sewage as pure water. *Desalination and Water Treatment*, **41**, 114–121.

Sandler S. (1999). Chemical and Engineering Thermodynamics, 3rd edn. Wiley, New York.

Setiawan L., Wang R., Li K. and Fane A. G. (2011). Fabrication of novel poly(amide–imide) forward osmosis hollow fiber membranes with a positively charged nanofiltration-like selective layer. *Journal of Membrane Science*, **369**, 196–205.

Shaffer D. L., Yip N. Y., Gilron J. and Elimelech M. (2012). Seawater desalination for agriculture by integrated forward and reverse osmosis: improved product water quality for potentially less energy. *Journal of Membrane Science*, **415–416**, 1–8.

She Q., Wong Y. K. W., Zhao S. and Tang C. Y. (2013). Organic fouling in pressure retarded osmosis: experiments, mechanisms and implications. *Journal of Membrane Science*, **428**, 181–189.

Skilhagen S. E., Dugstad J. E. and Aaberg R. J. (2008). Osmotic power – power production based on the osmotic pressure difference between waters with varying salt gradients. *Desalination*, **220**, 476–482.

Statkraft. (2014). Osmotic power – http://www.statkraft.com/about-statkraft/innovation/osmotic-power/

Straub A. P., Yip N. Y. and Elimelech M. (2014). Raising the bar: increased hydraulic pressure allows unprecedented high power densities in pressure-retarded osmosis. *Environmental Science & Technology Letters*, **1**, 55–59.

Sukitpaneenit P. and Chung T.-S. (2012). High performance thin-film composite forward osmosis hollow fiber membranes with macrovoid-free and highly porous structure for sustainable water production. *Environmental Science & Technology*, **46**, 7358–7365.

Sun G., Chung T.-S., Jeyaseelan K. and Armugam A. (2013). A layer-by-layer self-assembly approach to developing an aquaporin-embedded mixed matrix membrane. *RSC Advances*, **3**, 473–481.

Tang C. Y., Zhao Y., Wang R., Hélix-Nielsen C. and Fane A. G. (2013). Desalination by biomimetic aquaporin membranes: review of status and prospects. *Desalination*, **308**, 34–40.

Tang W. and Ng H. Y. (2008). Concentration of brine by forward osmosis: performance and influence of membrane structure. *Desalination*, **224**, 143–153.

The Barila Group, The Coca-Cola Company, The International Finance Corporation, McKinsey & Company, Nestle S.A., New Holland Agriculture, SABMiller plc, Standard Chartered Bank, and Syngenta AG. (2009). Charting Our Water Future, Economic Frameworks to Inform Decision-Making. 2030 Water Resources Group, Washington, DC.

Thelin W. R., Sivertsen E., Holt T. and Brekke G. (2013). Natural organic matter fouling in pressure retarded osmosis. *Journal of Membrane Science*, **438**, 46–56.

Thorsen T. and Holt T. (2009). The potential for power production from salinity gradients by pressure retarded osmosis. *Journal of Membrane Science*, **335**, 103–110.

Tibbetts J. (2002). Coastal cities: living on the edge. *Environmental Health Perspectives*, **110**, A674–A681.

Tiraferri A., Yip N. Y., Phillip W. A., Schiffman J. D. and Elimelech M. (2011). Relating performance of thin-film composite forward osmosis membranes to support layer formation and structure. *Journal of Membrane Science*, **367**, 340–352.

Valladares Linares R., Yangali-Quintanilla V., Li Z. and Amy G. (2011). Rejection of micropollutants by clean and fouled forward osmosis membrane. *Water Research*, **45**, 6737–6744.

Valladares Linares R., Yangali-Quintanilla V., Li Z. and Amy G. (2012). NOM and TEP fouling of a forward osmosis (FO) membrane: foulant identification and cleaning. *Journal of Membrane Science*, **421–422**, 217–224.

Valladares Linares R., Li Z., Abu-Ghdaib M., Wei C.-H., Amy G. and Vrouwenvelder J. S. (2013a). Water harvesting from municipal wastewater via osmotic gradient: an evaluation of process performance. *Journal of Membrane Science*, **447**, 50–56.

Valladares Linares R., Li Z., Yangali-Quintanilla V., Li Q. and Amy G. (2013b). Cleaning protocol for a FO membrane fouled in wastewater reuse. *Desalination and Water Treatment*, **51**, 4821–4824.

Valladares Linares R., Bucs S. S., Li Z., AbuGhdeeb M., Amy G. and Vrouwenvelder J. S. (2014). Impact of spacer thickness on biofouling in forward osmosis. *Water Research*, **57**, 223–233.

van't Hoff J. H. (1887). The function of osmotic pressure in the analogy between solutions and gases. *Proceedings of the Physical Society of London*, **9**, 307.

Wang H., Chung T.-S., Tong Y. W., Jeyaseelan K., Armugam A., Chen Z., Hong M. and Meier W. (2012). Highly permeable and selective pore-spanning biomimetic membrane embedded with aquaporin Z. *Small*, **8**, 1185–1190.

Wang K. Y., Chung T.-S. and Qin J.-J. (2007). Polybenzimidazole (PBI) nanofiltration hollow fiber membranes applied in forward osmosis process. *Journal of Membrane Science*, **300**, 6–12.

Wang K. Y., Ong R. C. and Chung T.-S. (2010a). Double-skinned forward osmosis membranes for reducing internal concentration polarization within the porous sublayer. *Industrial & Engineering Chemistry Research*, **49**, 4824–4831.

Wang R., Shi L., Tang C. Y., Chou S., Qiu C. and Fane A. G. (2010b). Characterization of novel forward osmosis hollow fiber membranes. *Journal of Membrane Science*, **355**, 158–167.

Werner C. M., Logan B. E., Saikaly P. E. and Amy G. L. (2013). Wastewater treatment, energy recovery and desalination using a forward osmosis membrane in an air-cathode microbial osmotic fuel cell. *Journal of Membrane Science*, **428**, 116–122.

Xie M., Nghiem L. D., Price W. E. and Elimelech M. (2013). A forward osmosis–membrane distillation hybrid process for direct sewer mining: system performance and limitations. *Environmental Science & Technology*, **47**, 13486–13493.

Xie M., Nghiem L. D., Price W. E. and Elimelech M. (2014). Toward resource recovery from wastewater: extraction of phosphorus from digested sludge using a hybrid forward osmosis–membrane distillation process. *Environmental Science & Technology Letters*, **1**, 191–195.

Yaeli J. (1992). Method and apparatus for processing liquid solutions of suspensions particularly useful in the desalination of saline water. US Patent 5098575 A.

Yangali-Quintanilla V., Li Z., Valladares R., Li Q. and Amy G. (2011). Indirect desalination of Red Sea water with forward osmosis and low pressure reverse osmosis for water reuse. *Desalination*, **280**, 160–166.

Yap W. J., Zhang J., Lay W. C. L., Cao B., Fane A. G. and Liu Y. (2012). State of the art of osmotic membrane bioreactors for water reclamation. *Bioresource Technology*, **122**, 217–222.

Yip N. Y. and Elimelech M. (2012). Thermodynamic and energy efficiency analysis of power generation from natural salinity gradients by pressure retarded osmosis. *Environmental Science & Technology*, **46**, 5230–5239.

Yip N. Y. and Elimelech M. (2013). Influence of natural organic matter fouling and osmotic backwash on pressure retarded osmosis energy production from natural salinity gradients. *Environmental Science & Technology*, **47**, 12607–12616.

Yip N. Y., Tiraferri A., Phillip W. A., Schiffman J. D. and Elimelech M. (2010). High performance thin-film composite forward osmosis membrane. *Environmental Science & Technology*, **44**, 3812–3818.

You S., Tang C., Yu C., Wang X., Zhang J., Han J., Gan Y. and Ren N. (2013). Forward osmosis with a novel thin-film inorganic membrane. *Environmental Science & Technology*, **47**, 8733–8742.

Zhang F., Brastad K. S. and He Z. (2011). Integrating forward osmosis into microbial fuel cells for wastewater treatment, water extraction and bioelectricity generation. *Environmental Science & Technology*, **45**, 6690–6696.

Zhang J., Loong W. L. C., Chou S., Tang C., Wang R. and Fane A. G. (2012). Membrane biofouling and scaling in forward osmosis membrane bioreactor. *Journal of Membrane Science*, **403–404**, 8–14.

Zhang M., Hou D., She Q. and Tang C. Y. (2014). Gypsum scaling in pressure retarded osmosis: experiments, mechanisms and implications. *Water Research*, **48**, 387–395.

Zhang S. and Chung T.-S. (2013). Minimizing the instant and accumulative effects of salt permeability to sustain ultrahigh osmotic power density. *Environmental Science & Technology*, **47**, 10085–10092.

Zhang S., Wang K. Y., Chung T.-S., Chen H., Jean Y. C. and Amy G. (2010). Well-constructed cellulose acetate membranes for forward osmosis: minimized internal concentration polarization with an ultra-thin selective layer. *Journal of Membrane Science*, **360**, 522–535.

Zhang Y., Pinoy L., Meesschaert B. and Van der Bruggen B. (2013). A natural driven membrane process for Brackish and wastewater treatment: photovoltaic powered ED and FO hybrid system. *Environmental Science & Technology*, **47**, 10548–10555.

Zhao Q., Chen N., Zhao D. and Lu X. (2013). Thermoresponsive magnetic nanoparticles for Seawater desalination. *ACS Applied Materials & Interfaces*, **5**, 11453–11461.

Zhao S., Zou L. and Mulcahy D. (2012a). Brackish water desalination by a hybrid forward osmosis–nanofiltration system using divalent draw solute. *Desalination*, **284**, 175–181.

Zhao S., Zou L., Tang C. Y. and Mulcahy D. (2012b). Recent developments in forward osmosis: opportunities and challenges. *Journal of Membrane Science*, **396**, 1–21.

Zhao Y., Qiu C., Li X., Vararattanavech A., Shen W., Torres J., Hélix-Nielsen C., Wang R., Hu X., Fane A. G. and Tang C. Y. (2012c). Synthesis of robust and high-performance aquaporin-based biomimetic membranes by interfacial polymerization-membrane preparation and RO performance characterization. *Journal of Membrane Science*, **423–424**, 422–428.

Zhong P., Fu X., Chung T.-S., Weber M. and Maletzko C. (2013). Development of thin-film composite forward osmosis hollow fiber membranes using direct sulfonated polyphenylenesulfone (sPPSU) as membrane substrates. *Environmental Science & Technology*, **47**, 7430–7436.

Zhu H., Zhang L., Wen X. and Huang X. (2012). Feasibility of applying forward osmosis to the simultaneous thickening, digestion, and direct dewatering of waste activated sludge. *Bioresource Technology*, **113**, 207–213.

Supplementary material

Chapter 2.1

Table 2.1.S1 Rejection (%) of trace metals in the simulated osmotic detention pond under different feed water conditions.

Batch Number	Neutral				Acidic				NOM Fouling			
	Cu	Zn	Cd	Pb	Cu	Zn	Cd	Pb	Cu	Zn	Cd	Pb
1	99.1	100	99.9	100	99.4	100	98.7	100	99.1	100	100	99.7
2	98.9	100	99.0	99.6	99.0	100	99.0	100	98.9	100	100	100
3	98.9	99.9	99.5	99.9	99.5	100	98.7	100	100	100	100	100
4	92.4	99.9	99.3	97.1	99.6	100	98.9	100	100	100	100	100
5	98.5	99.9	99.1	100	99.9	100	99.8	100	100	100	100	99.8
6	98.1	100	99.4	99.8	99.6	100	99.2	100	100	100	100	100
7	98.2	100	99.4	99.9	99.5	100	99.1	100	100	100	100	99.7

Table 2.1.S2 Rejection (%) of nutrients in the simulated osmotic detention pond under different feed water conditions.

Batch Number	Neutral			Acidic			NOM Fouling		
	P	NO_3^-	TN	P	NO_3^-	TN	P	NO_3^-	TN
1	100	88.4	85.4	98.7	86.2	84.7	98.6	93.9	85.1
2	100	89.0	84.7	99.5	82.4	71.1	99.1	91.1	80.4
3	100	88.2	84.0	97.1	79.7	75.1	99.7	80.5	72.9
4	100	86.0	82.8	98.5	79.1	75.7	98.9	68.5	71.1
5	100	87.8	72.5	99.2	71.4	78.1	99.3	65.5	69.0
6	100	86.4	71.9	99.5	68.7	80.6	99.7	65.1	69.0
7	100	82.9	75.3	98.8	66.9	81.6	98.5	52.1	64.7

Chapter 3.1

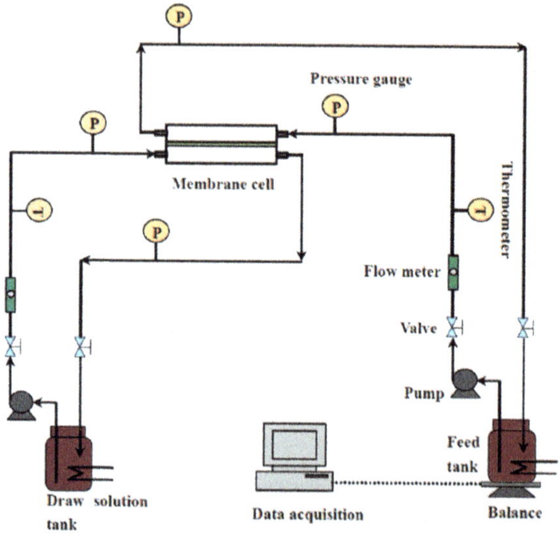

Figure 3.1.S1 Schematic diagram of the experimental set-up for SWFO.

Chapter 3.3

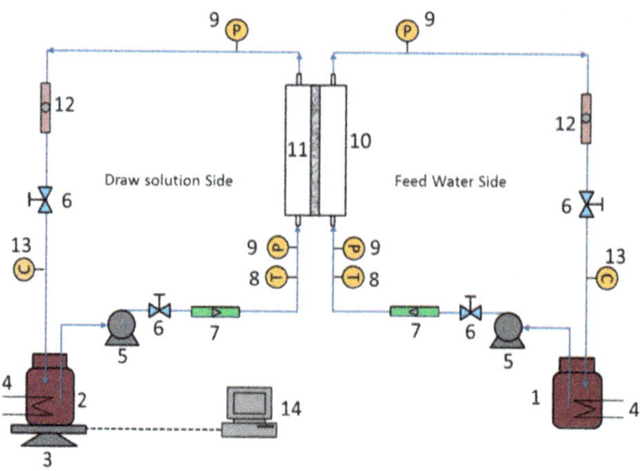

Figure 3.3.S1 The schematic of experimental FO set-up. (1, feed tank; 2, draw solution tank; 3, digital balance; 4, heat exchange coil; 5, pump; 6, valve; 7, flow meter; 8, temperature meter; 9, pressure gauge; 10, FO cell; 11, FO membrane; 12, pH meter; 13, conductivity meter; 14, data acquisition).

Index

W

Lightning Source UK Ltd.
Milton Keynes UK
UKOW06f2126020617
302556UK00003B/8/P

9 781780 408118